MW00826452

What Is a Bird?

What Is a Bird?

An Exploration of Anatomy, Physiology, Behavior, and Ecology

Consultant Editor Tony D. Williams
With a foreword by Dr Andy Clements

Princeton University Press
Princeton • Oxford

Published in the United States and Canada in 2020 by
Princeton University Press,
41 William Street, Princeton, New Jersey 08540
press.princeton.edu

ISBN: 978-0-691-20016-3

Ebook ISBN: 978-0-691-21187-9

Library of Congress Control Number: 2020940399

This book has been composed in Gotham and Sabon
Printed on acid-free paper

Conceived, designed and produced by
The Bright Press
Part of the Quarto Group
The Old Brewery, 6 Blundell Street, London, N7 9BH, England

Publisher: James Evans
Art Director: Katherine Radcliffe
Editorial Director: Isheeta Mustafi
Managing Editor: Jacqui Sayers
Project Editor: Emily Angus
Project Manager: David Price-Goodfellow, D & N Publishing
Design: JC Lanaway
Copy Editor: Susi Bailey
Illustrations: Robert Brandt
Commissioned by Jacqui Sayers

Cover Design: Layla Mac Rory

Photograph previous page: Grey parrot (*Psittacus erithacus*)

Printed in Singapore

10 9 8 7 6 5 4 3 2 1

Contributors

PREFACE, CHAPTERS 1, 4, 7 AND 9, AND CONSULTANT EDITOR

Tony D. Williams PhD

Department of Biological Sciences, Simon Fraser University, Canada

Dr Williams completed his PhD at the Natural Environment Research Council's Monks Wood Experimental Station and the University of Bristol in 1986. He has more than 35 years' research experience, with a focus on reproductive physiology, evolutionary biology and ecology, migration and ecotoxicology. Dr Williams is a fellow of the American Ornithological Society and the International Ornithologists' Union, and has published more than 270 scientific papers and two previous books.

CHAPTERS 1 AND 2

Julia A. Clarke PhD, FRSB

Jackson School of Geosciences, University of Texas at Austin, United States of America

Dr Clarke received her PhD from Yale University. She is the John A. Wilson Professor in Vertebrate Paleontology at the University of Texas at Austin and a Howard Hughes Medical Institute professor. Her work explores how new structures and functions arise in deep time, with a focus on the evolution of dinosaurs – including birds. She has an international field programme in palaeontology and has led interdisciplinary teams integrating data on living animals to ask new questions of the fossil record.

CHAPTER 2

Chad Eliason PhD

Grainger Bioinformatics Center, Field Museum of Natural History, United States of America

Dr Eliason received his PhD in integrated bioscience from the University of Akron and is currently a postdoctoral fellow at the Field Museum in Chicago. His research uses a combination of mathematical modelling and experimental work to study the evolution of form and function in birds, with a focus on feathers and their colour.

CHAPTERS 3 TO 6

Scott A. MacDougall-Shackleton PhD

Departments of Psychology and Biology, Western University (University of Western Ontario), Canada

Dr MacDougall-Shackleton earned his BSc and MSc in biology at Queen's University in Canada. He then completed a PhD in psychology at Johns Hopkins University and a postdoctoral fellowship in ecology and evolutionary biology at Princeton University. Dr MacDougall-Shackleton is currently chair of the Department of Psychology and a professor in the Department of Biology at Western University, Canada. He is also director of the Advanced Facility for Avian Research, where he investigates songbird communication, photoperiodism and seasonal changes in birds, as well as other aspects of avian behaviour.

CHAPTERS 4 TO 6

Scott McWilliams PhD

Department of Natural Resources Science, University of Rhode Island, United States of America

Dr McWilliams is a professor of wildlife ecology and physiology at the University of Rhode Island. He received his PhD in wildlife ecology from the University of California, Davis, and his MSc in animal ecology from Iowa State University. He conducted his postdoctoral training in wildlife ecology at the University of Wisconsin–Madison, and today studies the ecology and physiology of migratory birds.

CHAPTER 8

Elizabeth MacDougall-Shackleton PhD

Department of Biology, Western University (University of Western Ontario), Canada

Dr MacDougall-Shackleton completed her undergraduate and master's degrees in biology at Queen's University, Canada, and her doctorate in ecology and evolutionary biology at Princeton University. She is now a professor of biology at Western University, Canada, where she teaches undergraduate and graduate courses in evolution and behavioural ecology. Dr MacDougall-Shackleton's research addresses key stages in the lives of birds through the lens of host–parasite interactions and coevolution.

CHAPTER 10

Frances Bonier PhD

Department of Biology, Queen's University, Ontario, Canada

Dr Bonier completed her MSc in zoology at the University of Idaho and her PhD in zoology at the University of Washington. She is a field biologist with a research focus on the ways animals respond to diverse challenges, including parasites, urbanisation and climate change. Dr Bonier and her students employ an array of tools and approaches in their research, from measuring hormones, to using genetic tools and drawing on the expertise of citizen scientists around the world.

Contents

6

7

8

9

10

Foreword

Dr Andy Clements, Chief Executive,
British Trust for Ornithology

Scientists who understand birds are ornithologists. In that short sentence there are already two words that might put off non-specialist members of society who are interested in finding out more about birds. Science and ornithology sound technical and dry by comparison with, say, a dramatically coloured blue jay inspiring wonder in your backyard. But don't be put off. This book, brought to you by established scientists, describes in wondrous terms everything you would ever want to know about birds. And it does so in beautiful detail and with stunning images. Explaining how birds first originated to how they now undertake their awesome, almost magical, journeys of migration, these pages open a door to the forefront of our knowledge. And reading this book may draw you in to getting involved. Citizen science, whereby your observations of birds can become that body of science, is your means to participate in saving birds.

Aren't birds brilliant? This book answers that question with an emphatic, yes!

► A pair of tawny frogmouths (*Podargus strigoides*).

▼ Regarded as one of the more successful conservation achievements, there were only four known Mauritius kestrels (*Falco punctatus*) in 1974. There are today believed to be about 400 in a self-sustaining population.

Editor's preface: The world of birds

I remember being perhaps five or six and sitting watching birds in my back garden, when I noticed one particular bird flying back to the feeder and aggressively displacing all others. I wanted to know why this one bird was able to dominate others around it while feeding. This book is about 'why' questions like this: why do birds behave the way they do? Why are they built in a particular way? Why are they so diverse? But it is also about the 'how': how do red knots (*Calidris canutus*) manage to fly non-stop from Alaska to New Zealand? How do emperor penguins (*Aptenodytes forsteri*) breed in the depths of the Antarctic winter when temperatures drop to –40°C?

Red knot (*Calidris canutus*).

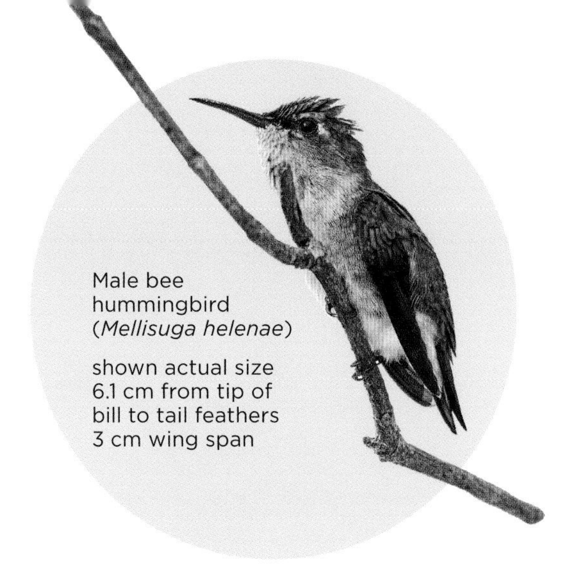
Male bee hummingbird (*Mellisuga helenae*)

shown actual size
6.1 cm from tip of bill to tail feathers
3 cm wing span

The answer to this last question involves explaining how birds work in terms of their anatomy and morphology (skeleton, muscle, brain), and their physiology (regulation of body temperature, cardiac performance, reproductive system). This lavishly illustrated book thus reveals the incredible diversity of birds while exploring all aspects of their biology – from their internal anatomy to their external morphology, from their fascinating behaviour and cognitive abilities to their ecology, and from their evolutionary history to their life history. In doing so, it provides a fascinating enquiry into, and celebration of, the rich complexity of birdlife.

Among living terrestrial vertebrates, the birds form the most abundant group in terms of species. These range from the tiny bee hummingbird (*Mellisuga helenae*), weighing less than 2 g, up to the flightless common ostrich (*Struthio camelus*), which tips the scales at more than 150 kg. We start by exploring what birds are and how this incredible diversity came about. Birds are living dinosaurs that have evolved the power of flight and have truly mastered the air. Many features of modern birds (class Aves) appear to reflect the demands of flapping or powered flight, although we now know that a number of these were present in their theropod dinosaur ancestors, at least some of which were not capable of powered flight.

The following three chapters consider the diversity of structure in birds, including the integument (the feathers that allow for flight, and the skin, beak and claws), the neuromuscular system that powers flight, and physiological adaptations that allow birds to live in a diverse range of habitats, from polar regions to deserts and the tropics, and even underwater. Powered flight is the most energetically demanding form of animal locomotion and requires a wing, an 'engine', control systems and a fuel source, among other things, all of which are described in Chapter 5.

Once birds evolved the ability to fly, they could move easily around the globe. Although some species spend their lives in one area as residents, others migrate thousands of kilometres between distant wintering and breeding areas – this is the topic of Chapter 6. The bar-tailed godwit (*Limosa lapponica*), for example, flies 11,000 km non-stop between its Alaskan breeding grounds to overwinter in New Zealand, while the Arctic tern (*Sterna paradisaea*) migrates 19,000 km between the high Arctic and Antarctica. This has required the evolution of a sophisticated ability to navigate and tell the time, without the benefits of the familiar tools we humans use, such as GPS, a compass or a watch. As birds spread into a multitude of different habitats, from the highest mountains to the total darkness of the deep oceans, they developed different ways of obtaining food, different social systems and different patterns of breeding. This diversity of foraging, social behaviour and reproduction is covered in Chapters 7, 8 and 9, including the answer to my childhood question of why some birds dominate at feeders.

Being diurnal, highly vocal and often brightly coloured, birds are among the animals most familiar to us, as explored in the final chapter. Records of our shared history with birds dates back to the impression of an owl created by an artist some 30,000 years ago at Chauvet Cave in France. We have domesticated birds, we feed birds in our gardens, we birdwatch and many of us contribute to citizen science projects to study birds. However, over time we have also lost iconic species such as the passenger pigeon (*Ectopistes migratorius*) and the great auk (*Pinguinus impennis*), and birds continue to face major human-induced challenges from climate change, habitat loss and pollutants, and 'incidental mortality' from artificial lights, power lines and even domestic cats.

Hopefully, this book will encourage people to support or contribute to global efforts to conserve birds, so that we can continue to be inspired by their beauty and diversity.

Introduction

Birds are living dinosaurs, but what are their origins? The first birds evolved from other tiny dinosaurs around 165–150 million years ago. We now know that many earlier dinosaurs evolved feathers and even a form of aerial locomotion, but their wings were anatomically distinct from those of living birds and they had different ways of moving. Modern birds started diversifying in the Late Cretaceous prior to 66 million years ago, and most major groups emerged about 50 million years ago during a period of rapid diversification. The speciation events that gave rise to songbirds (Passeriformes), comprising half of all birds (around 5,000 species), appear to have been even more recent in evolutionary terms.

Today, birds are the most species-rich group of living land-dwelling vertebrates, with around 10,000 species occupying every continent and almost every habitat on Earth. But how did this diversity come about, and how do scientists classify the seemingly bewildering variety of avian species?

◀ Pied avocet (*Recurvirostra avosetta*).

What is a bird?

Birds are flying animals. They are not the only animals to take to the skies: bats and insects have both evolved powered flight, and extinct pterosaurs, non-bird dinosaurs, flying squirrels and even some snakes all moved or move aerially in a variety of ways, including gliding. However, birds have truly mastered the air. They show incredible diversity yet share a similar basic design, with a beak, feathers, a head and body with two wings and two legs, and a short bony tail. Birds also share other arguably less obvious traits, such as relatively bigger brains, unique ways of breathing and complex vocal communication.

Adaptations for flight

Modern birds (Aves) possess many features that appear to reflect the demands of flapping or powered flight, suggesting that the need to reduce body mass in order to take to the air has dominated all aspects of their anatomy, physiology and behaviour. Birds have feathers, strong but very light structures that are capable of lifting them into the air, they have toothless bills or beaks that are much lighter than the jaws and teeth of mammals, and even their skin is thinner than that of mammals.

Flight can appear to dominate the internal design of birds too. Their gas-exchange system is more efficient than that of mammals and includes substantial air sacs that extend into hollow, air-filled (pneumatic) bones. Weight reduction even extends to the reproductive system. Most female birds only have one oviduct, through which the developing egg passes, rather than the two equivalent fallopian tubes of female mammals. Males in most species lack a penis, instead having a simple cloacal protuberance. And birds also lack a urinary bladder.

Retention of ancestral reptilian egg-laying, rather than evolution of viviparity (bearing live young, as seen in mammals and some reptiles), might also be associated with the demands of flight. It has even been suggested that the smaller, more compact genome of birds – the total amount of genetic material packed into the nucleus of individual cells – is a weight-saving adaptation (the trait is shared by bats), although alternative hypotheses link it to metabolic shifts and generation time of birds.

◄ Male (left) and female (below) mallards (*Anas platyrhynchos*), showing upper and lower wing feathers.

However, although it may appear that all these traits are adaptations for flight, it is becoming increasingly clear that many (e.g. egg-laying, small body sizes and genomes, feathers, aspects of the respiratory system and pneumatic bones) were present in the hollow-boned, three-toed bipedal theropod dinosaur ancestors of modern birds, at least some of which were not capable of powered flight.

A diversity of body shapes in relation to the basic body plan of a bird (house sparrow, centre)

Evolution of birds

Birds seemed incontrovertibly unique relative to all other known extinct and extant species, so even locating them on the tree of life was challenging. However, less than 100 years after Swedish taxonomist Carl Linnaeus named his 'Class Aves', the first evidence of the affinities of birds with certain extinct reptile species (Dinosauria) was noted by the British anatomist Sir Richard Owen and biologist Thomas Huxley, contemporaries of naturalist Charles Darwin. Some dinosaur fossils also had hollow bones, and others had distinctly bony features that were otherwise seen only in birds.

Mosaic evolution

Fossil discoveries in the nineteenth century seemed to show a mix of characters that indicated close genealogical relationships between birds and reptiles. *Archaeopteryx lithographica* was one example: fossils of this species from rocks in southern Germany dating from the Jurassic period 150 million years ago had some features typical of reptiles and others that were known only in birds. Thomas Huxley and other biologists noted that while *Archaeopteryx* had teeth and a long bony tail, it also had feathered wings like modern birds. Huxley coined the now widely used term 'mosaic evolution' to describe what he saw: a pattern indicating that different features must evolve at different rates. Some features of *Archaeopteryx* evolved more slowly and were known in other fossil birds, such as the long bony tail and big claws with separate digits on wings. However, *Archaeopteryx* had other features such as feathers that showed a much more limited, exclusive distribution and therefore must have arisen more recently.

Soon after the publication of Charles Darwin's *On the Origin of Species* in 1859, more fossil finds were reported by early American palaeontologist Othniel Charles Marsh from the chalk rocks of Kansas. These reptile fossils showed a strongly keeled sternum and bony wings that were much more similar to those of living birds than were those of *Archaeopteryx*, but like that species they retained teeth in their jaws. Darwin wrote to Marsh to congratulate him on discoveries that were key to illuminating the great transitions in the history of life, including the evolution of birds, predicted by his theory.

► The Berlin specimen of *Archaeopteryx lithographica*, discovered in the late nineteenth century. This fossil provides key evidence of the affinities of birds within Reptilia.

Shared features

Most textbooks emphasise that every aspect of bird biology is shaped by flight, as noted above. When we take the long view, however, we see that many of the features that characterise modern birds have a much deeper evolutionary history stretching back far before the origin of flight. Birds lay eggs, but this feature is inherited from their reptilian ancestors – all extinct dinosaurs found so far laid eggs. Dinosaurs, including birds, evolved a fully 'erect gait', with the hindlimbs always under the body. Dinosaurs and modern birds also share an expanded area of muscle origin in the hip and a relatively more elongate neck.

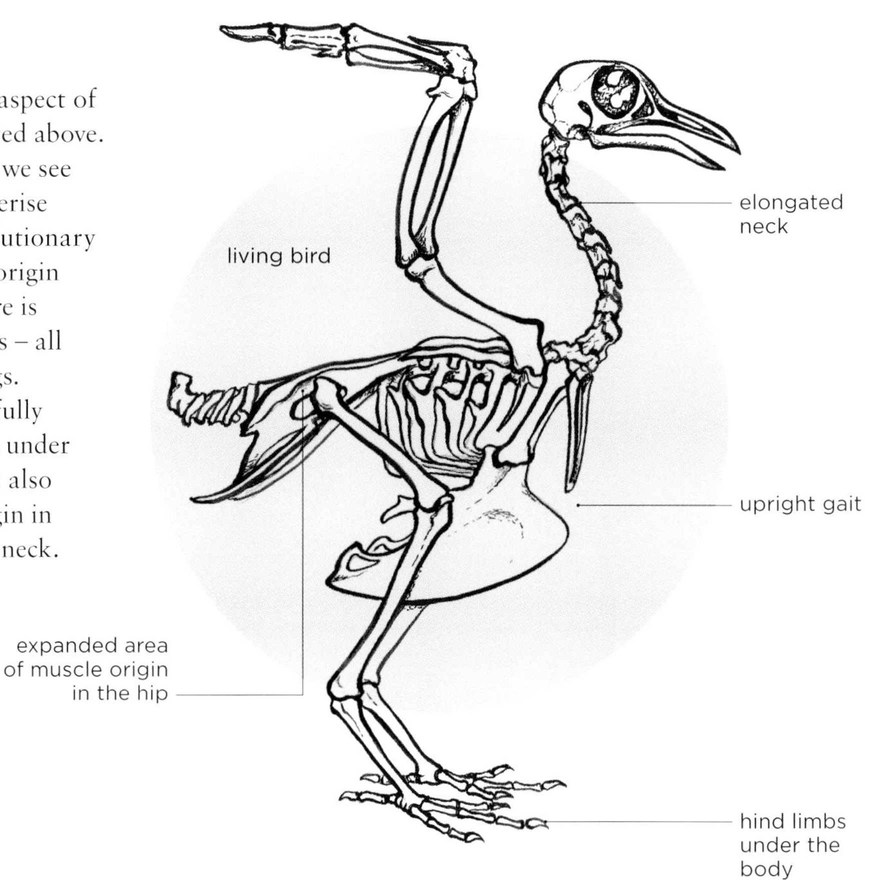

► The skeleton of a feral pigeon (*Columba livia*) contains evidence of adaptations that developed deep in dinosaur evolution. These include an upright gait, with the hindlimbs under the body.

◄ Reconstruction of the feathered dinosaur *Caihong juji*, showing iridescent and black coloration inferred from fossilised evidence of melanin-containing organelles analysed against the shapes of those associated with colours in living birds.

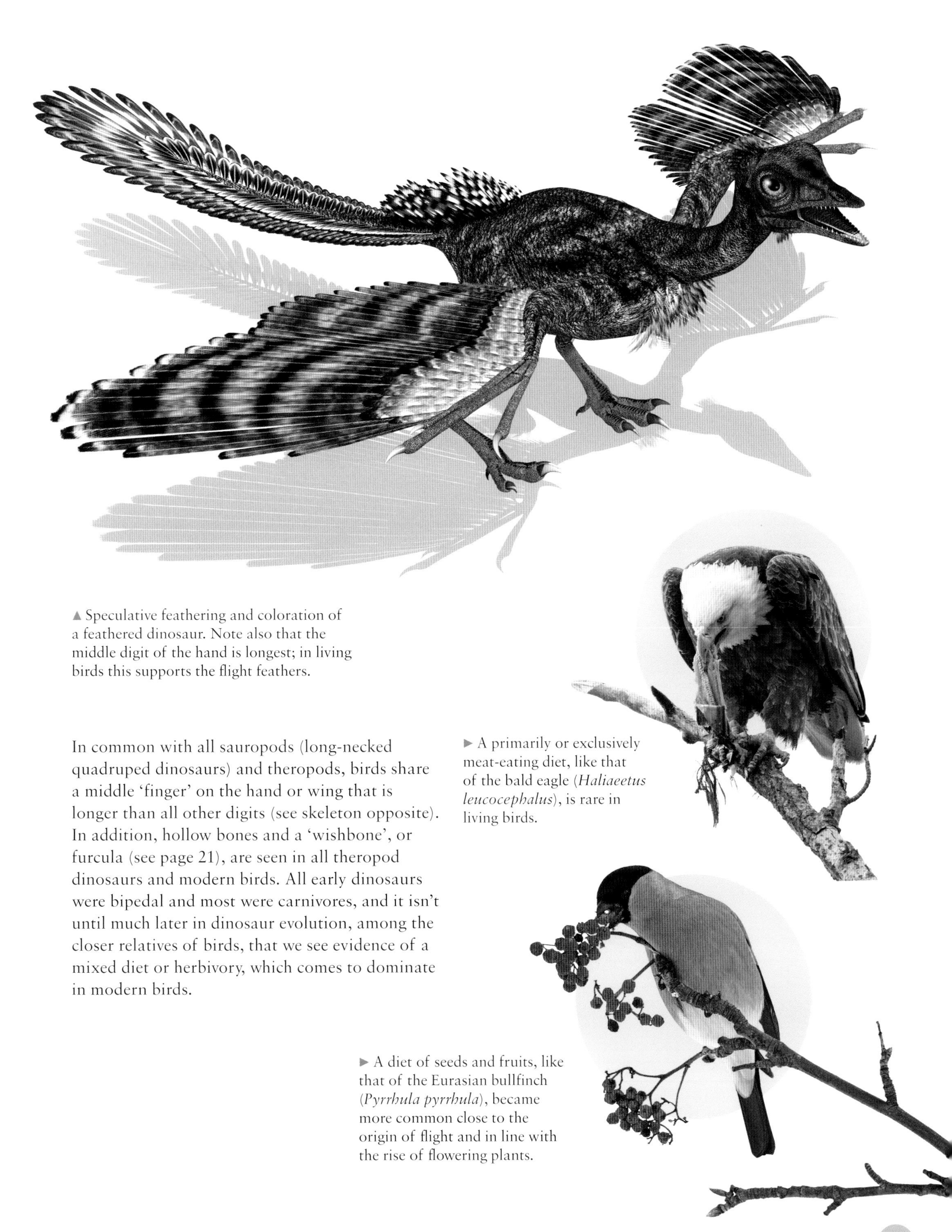

▲ Speculative feathering and coloration of a feathered dinosaur. Note also that the middle digit of the hand is longest; in living birds this supports the flight feathers.

► A primarily or exclusively meat-eating diet, like that of the bald eagle (*Haliaeetus leucocephalus*), is rare in living birds.

► A diet of seeds and fruits, like that of the Eurasian bullfinch (*Pyrrhula pyrrhula*), became more common close to the origin of flight and in line with the rise of flowering plants.

In common with all sauropods (long-necked quadruped dinosaurs) and theropods, birds share a middle 'finger' on the hand or wing that is longer than all other digits (see skeleton opposite). In addition, hollow bones and a 'wishbone', or furcula (see page 21), are seen in all theropod dinosaurs and modern birds. All early dinosaurs were bipedal and most were carnivores, and it isn't until much later in dinosaur evolution, among the closer relatives of birds, that we see evidence of a mixed diet or herbivory, which comes to dominate in modern birds.

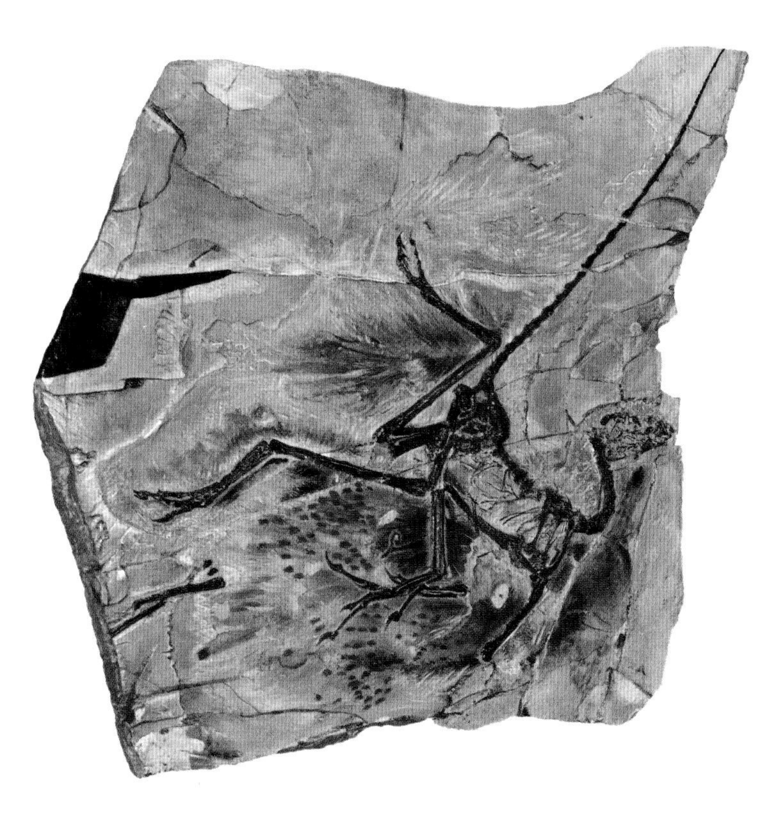

◄ Fossil specimen of *Anchiornis huxleyi*, showing preserved spangles on the flight feathers of the wing.

Closest living relatives

DNA, morphological and life-history data indicate that crocodilians are the nearest living relatives of birds in a group called Archosauria, which also includes all extinct dinosaurs and pterosaurs. Birds and crocodilians share a huge number of unique beta-keratin genes – a key component of feathers. They also share a four-chambered heart (evolutionarily convergent with the four-chambered heart in mammals) and the rotation of the hindlimbs under the pelvis for a more upright stance. While most reptiles mix deoxygenated and oxygenated blood fresh from the lungs in an undivided heart, crocodiles and birds separate these. Crocodilians do, however, have a novel valve that allows recycling of blood that is largely deoxygenated during breath-holding in deep dives.

Most people think of extant crocodilians as having a sprawling lizard-like gait that is dissimilar to the bipedality or two-leggedness of birds. Unlike lizards, however, crocodilians can also swing their hindlimbs under the body, allowing a fast-moving gait called the crocodilian high walk. In this, the legs move largely parallel to the midline of the body, a bipedal form of locomotion also seen in birds and humans; in other reptiles the legs describe wide arcs out to the right and left. Both crocodilians and birds also use vocalisations in communication between young and adults, as well as in signalling to other adults for mate attraction and territorial defence. As discussed on the following pages, when we combine evidence from fossils with the study of living animals we can learn more about the evolution of features like feathers and behaviours.

DINOSAURS ON YOUR DINNER TABLE

We can explore the evidence that birds are living dinosaurs by examining any roast chicken or Thanksgiving turkey. When we pull the wishbone to make a wish, we use a bony feature that is more than 200 million years old and shared with theropod dinosaurs like *Coelophysis*. As we carve the breast we come to the bony sternum, which is also seen in extinct bipedal non-flying dinosaurs with teeth and unusual feathers, including *Microraptor*, *Oviraptor* and *Troodon*. However, the muscles around the top of the chicken drumstick are ancient indeed. The pelvis and enlarged areas for muscle origin in birds go back to the oldest known dinosaurs.

► The earliest scales, filaments or feathers preserved in the geologic record are from the Late Jurassic. Filamentous feather precursors may have originated nearly 100 million years before the origin of flight, but very few fossil deposits have been found from this period.

Fossil evidence for the evolution of body coverings in dinosaurs including birds

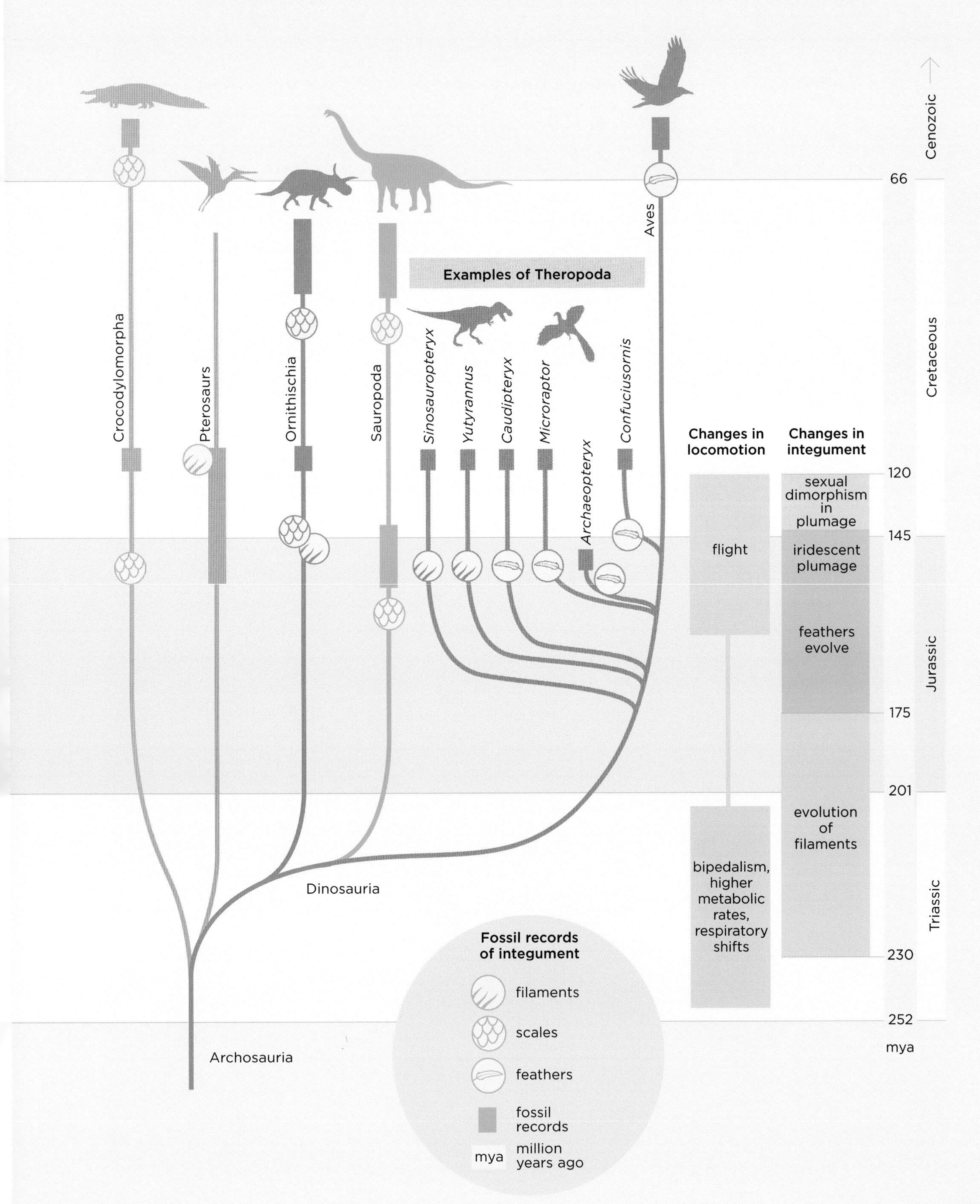

Evolution of feathers and flight

Feathers are branched, hierarchically organised structures made from the protein beta-keratin. In living birds, typical feathers have a central rachis with branching barbs and barbules, the latter bearing interlocking hooklets that are critical for flight. Feathers allow birds to use flight to forage and scavenge over large areas, pursue prey, escape predators and travel extreme distances during migration. They can also be highly coloured and otherwise modified for important functions in communication and signalling, and for camouflage, insulation or waterproofing.

Land animals have diverse structures covering their bodies, including hair, skin, scales and feathers. A key question facing evolutionary biologists in the nineteenth and early twentieth centuries was how the lightweight branched feathers of birds developed, given the scales and skin of their reptile relatives. Hypotheses of feather origins tried to explain how flat scales might have become elongated and divided to form flat, branched feathers. Other hypotheses imagined a membrane-covered bat-like wing used in gliding as an intermediate before the evolution of feathers.

A discovery that makes feathers fly

Fossils first discovered in northeastern China in 1996 have profoundly changed our understanding of the evolution of feathers and flight. It is now clear that branched bird feathers were preceded in the history of dinosaurs by the development of bristles and superficially 'hair-like' filaments.

▼ Pterosaurs flew on thin membranes supported by their finger digits. They are dinosaur cousins that evolved flight independently from birds.

Some proposed stages of feather evolution

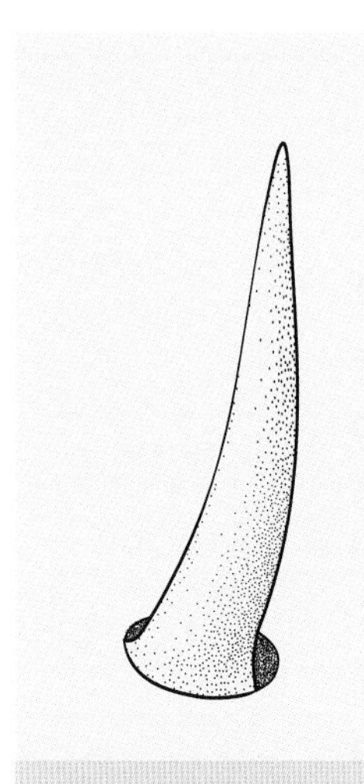

Stage 1

A simple hollow filament, the first feather precursor in the fossil record, may have been present in the Triassic ancestors of both dinosaurs and pterosaurs (see pages 20–1). It links to the earliest part of feather development.

Stage 2

Filaments diverging from a shared base. Stuctures interpreted as fitting this proposed stage are reported from the Late Jurassic and Early Cretaceous.

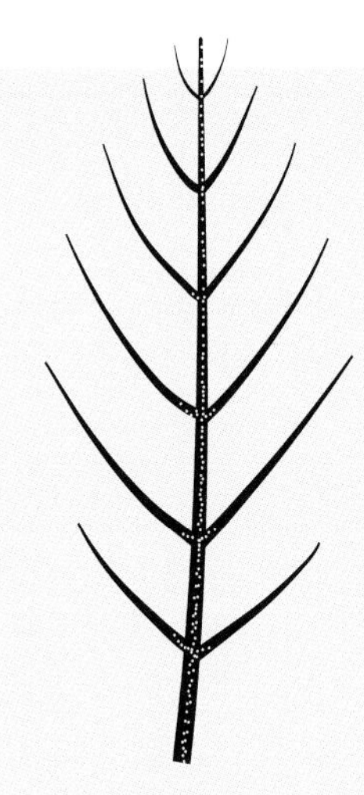

Stage 3

Branched structure with a central rachis and barbs. This stage is predicted from feather development. Representatives in the fossil record are debated.

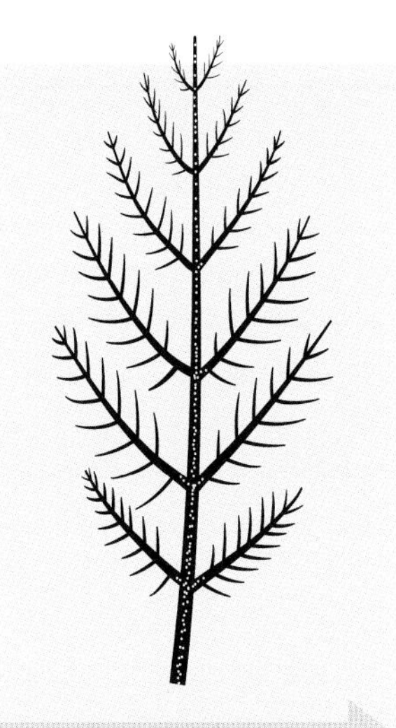

Stage 4

Complete basic structure of a living bird feather, with rachis, barbs and barbules. This is first seen in the fossil record about 150 million years ago (see pages 20–1).

▲ Branched feathers evolved from simple hollow filaments that are found in many different groups of dinosaurs and in pterosaurs. Some subsequent stages in the development of feathers have been inferred from models and are debated.

These filament structures are present in the two main groups of dinosaurs, Ornithischia and Saurichia, as well as in the closest extinct relatives of dinosaurs, the pterosaurs, which quite independently evolved flight via membranes extended from the digits of the forelimb. Although these bristles and filaments are sometimes referred to as protofeathers, they apparently arose more than 100 million years earlier than branched feathers, and their acquisition may be related to changes in posture and physiology that evolved early in the origin of all archosaurs, including dinosaurs and pterosaurs, during the Triassic. But why might bristles and filaments have arisen in dinosaurs and why, much later, would feathers? How does a bristle relate to a feather and how might either evolve from structures present in reptiles? Answers to these questions have come from studying living animals.

Research into the early development of feathers has shown that they form a hollow cone shape from a round placode or zone of growth (see drawing above). This cone or tiny hollow, stump-like filament is segmented into barbs that may fuse at one side of the disc or plate to form a central rachis. Thus, feather development in living birds does not involve a stage with a flat, scale-like sheet, but rather one with a round, tubular form. If formation of the feather is truncated at an early stage, a hollow filament would be produced. Aspects of the early growth plates of feathers and scales are shared and homologous between birds, but in their reptile relatives the flat plate does not project or split. Other work has shown that most genes so far linked to feather development are ancient and are shared with all other land vertebrates. Many of these same genes are deployed in the development of hair, which is a very different structure. However, beta-keratin protein genes show an amazing proliferation within the Archosauria, and some examples of these are unique to crocodilians and birds, while many others are found only in birds.

Why did feathers evolve?

Some hypotheses for the origin of the filaments that precede the acquisition of feathers in archosaurs include insulation. This would imply higher metabolic rates in this group than seen in other reptiles and would be associated with the maintenance of body temperatures above ambient (endothermy). Hypotheses for the origin of true feathers centre on the creation of more complex patterns on the body surface, perhaps associated with visual communication and signalling (see page 69), and feathers used for these purposes are diverse in structure and coloration mechanisms. The discovery of fossil evidence for coloration involving the most common pigment in living birds, melanin, has allowed the reconstruction of colours present in fossil feathers. The earliest evidence for large colour patches (perhaps used in sexual signalling) and two different forms of iridescence are seen in dinosaurs about 150 million years ago. This new evidence has informed hypotheses for why feathers might have arisen before they were used in flight.

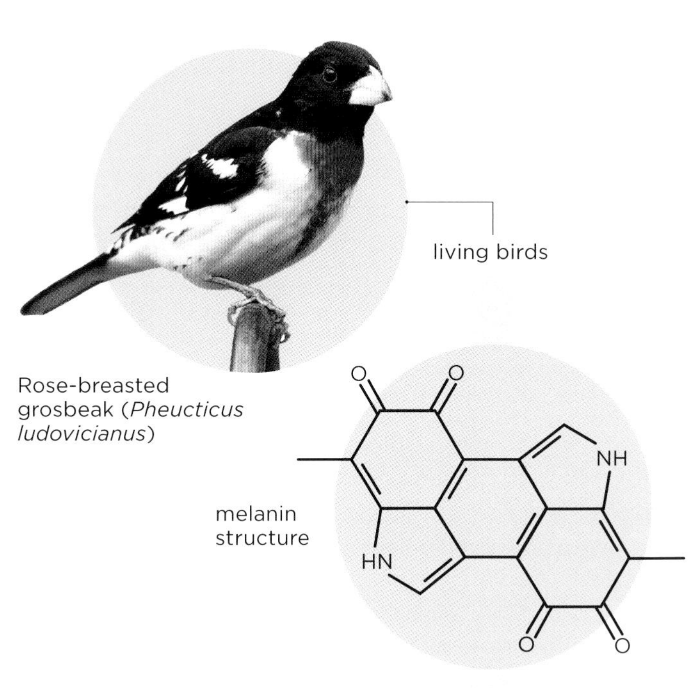

Rose-breasted grosbeak (*Pheucticus ludovicianus*)

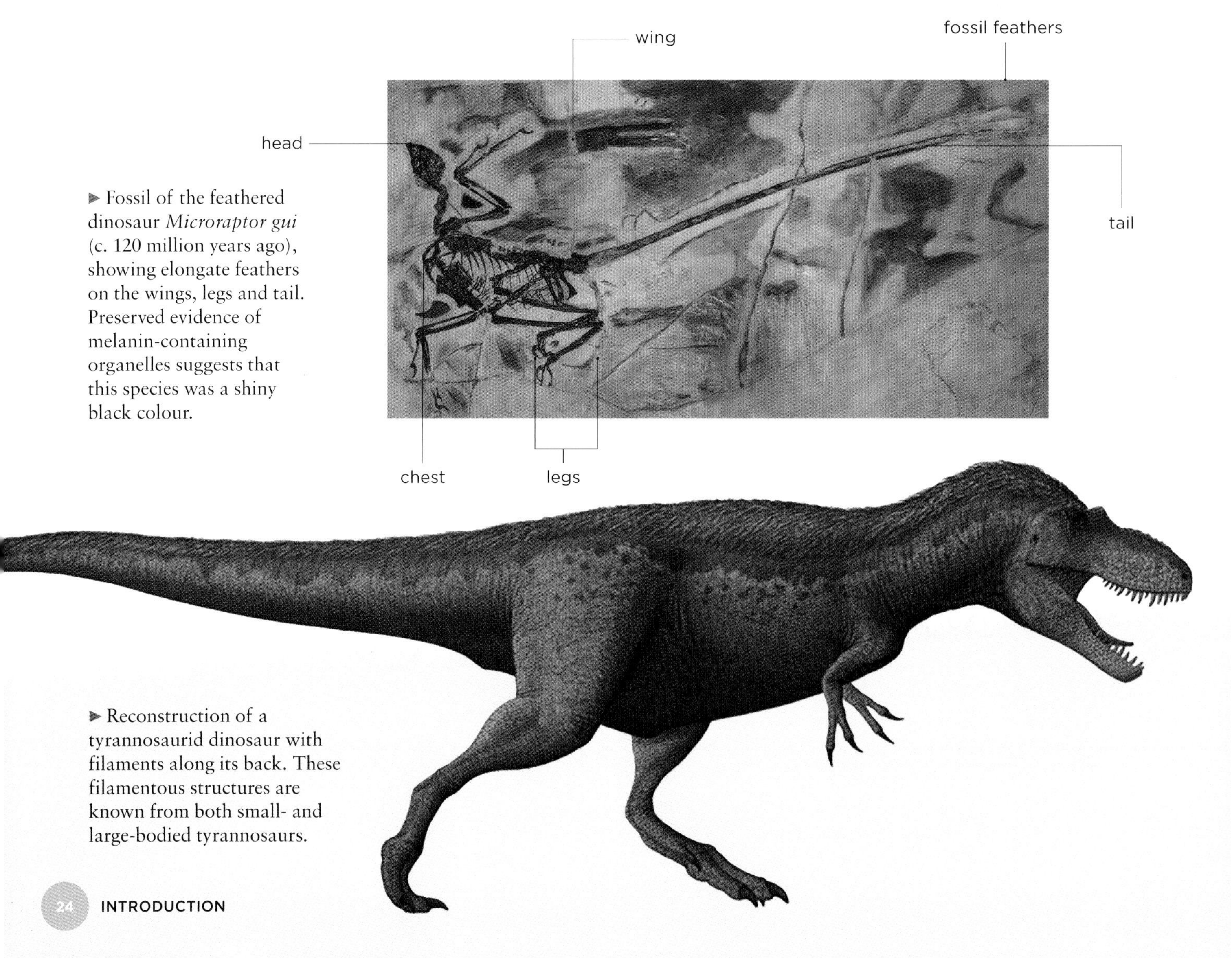

▶ Fossil of the feathered dinosaur *Microraptor gui* (c. 120 million years ago), showing elongate feathers on the wings, legs and tail. Preserved evidence of melanin-containing organelles suggests that this species was a shiny black colour.

▶ Reconstruction of a tyrannosaurid dinosaur with filaments along its back. These filamentous structures are known from both small- and large-bodied tyrannosaurs.

Recent studies on the ways living birds move have revealed new things about how natural selection could favour the development of 'half-wings': wings comprising elongate branched feathers that are too small to generate enough lift for gliding or flying. When new structures arise in evolution they often have to confirm a benefit in terms of natural selection in an intermediate state. Scientists first found that many young birds use their half-wings to gain traction when running up steeply inclined and even near-vertical surfaces to escape a perceived threat. This recently discovered method of moving is termed wing-assisted incline running (WAIR) and has been observed in both adult and young birds from many major groups of Aves. These observations led to the hypothesis that small dinosaurs might also have first used their proto-, or half-, wings in escape.

◀ A major question in the evolution of flight concerns how wings might have first been used. The wings of many early feathered dinosaurs were too small to generate sufficient lift for sustained flight.

Another hypothesis for the evolution of flight proposes that small, half-winged feathered dinosaurs moved to living in the trees, and that from jumping from branch to branch they moved to gliding, and from there to active flight. This hypothesis requires an archaic way of moving that is not seen in living birds: no bird species glides that does not also flap. These WAIR and gliding hypotheses also differ in the proposed natural selective pressures driving the origin of flight. In one case the pressure is escape from land-bound predators, while in the second it is movement among trees during foraging. On top of these open questions, close relatives of the first active flapping birds show evidence of aerial behaviours not seen in living birds. For example, *Microraptor* had an elongate 'wing' on its hindlimb, and the strange *Epidexipteryx hui* apparently had some extended membranes on its wing but no long wing feathers. While neither of these species is a very close cousin or direct ancestor of living birds, they illustrate that, in the Late Jurassic and Early Cretaceous, many different feathered dinosaur species developed new ways of moving in the air.

Wing-assisted incline running in a horned screamer (*Anhima cornuta*)

▲ All studied species of living birds use wing-assisted incline running (WAIR). This behaviour involves flapping while ascending small or large inclines on foot. Even the 'half-wings' of baby birds, which cannot support them in flight, are used to provide increased traction in this common escape behaviour. In WAIR, flapping the wings or half-wings can enable birds to climb nearly vertical surfaces. Its description provided a key new model for how the flight stroke and protowings might have evolved.

Birds in the making

Although feathers are one of the most obvious characteristics of birds, many other aspects of a bird's body represent remarkable innovations. Some of these traits evolved in Dinosauria before the evolution of flight, while others arose in response to this radical new form of movement, including novel ways of breathing. Major shifts in tongue structure also occurred, along with the development of a new vocal organ deep in the chest, which is further modified in some living birds to produce song. Even changes in genomes (the DNA content of individual cells) are associated with the evolution of birds.

Respiratory and metabolic changes

Bird lungs are unique and very different from those of all other living animals. They do not expand or contract as in mammalian lungs, but are ventilated by the passage of air from sacs distributed across the bird's body. Other features of the lungs are uniquely shared with crocodilians, and there is evidence that components of this air sac system of ventilation were present in other dinosaurs. It is thought that the number of air sacs found in avian ancestors increases closer to the origin of flight, but since other features of the system – especially skeletal indicators of the posterior air sacs – have a limited fossil record, this story is still unfolding.

In the evolution of avian flight, the continuous development of respiration in warm-blooded animals is thought to be key. Higher metabolisms, like those in flighted animals, require more oxygen intake and more breath cycles. When exactly higher metabolisms evolved in dinosaurs or in the archosaur ancestors of birds is debated. Palaeontologists primarily use evidence in the cross sections of bones to investigate signatures of growth rate, which is linked to metabolic rate. There are indications of very early shifts toward increases in metabolism close to the evolutionary split between crocodiles and birds. Other evidence suggests that shifts in bone growth and metabolic rate took place later in dinosaur evolution, close to the origin of flight. Some syntheses suggest that metabolic rates may have had several key phases of increase in Dinosauria.

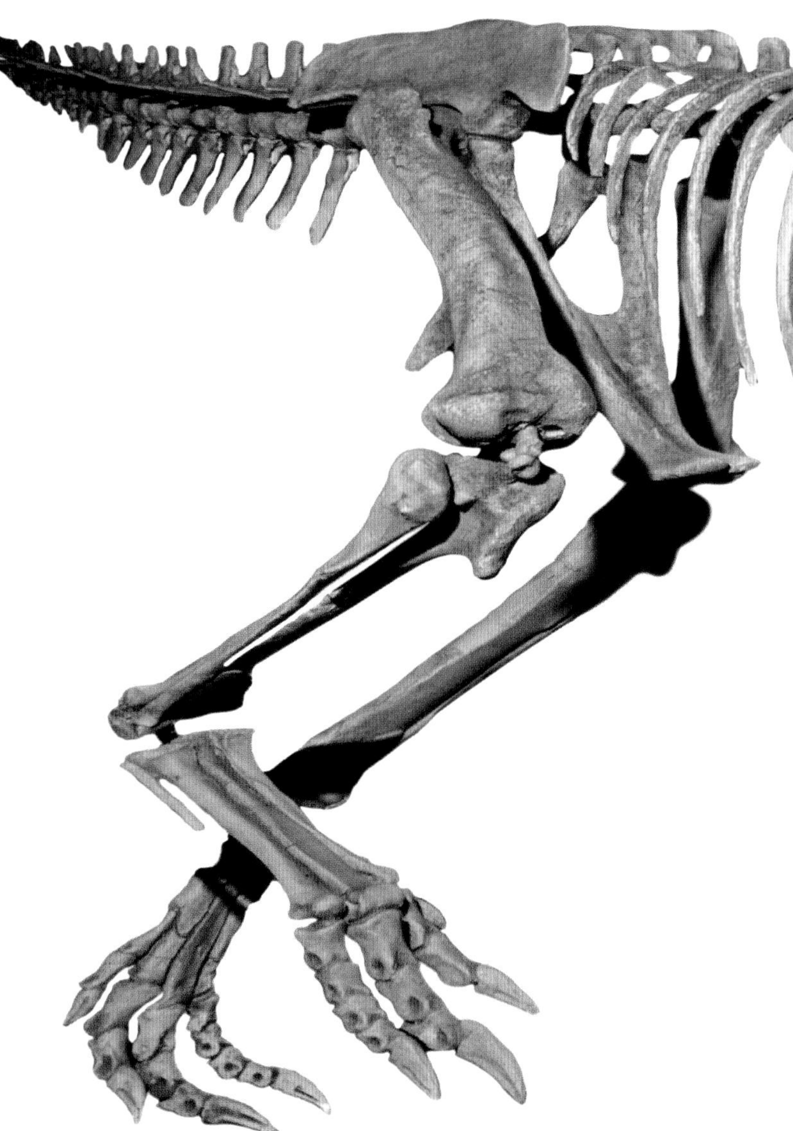

Skeletal changes

Many skeletal changes are also associated with the evolution of active powered flight. Even before archosaurs took to the air, an expanded bony sternum evolved, to which the pectoralis muscles that power the active downstroke of modern birds attach. This feature is seen in dinosaurs like *Oviraptor*, which had wings that were far too small to power flight. At around the same time, more than 160 million years ago, the shoulder girdle changed and the forelimbs became more elongate. After flight evolved, changes to the forelimb muscles continued and the sternum of early flyers developed a ridge, indicating that the pectoralis muscles became larger. Closer to the time the common ancestor of all living birds emerged in the Late Cretaceous, a true keel on the sternum finally developed, suggesting the presence of the large muscles needed to confer a powerful downstroke seen in living birds. The hand bones gradually fused after flight was achieved, providing a rigid attachment site for the primary feathers, and the shape of the upper arm bone changed to allow a larger surface for rotation in the shoulder socket.

Late in bird evolution and more than 50 million years after the origin of flight, the close ancestor of living birds lost its teeth. The loss of teeth occurred many times in other theropod dinosaurs, and is typically related to the acquisition of a fruit or plant diet.

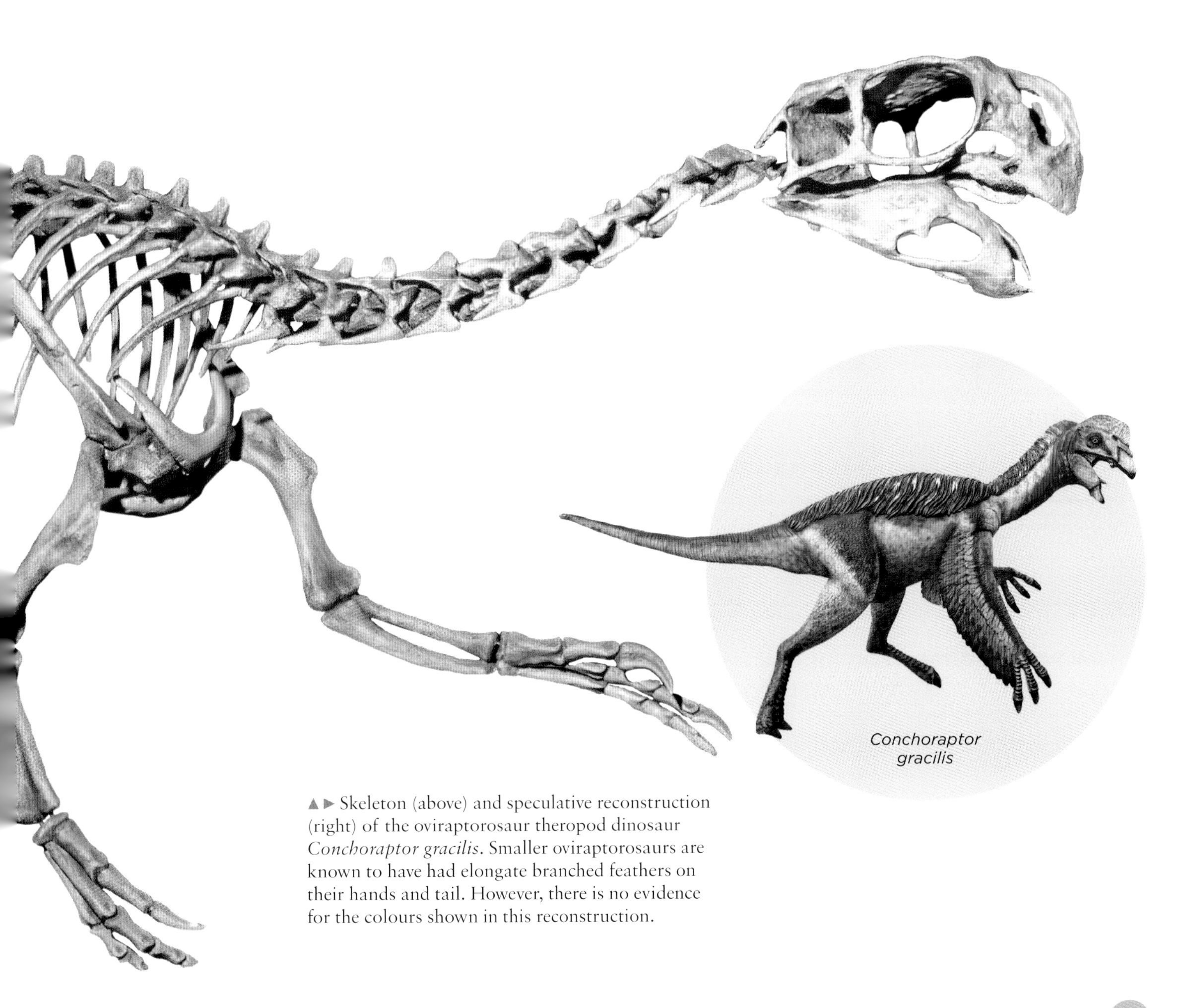

▲► Skeleton (above) and speculative reconstruction (right) of the oviraptorosaur theropod dinosaur *Conchoraptor gracilis*. Smaller oviraptorosaurs are known to have had elongate branched feathers on their hands and tail. However, there is no evidence for the colours shown in this reconstruction.

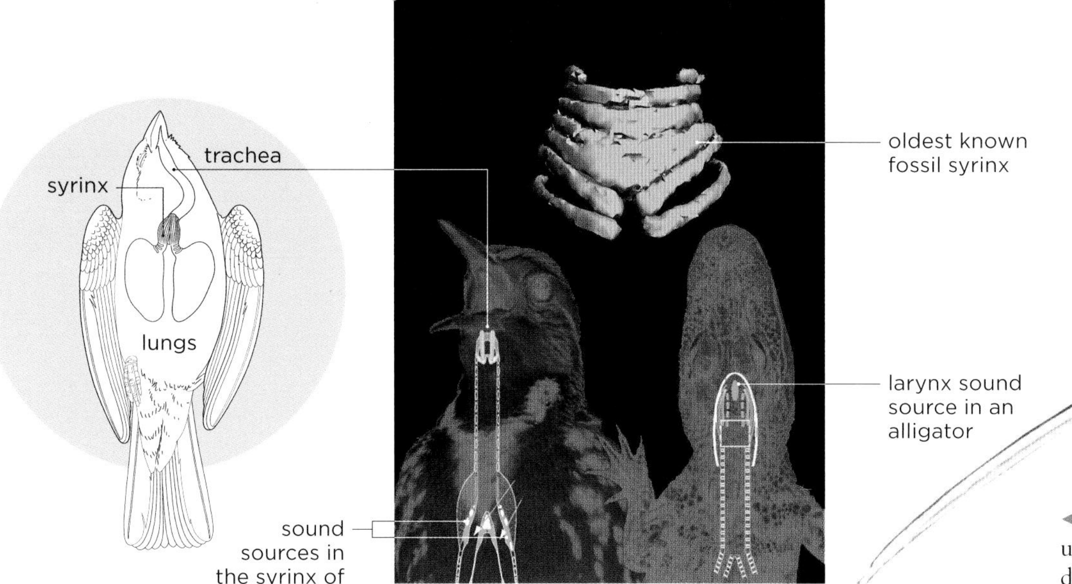

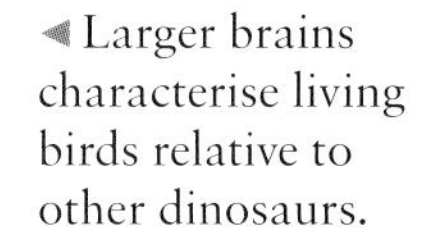

◄ Birds sing with a unique vocal organ deep in their chest, called a syrinx.

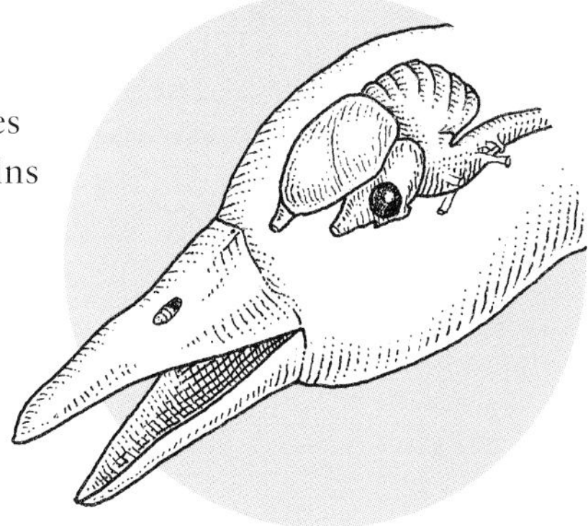

◄ Larger brains characterise living birds relative to other dinosaurs.

► Bony tongues, like the extreme example seen in this Andean emerald (*Amazilia franciae*), are highly elaborate in living birds.

Bird brains

While crows and parrots have the largest brain sizes relative to all other birds, all birds have bigger brains relative to their reptile relatives and they are also shaped differently, with a much larger forebrain. The shapes and sizes of the brains of extinct dinosaurs closely related to Aves can be examined by creating endocasts, or moulds of the inside of the skull. From these it can be seen that the forebrain is relatively larger in theropods than in other dinosaurs, but still smaller than in living birds. The brain of *Archaeopteryx*, while displaying the 'stacked' shape seen in the brains of living birds, is similar to those of other close cousins in extinct Dinosauria. The large forebrain seen in birds, which is involved in song learning in passerines, parrots and hummingbirds, evolved completely independently from the large forebrain of mammals and differs markedly in structure. Comparisons of the brains of passerines and primates have shown that there is a much higher neuronal density in the former – in other words, the bird brains may pack in more neural connections even though they are smaller.

Singing from the heart

While humans, frogs, crocodiles and all other land-dwelling vertebrates produce sound using a vocal organ called the larynx, situated at the end of the windpipe, birds evolved a completely different sound-producing organ deep in the chest where the windpipe branches to the lungs. This is called the syrinx and is situated right next to the heart; birds truly sing from the heart! While crocodiles and living birds both use sound to communicate in similar situations, they produce it in very different ways. Communication among young before hatching, with young and adults after hatching, and among adult males and females is seen in living archosaurs and was presumably also present in many extinct dinosaurs. However, the oldest known fossil evidence of a syrinx is from a Late Cretaceous bird species (see photo above) and there is so far no evidence of its origin in dinosaurs. As with increases in forebrain size, the syrinx may thus be a feature that evolved long after birds evolved flight.

Food for thought

Flight in birds is also linked to the subsequent evolution of new ways of acquiring food. While carnivorous or piscivorous birds swallow their prey whole and cannot chew, in common with their reptile relatives, birds that feed on nectar, seeds or fruits do so using a complex bony tongue. In the case of hummingbirds, the tongue bones unwrap from the surface of the skull to extend a long, narrow tip into a flower. Based on the existing fossil record, the diversification of the tongue bones took place in theropods only after the evolution of flight. It is also interesting to note that the diversification of flowering plants with true fruits and flowers (Angiospermae) occurred around the same time as the evolution of the first birds.

Classification and evolutionary history

The Linnaean system of taxonomy uses a hierarchy in which organisms are classified based on shared characteristics. Phylogenetic analysis uncovers relationships between taxonomic levels and their deeper evolutionary history, resulting in an evolutionary tree or phylogeny of evolutionary connectedness among groups – similar to a genealogical family tree in humans. Our understanding of the evolutionary history of birds has been revolutionised in the last decade and continues to be a hot area of research, due in part to the application of DNA techniques and the discovery of new fossils.

All living birds have a local, colloquial or common name, but this can vary by country, region, language or dialect. To avoid confusion, therefore, each distinct type of bird, or species, is given a binomial scientific name composed of both a genus and a species term. For example, the common starling belongs to the genus *Sturnus* and has the species name *vulgaris*, and is hence *Sturnus vulgaris*. The Linnaean system prescribes inclusion of each species in a genus, which is then nested within all higher levels of taxonomic classification, with families and orders making up the class Aves, the overarching taxon that includes all birds.

Defining a species

Different species within a named taxon are more closely related to one another, sharing common traits, than they are to species in other taxa. The species is therefore a fundamental unit of biological diversity, although scientists are still debating exactly how to define a species. In general, a species comprises individuals that share many common traits – in other words, they look or behave alike, they interbreed with members of the same species to produce fertile offspring and rarely breed with members of other species, and they share a common evolutionary history. Ornithologists often describe subspecies, the taxon below the level of species, based on smaller plumage differences among populations. For example, the North American dark-eyed junco (*Junco hyemelis*) has between eight and 15 subspecies; the variation in numbers is a consequence of the furious debate that takes place between 'lumpers' (who prefer fewer subspecies) and 'splitters' (who prefer more)! These issues can have conservation implications, as most laws cover only species and not subspecies.

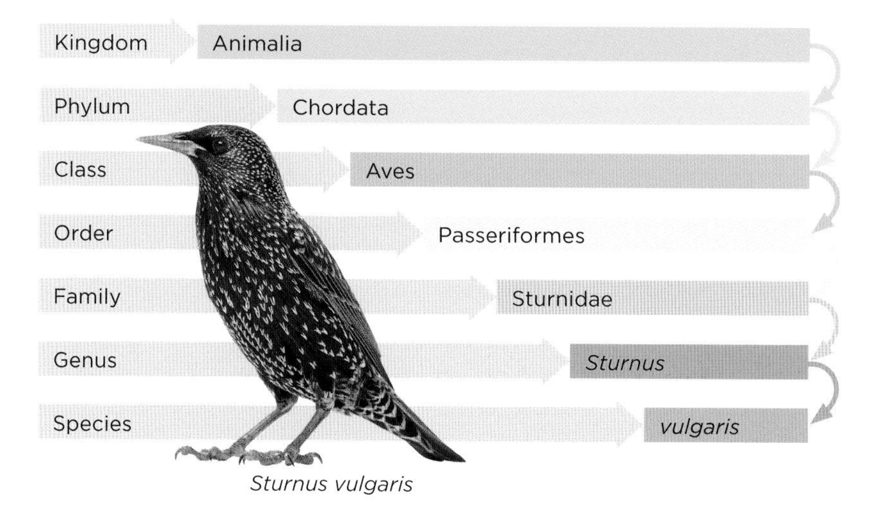

► An example taxonomic classification showing the common starling, with the binomial scientific name *Sturnus* (generic name) and *vulgaris* (species name).

Bird taxonomy

Phylogenetic analysis suggests that the radiation of all living bird species can be divided in two, with this evolutionary split taking place during the Cretaceous, some 110 million years ago. The taxon Palaeognathae includes the flightless or weakly flying ratites (e.g. ostriches, kiwi, tinamous), while the large, diverse taxon Neognathae contains all other birds. In turn, the Neognathae is split into two superorders: the Galloanserae, comprising the order Galliformes (landfowl: chickens and relatives) and the order Anseriformes (waterfowl: ducks, geese and relatives); and the Neoaves, comprising all other modern birds. Phylogenetic relationships among the Neoaves have been the subject of extensive work in recent years, which has involved integrating new fossil discoveries with large amounts of genomic data using increasingly complex statistical analysis. However, these analyses have resulted in conflicting estimates of phylogeny and differences in relationships between groups, perhaps because of the presumed rapid diversification of birds that took place around 50 million years ago. This makes it difficult to resolve precise connections, and as such this remains a very active area of bird research.

▼ A simplified phylogeny of birds (class Aves), showing the major clades or groups that are believed to have evolved from a single common ancestor.

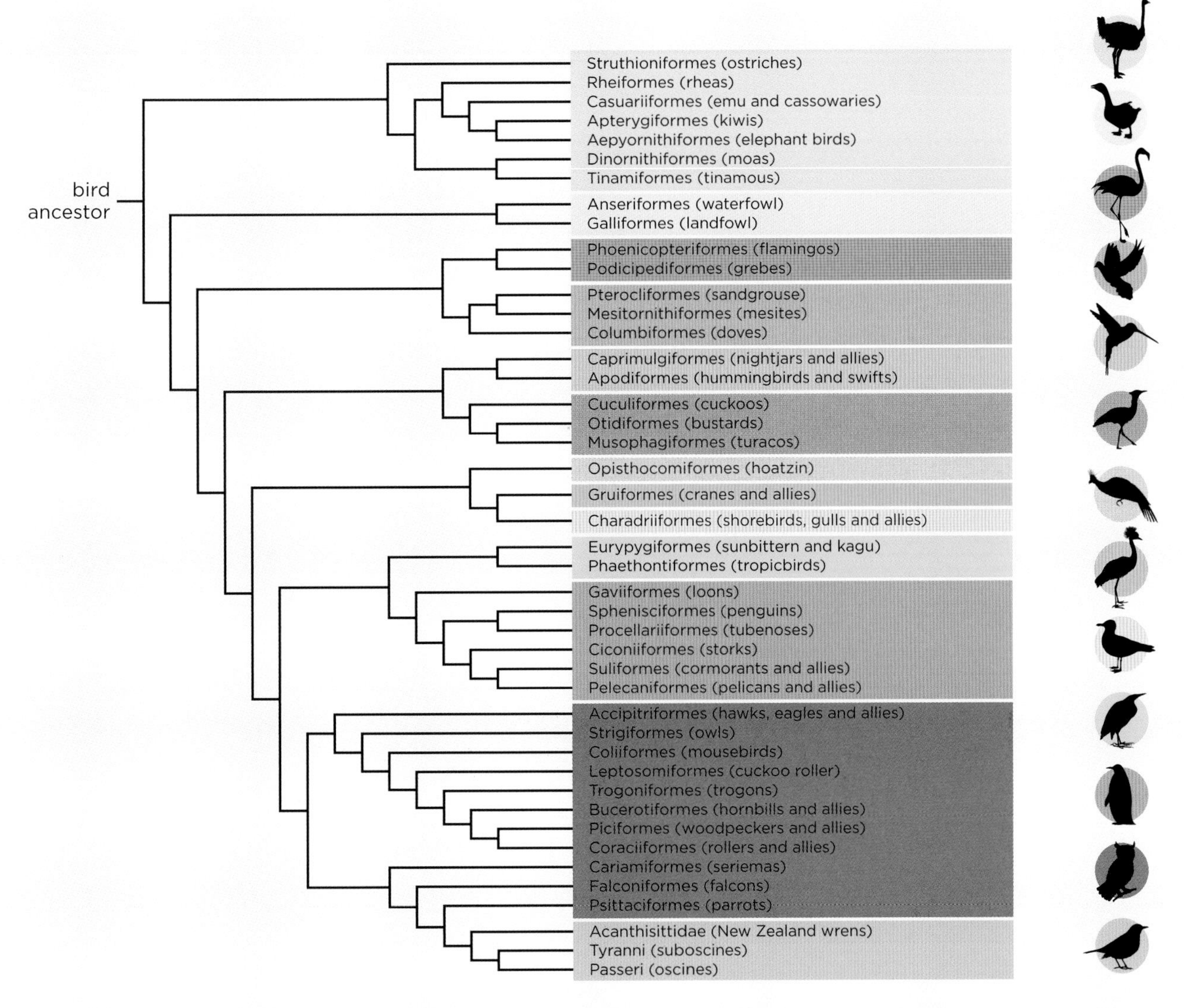

Diversity of extinct and living birds

Although all birds share a basic body plan, the shape and relative size of each component of the plan has been modified through natural selection to generate an incredible diversity of form and function. For example, bills may be longer or twisted, legs may be longer (as in wading birds) or shorter (as in penguins), and feet may be feathered, lobed or webbed. Wings may be long and narrow, short and very broad, or even completely reduced and non-functional. All of these adaptations arose in particular species to fit their particular lifestyles and habitats.

Diversity of size and shape

The smallest living bird, in terms of body mass, is the bee hummingbird (*Mellisuga helenae*; see page 11), which weighs less than 2 g, equivalent to half a teaspoon of sugar. The largest living birds capable of flight, including the wild turkey (*Meleagris gallopavo*) and mute swan (*Cygnus olor*), are more than 6,000 times heavier, averaging 12–13 kg. Flightless birds can be heavier still, with the common ostrich (*Struthio camelus*) weighing more than 150 kg. Wandering albatrosses (*Diomedea exulans*) have the largest wingspan of any living bird, averaging 3 m, compared with just 3 cm in the bee hummingbird. However, modern birds are dwarfed in size by some extinct birds. The heaviest bird still capable of flight is thought to have been *Argentavis magnificens*, similar to modern-day condors. Known from Miocene fossil beds in Argentina dating back 6 million years, it had a wingspan of up to 5–6 m and an estimated body weight of more than 70 kg. The largest extant penguin, the emperor penguin (*Aptenodytes forsteri*), stands 1.2 m tall and weighs more than 20 kg, but extinct penguins like *Icadyptes salasi* and *Kumimanu biceae*, which lived 55–30 million years ago, were

Aerodynamics of bird flight

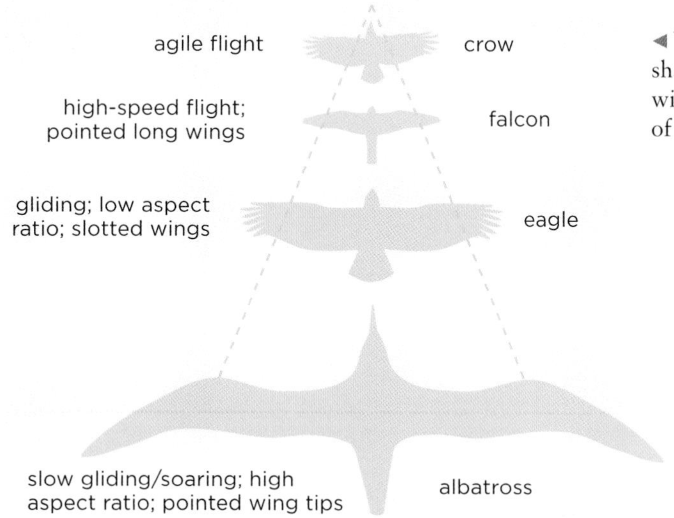

◄ Variation in wing shape associated with different types of bird flight.

A huge extinct penguin

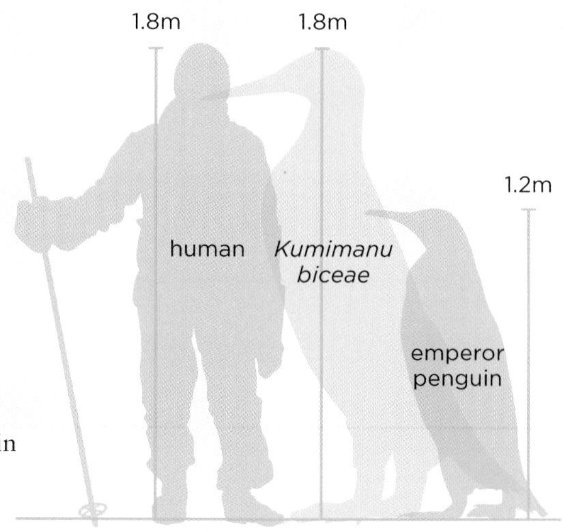

► *Kumimanu biceae*, the largest extinct penguin known from the fossil record, compared with the extant emperor penguin (*Aptenodytes forsteri*) and a person.

▶ A grey-headed albatross (*Thalassarche chrysostoma*) soaring over the ocean waves.

as tall as a human (1.8 m) and weighed about 60 kg. Extinct elephant birds (*Aepyornis* spp.) from Madagascar, the last of which died out only about 300 years ago, were probably the largest birds ever, exceeding 3 m in height and weighing 500 kg.

Modern birds are equally diverse in terms of body shape, with highly variable wing, bill, leg and feet structures. Wings are generally classified into four main types, based on length, breadth and pointedness. Many forest- and scrub-dwelling passerines have elliptical wings and flapping flight, allowing fast but short flights, rapid take-off and high manoeuvrability for evading predators. Seabirds such as albatrosses and shearwaters have long, narrow wings suitable for active soaring without flapping, allowing them to use the power of wind currents above the ocean waves. Eagles, vultures and pelicans have large, long, broad wings, with primary feathers spread out at the tips, forming slots, and are adapted for passive soaring on thermals above land. Finally, high-speed wings are medium length and narrow, facilitating rapid flapping and gliding for more sustained, fast flight, as seen in swallows, falcons, shorebirds and ducks.

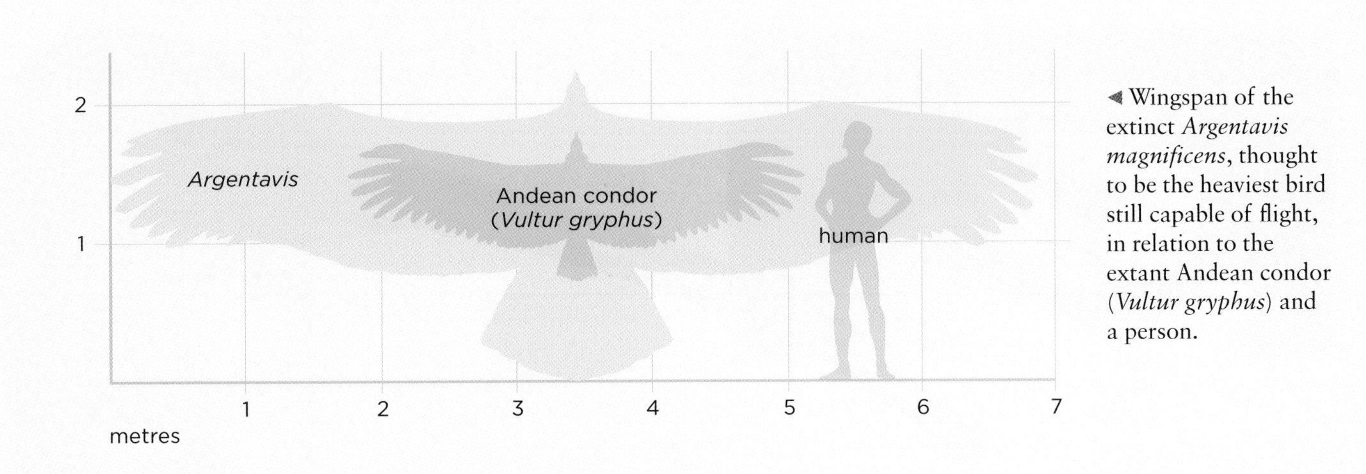

◀ Wingspan of the extinct *Argentavis magnificens*, thought to be the heaviest bird still capable of flight, in relation to the extant Andean condor (*Vultur gryphus*) and a person.

Bills, beaks and feet

Bird bills (or beaks) can be greatly elongated, as in shorebirds and hummingbirds, or flattened and broad (as in shoveler ducks and spoonbills). They can be curved downward (as in curlews), upward (as in avocets), sideways (uniquely and always to the right, as in the New Zealand wrybill, *Anarhynchus frontalis*) and even crossed (as in the aptly named crossbills). The beak of the South American sword-billed hummingbird (*Ensifera ensifera*) is actually longer than the rest of its body. Typically, the upper and lower parts of the bill (the mandibles) are the same size, but in skimmers (*Rynchops* spp.) the lower mandible is longer and enlarged to allow the birds to scoop up prey from just below the water surface. Flamingos also have a larger, stronger lower bill and a smaller, markedly bent upper bill; they feed with their head upside down, using their bill to filter prey from the water. Most of these bill differences relate to foraging, as they allow birds to capture and handle a wide variety of different prey or food items.

Internally, many bird bills contain large numbers of sensory cells (electroreceptors and mechanoreceptors), which also aid in the detection of prey. However, bills are not just for feeding – they are also used for preening, fighting, nest-building, feeding young and even manipulating tools. Bills can be brightly coloured, especially in males during breeding, and are used in courtship displays to attract females. In addition, they are highly vascularised (i.e. they have a good blood supply), and recent work shows that they can play a key role in heat exchange and regulation of body temperature in some species.

◀ The keel-billed toucan (*Ramphastos sulfuratus*) uses its colourful bill for display but also for thermoregulation.

Pied avocet (*Recurvirostra avosetta*)

Dunlin (*Calidris alpina*)

Wrybill (*Anarhynchus frontalis*)

◀▼ Variety in the shape of shorebird beaks. The curlew uses its long, downcurved bill to extract worms and other invertebrates from deep in the mud of saltmarshes.

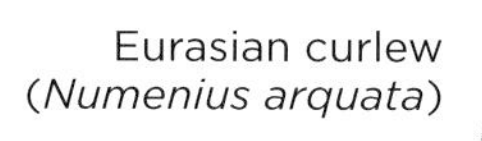

Eurasian curlew (*Numenius arquata*)

Birds are digitigrade animals, meaning that they walk on their toes rather than using the entire foot. However, the anatomy of bird legs and feet is diverse, again representing adaptations for multiple functions beyond simple movement. Many birds (e.g. storks, cranes and waders) have elongated legs, allowing them to wade through relatively deep water in estuaries and wetlands to catch prey. Conversely, penguins have short feet that are positioned well back on the body so that they can toboggan along the ice, pushing with their feet. Bird feet are covered with heavily scaled skin and typically have four toes with claws at the end of each. Woodpeckers have two toes pointing forward and two backward, allowing them to grip bark to climb trees. Passerines have three toes pointing forward and one back, allowing them to grip and perch on branches – even when asleep. Birds of prey (e.g. ospreys, sea eagles and owls) have powerful feet with sharp claws or talons for catching, killing and carrying prey, and many waterbirds (e.g. ducks, geese, swans and seabirds) have webs of skin between their toes so that they can use their feet to propel themselves through the water. Elongated toes or lobbed toes – as seen in jacanas, coots and moorhens – distribute a bird's weight, enabling it to walk on lily pads and other aquatic vegetation.

Variation in feet and the arrangement of toes in birds

▲ (Left) Arctic terns (*Sterna paradisaea*) fly more than 80,000 km a year between the Arctic and Antarctic. (Right) The greater roadrunner (*Geococcyx californianus*) has the fastest running speed for a bird that can also fly.

Diversity of movement

Wings, and the feathers they support, are typically used for flapping or soaring flight. They have allowed birds to master the air and range widely over large areas of the Earth, something that is impossible for a terrestrial mammal that relies on walking. Some bird movements are truly phenomenal. Arctic terns (*Sterna paradisaea*) weigh less than 125 g yet fly more than 80,000 km each year between their high Arctic breeding grounds and wintering areas in the Southern Ocean, sometimes covering more than 500 km a day. Common swifts (*Apus apus*) have been recorded spending ten months of the year in the air without landing – even sleeping and mating on the wing. Bar-headed geese (*Anser indicus*) and cranes are renowned for migrating over the Himalayas, although recent studies show that they sensibly follow low passes and valleys. Having evolved flight, numerous taxa have then given up on it, most notably in the ratites (e.g. ostriches, rheas and emu, *Dromaius novaehollandiae*) and penguins. Loss of flight also represents a major evolutionary transition associated with island species: more than 1,000 independent lineages of island birds have lost the power of flight, including rails, parrots, pigeons, owls, waterfowl and passerines. This is most common on islands with few or no natural predators, and many of these flightless 'endemics' have become extinct following the introduction of invasive predators such as domestic cats, weasels and pigs.

◀ Emperor penguins (*Aptenodytes forsteri*) are supremely adapted for diving deep underwater, reaching depths of 560 m.

Despite the clear advantages of flight, most birds routinely use terrestrial locomotion, hopping or walking to move around on the ground or among the branches of trees and shrubs. Although it is still capable of flight, the aptly named greater roadrunner (*Geococcyx californianus*) from the deserts of the southwest USA can run at up to 30 km/h – likely the fastest running speed for a flying bird. However, the flightless ratites hold the avian running record, with common ostriches sprinting at speeds of up to 72 km/h to escape predators (in comparison, human sprinters can reach speeds of only 40 km/h).

A wide range of aquatic birds swim on the water surface or dive beneath it to catch aquatic insects, fish, squid or a variety of other prey. Most have streamlined bodies to decrease water resistance while swimming, as well as highly waterproof feathers, and they propel themselves through the water with their feet (ducks, grebe) or their wings (penguins, auks). Most diving birds exploit only relatively shallow water, especially those that dive from flight, such as gannets. However, penguins are capable of frequent, prolonged deep dives using wings that have been modified into stiff flippers. Emperor penguins can reach depths of 560 m on dives lasting 22 minutes, taking the bird into total darkness.

Natural and sexual selection

Why do avocets have upturned bills? Why do spoonbills have spoon-shaped bills? Why do birds-of-paradise have such dazzling plumage and extreme display behaviours? All of the marvellously diverse features and behaviours showcased in this book are examples of adaptations - traits that increase the chances of individuals having higher reproductive or survival rates. Adaptations increase the 'fitness' of individuals, and species, to their environment, allowing them to contribute more copies of their genes to the next generation via their offspring. They arise through two main processes: natural selection and sexual selection.

Natural selection

Natural selection is the evolutionary process whereby some individuals in a population with favourable traits (e.g. longer bills) are more successful at reproduction or have better survival – Darwin's 'survival of the fittest'. These individuals have higher fitness through higher lifetime reproductive success, and this leads to these traits increasing over evolutionary time, from generation to generation. For example, as ratites evolved toward flightlessness they would initially have been increasingly at risk from predators. But those individuals that could run slightly faster than others of their kind were more likely to evade predators and survive to breed. If running speed has a genetic basis, then this trait would be inherited by the offspring of the fastest birds and, over time, running speed would increase in the population. Any trait – morphological, behavioural or physiological – can evolve if it varies among individuals, it is heritable (i.e. it can be transmitted from parents to offspring due to underlying genetic variation) and if the expression of the trait influences fitness, so that the trait is under selection.

Darwin's finches, studied by evolutionary biologists Peter and Rosemary Grant since the 1970s on the Galapagos Islands, are a classic case of natural selection in the wild. As just one example, a severe drought in 1976–1977 led to more than 85 per cent of medium ground finches (*Geospiza fortis*) dying of starvation. Surviving finches had larger bills than those that died – most likely because, during the drought, the few remaining seeds were larger and harder, and larger bills gave these birds a selective advantage. Following the drought, the surviving larger-billed birds mated and the increased bill size was inherited by subsequent generations.

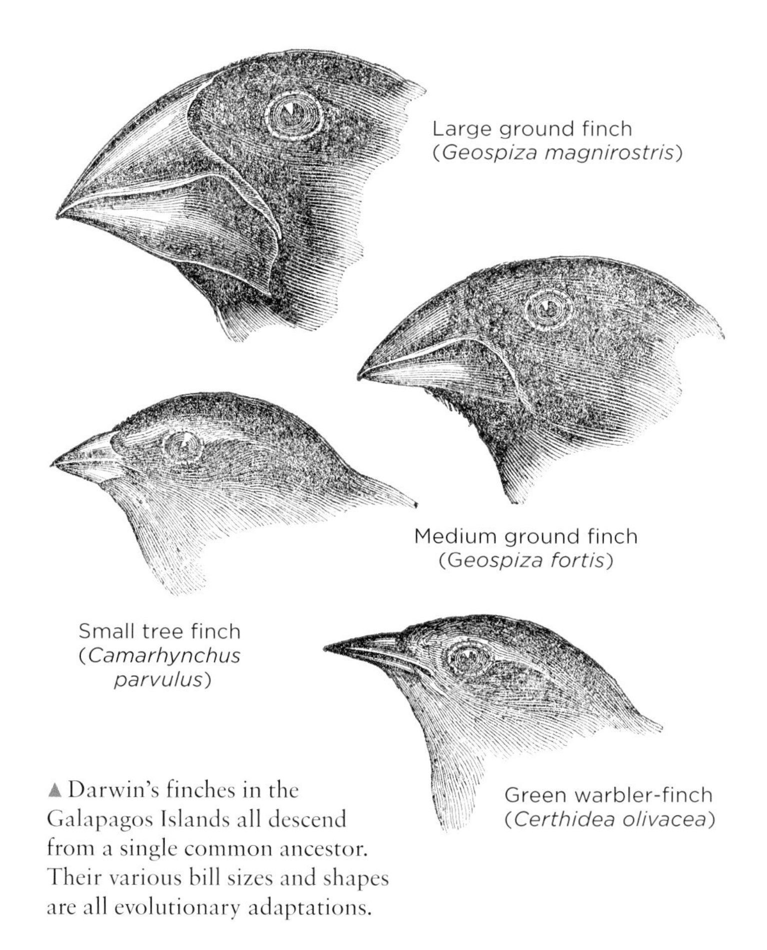

▲ Darwin's finches in the Galapagos Islands all descend from a single common ancestor. Their various bill sizes and shapes are all evolutionary adaptations.

Long-term studies of birds, such as those on great tits (*Parus major*) and Eurasian blue tits (*Cyanistes caeruleus*) dating back to the 1930s in the Hoge Veluwe in the Netherlands and Wytham Wood in the UK, provide evidence that numerous traits are related to fitness in birds, they are heritable and they respond to selection. Importantly, fitness is highly skewed – in other words, there is large variation in how many offspring individual birds produce. In fact, in most species many individuals produce no offspring at all in their lifetime, and 5–25 per cent of individuals in a population produce more than 80–90 per cent of the offspring recruited to the next generation. Fitter individuals (those with higher lifetime reproductive success) tend to breed earlier than other, less fit individuals, they lay more eggs and they have a longer lifespan – all potential components of overall 'individual quality'. In recent decades, climate change has become a key selective force acting on bird populations: warmer springs have led to earlier laying, and there have been shifts in migration timing and also in range, with more southerly breeding species expanding their distribution poleward. On the Antarctic Peninsula, the northern edge of the breeding range of ice-dependent Adelie penguins (*Pygoscelis adeliae*) is shifting further south, as sea-ice distribution changes, with concomitant increases in less ice-dependent, more subantarctic Gentoo penguins (*P. papua*).

▼ Effects of a projected 2°C increase in global temperature on the breeding and wintering distribution of the common loon or great northern diver (*Gavia immer*).

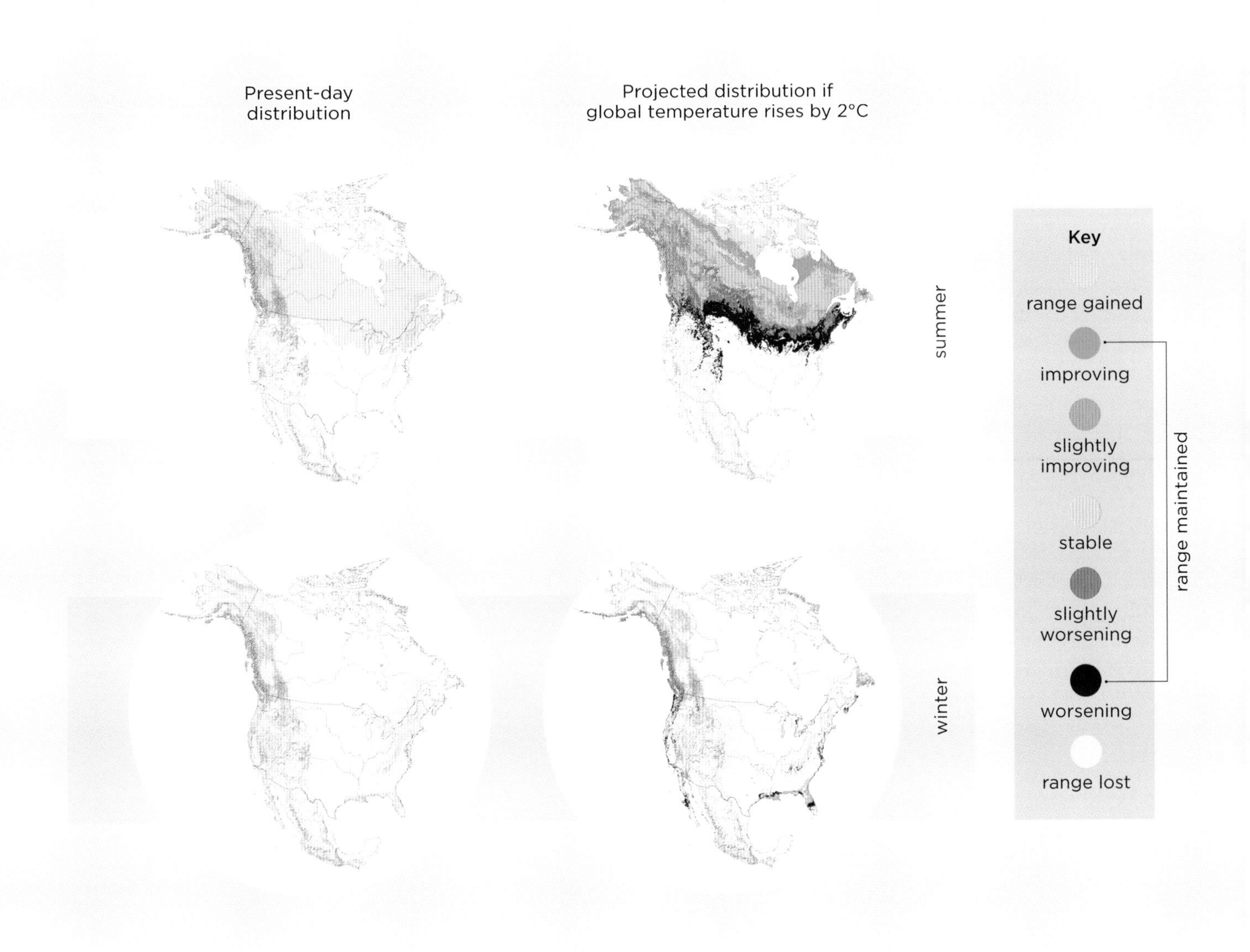

◄ Wilson's bird-of-paradise (*Diphyllodes respublica*) has ornate plumage as well as turquoise patches of blue skin, which it shows off in an elaborate courtship dance.

► The iridescent plumage and eyespots on the train or tail fan of the male Indian peafowl (*Pavo cristatus*) is an extreme example of a sexually selected trait.

Sexual selection

Sexual selection is a specific form of selection driven by access to mating opportunities. In most animals there is 'sexual conflict' between males and females when it comes to reproduction. Male reproductive success is limited by access to females: a male can increase his success by finding and fertilising as many different females as possible. Female reproductive success is limited by resources – in part because females generally incur greater costs of reproduction. Females need to mate with a only single male, but he should be a 'good-quality' male. So for females, choosing the right mate (mate choice) is important. Thus, males compete for females and females choose (good-quality) males. This 'sexual selection' gives rise to exaggerated sexually selected characters or traits when these traits increase the ability of individual males to compete for, and fertilise, more females, or if they increase the attractiveness of the male to females. There is abundant evidence that females prefer males with the most elaborate, complex songs, plumage traits or courtship displays. If females prefer males with longer tails, as is the case in the African long-tailed widowbird (*Euplectes progne*), then males with that feature will mate most often, and females choosing long-tailed males will produce more 'sexy' sons that have the genes for long tails inherited from their fathers. These sons with longer tails will then also be preferred by females and will mate more. As a consequence, tails will become longer and longer, or more and more elaborate – a process known as runaway selection.

Sexual selection involves traits solely concerned with increasing male mating success (rather than traits that contribute directly to survival via natural selection). Male–male competition can involve direct aggression (fighting) but is often ritualised through elaborate song or plumage displays, such as the large, colourful train of male Indian peafowls. This is why, in many species, males have more colourful plumage than females. Scientists refer to sexually selected traits as 'honest' signals: they can be used reliably by females during mate choice to assess male quality. Why is this? One idea is that males with elaborate traits have 'good genes', not just for the sexually selected traits themselves but for other traits, such as immune function. Alternatively, if elaborate traits are costly to develop and maintain, then only high-quality males can support the costs of these traits (this is known as the handicap hypothesis). Such high-quality males might be more likely to provide resources to females, they might be healthier or they might make better parents. If sexually selected traits are 'costly', this also prevents poor-quality males from cheating, hence the signal is 'honest'.

The cost of maintaining sexually selected traits also suggests why runaway selection might slow down, or even stop. As traits become more and more costly, the benefits in terms of reproductive success might be outweighed by negative effects of the traits on survival (natural selection). For example, birds with ever more elaborate plumage for courtship might have compromised flight ability and increased risk of predation.

► Many species have sexually dimorphic plumage, as in this king quail (*Excalfactoria chinensis*), where males have brighter, more colourful plumage and females duller, browner plumage.

Annual cycles

As with most other animals, birds divide their year between a breeding and non-breeding state, with breeding occurring during a limited period when conditions for rearing chicks are optimal. Some birds live in the same area year-round or make just small movements (residents), while others move between often distant breeding and wintering areas (migrants). All birds need to replace their feathers or plumage each year, so periods of moult occur, commonly twice a year, before and after breeding. Interactions between different phases of the annual cycle are common, generating connectivity and carry-over effects between different life-history stages.

Common tern (*Sterna hirundo*)

Reproduction

An individual's overall fitness relies on successful integration of multiple life-history stages through the complete annual cycle, including non-breeding (overwintering), breeding, moulting and migrating. The breeding season comprises phases of courtship, pair formation, egg-laying, incubation and rearing of chicks. In most species breeding is restricted to a limited part of the year, typically two to three months during late spring or early summer, when environmental conditions are optimal for successful reproduction. In less seasonal, tropical environments, some species can breed year-round, although individual birds will still alternate breeding and non-breeding periods. Smaller species, with shorter breeding cycles, can be multiple-brooded, producing up to five clutches in a single breeding season, although many small birds are single-brooded. While larger species tend to make a single breeding attempt each year, many birds will lay replacement clutches if there is early breeding failure.

▼ A bird's annual cycle comprises several life-history stages (reproduction, migration), with linkages or carry-over effects between the stages.

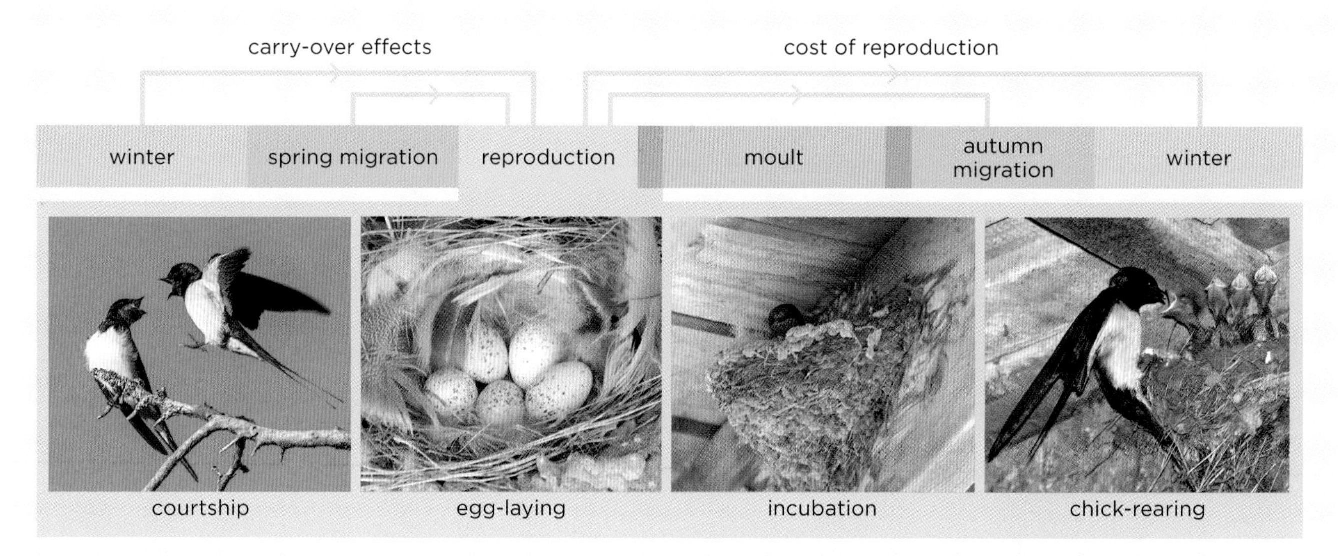

King penguins (*Aptenodytes patagonicus*) and wandering albatrosses take almost 12 months to rear their single chick, so they commonly breed in two out of every three years, or every other year (biennially).

Annual migration

Resident species live in the same area year-round, wintering and breeding in the same location, or make just small movements. In contrast, migratory species make return trips between often distant breeding and wintering (non-breeding) areas, with spring and autumn migrations linking non-breeding and breeding stages of the annual cycle. In the northern hemisphere, Arctic and boreal breeding species typically migrate south, overwintering in Africa, Central and South America, and south Asia. In contrast, few species breeding in the southern hemisphere migrate between continents and only a very few cross the equator. Only a few species move routinely between high-latitude breeding areas in the northern hemisphere and wintering areas in the southern hemisphere, or vice versa. Arctic terns are an exception, migrating more than 19,000 km between the high Arctic and Antarctic, and in the process experience two summers each year and likely more daylight than any other bird. Bar-tailed godwits (*Limosa lapponica*) make a single non-stop, eight-day flight across the Pacific Ocean from Alaska to New Zealand, covering a distance of more than 11,000 km.

▲▲ Adult king penguins (*Aptenodytes patagonicus*) incubating eggs while fully grown chicks from the previous breeding attempt (brown) still attend the colony.

▲ Bar-tailed godwits (*Limosa lapponica*) and red knots (*Calidris canutus*) at a stopover site in New Zealand during their transoceanic migration.

Partial migration and dispersal

Birds display many other forms of migratory movement. Partial migration occurs when a breeding population contains both migratory and resident individuals. In this case, some individuals migrate after breeding, whereas others remain on the same breeding area year-round. More than a third of birds that breed in Europe have been classified as partial migrants, and many familiar North American birds – including blue jays (*Cyanocitta cristata*) and American robins (*Turdus migratorius*) – as well as Amazonian birds such as tropical kingbirds (*Tyrannus melancholicus*) are partial migrants, as are ducks, shorebirds and *Accipiter* hawks. Partial migration can include altitudinal migration, in which a short-distance migration occurs from lower-altitude non-breeding areas to higher altitudes and back. This is especially common in the tropics, as seen in white ruffed manakins (*Corapipo altera*), which breed at 400–900 m in wet montane forests in Costa Rica from April to June. Many individuals migrate downhill to lowland forests asynchronously during the second half of the year, and then migrate back uphill in February and March. Partial migration may represent an evolutionary transition between residency and complete migration, but the mechanisms driving individual differences in migratory tendency remain poorly studied and controversial.

Finally, natal dispersal represents another form of movement, whereby young birds move from their natal site, where they were hatched, to their subsequent adult breeding site. Breeding dispersal, which is defined as movement from one breeding site to another, is far less common. Generally, once adult birds have established a breeding site they show a high degree of site fidelity, returning to the same area – or even the same territory or actual nest site – in successive years, even if they migrate long distances between breeding seasons.

▼ The white-ruffed manakin (*Corapipo altera*) undergoes an altitudinal migration in the wet montane forests of Costa Rica.

Feather replacement

Given the importance of feathers for flight and the fact that they wear or get broken, it is essential that birds maintain feather quality. Most birds do this by replacing all or part of their plumage every year through moulting. Birds display extensive variation in moult cycles, which can often be quite complicated (along with the terminology to describe it!). Many young birds first develop a juvenile or subadult plumage that is very different from the adult 'definitive' plumage, and attain full adult plumage over one to several annual moults in their first few years of life. Albatrosses can take seven or eight years to obtain their definitive plumage.

Once birds reach adulthood, they typically have annual cycles of moult that replace major parts of their plumage, although these vary among species in their frequency and timing, and in which feathers are replaced. Most adult birds moult twice a year. A prenuptial (or pre-alternate) moult is often a partial moult and occurs before breeding, with birds developing nuptial or breeding (alternate) plumage in spring and summer. A second, postnuptial (or pre-basic) moult is a complete moult and occurs after breeding, resulting in winter or 'basic' plumage over autumn and winter. However, many species of birds have no distinct breeding plumage, appearing the same year-round.

▲ A male northern red bishop (*Euplectes franciscanus*) moulting from its bright breeding plumage into its duller winter, or 'basic', plumage.

▲► A silver gull (*Chroicocephalus novaehollandiae*) in adult breeding plumage (above) and subadult juvenile plumage (right).

◀ Female American yellow warblers (*Setophaga petechia*) appear to start yolk formation a few days before arrival on their breeding grounds, while still on migration.

▼ The quality of wintering habitats used by American redstarts (*Setophaga ruticilla*) can have 'carry-over effects' on pre-migratory condition, and timing of arrival on the breeding grounds.

Carry-over effects

It might be expected that the energetic costs of breeding, moulting and overwintering would lead to minimisation of the overlap between the successive stages of the annual cycle. In contrast, we now know that life-history stages do overlap in many birds. In some species, reproduction (gonadal development) is initiated while the birds are still completing their spring migration. Male garden warblers (*Sylvia borin*), for example, have enlarged testes during their spring migration, and female macaroni penguins (*Eudyptes chrysolophus*) and perhaps also females of some Neotropical migrant songbirds initiate yolk (egg) formation during their return to their breeding sites. A lengthy period of post-fledging parental care is common in birds, and can often be longer than the nestling period. This generates overlap among breeding attempts, where individuals are still feeding chicks in their first brood while preparing for their second breeding attempt (e.g. forming eggs in females). In addition, many birds start moulting before the end of their breeding season while still caring for their chicks. In pied flycatchers (*Ficedula hypoleuca*) in Sweden, 67 per cent of males and 41 per cent of females start moulting before their last chicks fledge, and some white-crowned sparrows (*Zonotrichia leucophrys*) and giant petrels (*Macronectes* spp.) moult during incubation. Individuals that start moulting while breeding can have reduced chick feeding rates and produce fewer, lower-quality chicks, although some studies have found no negative effects of overlap.

Carry-over effects are also common between successive life-history stages, even if these don't overlap directly. Habitat quality, or a bird's activity, on the wintering grounds can have important carry-over effects that influence the subsequent breeding period. For example, American redstarts (*Setophaga ruticilla*) that used high-quality winter habitats have higher pre-migration body condition, earlier timing of departure and earlier arrival on the breeding grounds, all of which are associated with higher breeding productivity. In migratory passerines and geese, body condition (fat stores) prior to migration or at staging areas during migration are positively correlated with timing of egg-laying and reproductive success.

Connectivity

Early studies following banded or ringed birds were invaluable in providing information on linkages between breeding and wintering areas in migratory birds. For example, studies of black-tailed godwits (*Limosa limosa islandica*) revealed 'seasonal matching', whereby individuals occupying higher-quality breeding sites in Iceland also use higher-quality winter sites in western Europe, with important consequences for fitness. Recent technological advances in wildlife tracking (e.g. satellite-linked ground positioning systems, or GPS, and landscape-scale arrays of automated radiotelemetry stations such as Motus) have greatly increased our ability to follow individual birds year-round and identify linkages, or connectivity, between breeding, moulting and wintering locations. Since 2012, detection data from the Motus network has provided detailed data on migration routes tracking some 20,000 individuals from more than 190 different species. Geolocators are low-cost devices that determine a bird's position based on recorded time of sunrise and sunset (day length), which varies systematically with latitude. Studies using this technology are helping identify previously unknown migration routes and wintering locations for species of conservation concern.

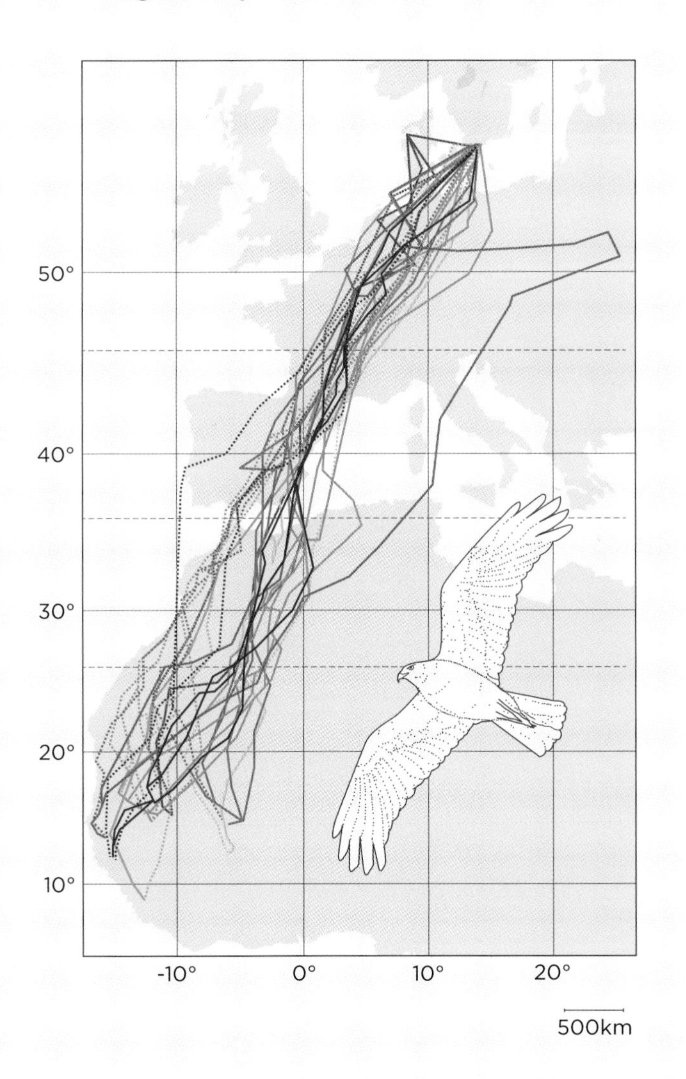

◄ Migration routes of autumn (solid lines) and spring (dotted lines) journeys from satellite tracking of seven adult western marsh harriers (*Circus aeruginosus*).

▲ A northern gannet (*Morus bassanus*) in flight with a tracking device fixed to its tail feathers.

▼▼ Red knots (*Calidris canutus*) arriving in autumn on their wintering grounds in eastern England.

Life-history variation

The lives of birds comprise a number of life-history traits that describe how the lives are organised – for example, how quickly offspring grow, at what age they first reproduce and how long they live. Life histories are commonly arranged along a fast–slow continuum: species with high rates of reproduction have short lifespans, and vice versa. This inverse relationship is an example of a trade-off. Trade-offs can also occur between different physiological systems, reproductive traits or phases of the annual cycle.

Life-history traits

Reproduction and survival are directly influenced by life-history traits, and birds have evolved a range of different ways to link these. The main life-history traits include offspring size (egg size or size of chicks at hatching), duration of the nestling period and growth rates, age at reproductive maturity, reproductive rates, age-specific patterns of survival and lifespan. Many of the traits correlate broadly with body mass (i.e. they 'scale' with mass), such that larger birds typically having longer times to independence, delayed sexual maturation, lower reproductive rates and longer lifespans compared to smaller birds.

Egg (or offspring) size varies markedly among species, from 1.4 kg in the common ostrich to less than 1 g in the bee hummingbird, and it also varies up to twofold among individuals within species. In general, egg size is a strong predictor of size of chicks at hatching. However, some precocial species produce very energy- or lipid-rich eggs and their chicks hatch at a more advanced state of developmental maturity (see page 309). Egg mass scales positively with body mass – larger birds lay absolutely larger eggs – but relative egg mass declines as a proportion of body mass as this increases. The number of eggs birds lay (their clutch size) varies from one in many seabirds, to ten or more in some small passerines, ducks and

◄▲ Eurasian skylarks (*Alauda arvensis*) have altricial chicks, born blind, sparsely feathered and dependent on their parents for food (left), but they fledge, or leave the nest, before they are capable of flying (above).

gamebirds. Life-history theory considers that the number and size of eggs or offspring produced in a single breeding event in different species has evolved to optimise fitness. So, in general, smaller, short-lived birds produce more, smaller eggs, and larger, longer-lived birds produce fewer, or even a single, larger egg.

Rates of chick growth, time to nest departure (fledging) and final size are all interrelated, but in complex, species-specific ways. Many birds reach somatic maturity (close to adult size), but not physiological maturity, at fledging, while others complete their growth as juveniles. However, some small passerines leave the nest at only eight days of age, before they are fully capable of flying or self-feeding. Rapid growth and/or early nest departure can be associated with high rates of nest predation, with chicks 'escaping' to a safer environment outside the nest. In contrast, chicks of larger species can spend months in the nest – wandering albatross chicks, for example, fledge at 11–12 months of age (see page 310).

▲▲ African jacana (*Actophilornis africanus*) have precocial chicks that can leave the nest as soon as they hatch and find their own food.

▲ A common tailorbird (*Orthotomus sutorius*) feeding altricial chicks at the nest.

Many small birds breed first in the spring or early summer following the year in which they were hatched, when they are around 12 months of age. In contrast, some penguins and albatrosses show long-deferred sexual maturation, breeding first at six to 12 years of age. Typically, females start breeding at a slightly younger age compared with males within species. Most birds breed annually once they reach sexual maturity, but some larger birds skip years of breeding – in common eider ducks (*Someteria mollissima*), up to 20 per cent of birds defer breeding in some years. Additionally, some smaller species can produce up to five clutches in a single year, whereas larger species will produce only a single clutch. Average lifespan varies from one year in many small songbirds, to 20–30 years in large seabirds, and juvenile survival is always lower than adult survival. There is accumulating evidence for senescence (decreases in reproductive performance with age) in some long-lived birds.

Trade-offs

The various life-history traits described above are linked by trade-offs: negative relationships between pairs of traits. Major trade-offs include that between current reproductive investment and future fecundity (fertility) and/or survival (known as the cost of reproduction), and that between the number of eggs or offspring and their size. Trade-offs must have a physiological basis, although this remains poorly understood. The standard explanation for trade-offs invokes competition for finite resources: if animals have limited resources, investment in one trait has to be associated with decreased investment in the related trait. For example, increased investment in producing offspring might increase reproductive success, but it will do so only at a cost somewhere else (e.g. decreased survival). Similarly, theory predicts that species or individuals that invest in a large number of eggs must lay small eggs, whereas those that lay few eggs can invest in larger eggs. Although some researchers have suggested trade-offs are ubiquitous, they should, in fact, be apparent only when resources are limited – that is, when birds are food stressed. It is also difficult to detect trade-offs through simple correlations of traits, because some individuals have access to more resources than others, which generates positive (not negative) relationships. Trade-offs may also occur independently of resource allocation per se, due to direct 'conflict' between different physiological or hormonal regulatory systems, such as when a single hormone has both positive and negative effects on multiple physiological systems.

Trade-offs can also occur among different stages of the annual cycle, especially when these overlap. For example, many birds initiate their moult toward the end of the breeding season, before they fledge their last chicks and reproduction is complete. This can generate costs, either through reduced feeding rates of offspring or reduced quality of newly moulted feathers, affecting the adult's thermoregulation and survival in the subsequent winter. More intriguingly, some birds appear to face a trade-off between spring migration and reproduction when they initiate gonadal development during migration and before they arrive on their breeding grounds – although any costs here have yet to be identified. Finally, trade-offs can occur more broadly, such as males with elaborate plumage used during courtship might incur a decrease in flight performance, which could lead to a greater risk of predation – a trade-off between investment in reproductive success and survival.

◄ A male American yellow warbler (*Setophaga petechia*) feeding his female partner at the nest.

▼ Sexually dimorphic plumage in female (left) and male (right) common eider (*Somateria mollissima*).

CHAPTER 2

Feathers, Beaks, Scales and Claws

They say that feathers make the bird. But as we learned in the previous chapter, feathers were present in a wider group of dinosaurs. Furthermore, avian body coverings, or integuments, are complex and comprise much more than just branched feathers. In birds, the integumentary system includes the skin and accessory structures, these ranging from scales and claw coverings, to complexly coloured feathers. All, however, are made of the protein beta-keratin, which is also seen in the birds' reptile relatives. Feathers function in making up the flexible, light aerofoils in the wings of birds, but they also serve as sensory structures, act as insulation for both chicks and adults, are used to attract mates and provide camouflage.

Birds are arguably the most brilliantly coloured terrestrial vertebrates, with their feathers and other skin structures containing many novel pigments. But these remarkable animals also create structural colours with tiny light-refracting surfaces, producing blues, glossy blacks and the bright, iridescent hues that light up our everyday environments. Given their many functions, feathers must be maintained through preening and replaced via systematic shedding and regrowth during the moulting period.

◀ Scarlet macaw (*Ara macao*) feathers.

An array of strange structures

The structures in bird skin are much more diverse than they appear at first glance. Aside from feathers, the skin itself, the scales on the top and bottom of the feet, claw coverings on the wings and toes, bristle structures and beaks all vary across birds depending on their ecology. Together, these features are called integumentary structures.

Beta-keratin

In birds, as in their reptile relatives, integumentary structures are made of a protein called beta-keratin. By contrast, integumentary structures in mammals, including hair and fingernails, are made of alpha-keratin. Alpha- and beta-keratin are very different in composition. If you put a feather under a microscope at the highest magnification it will look smooth, whereas mammalian hair and fingernails have a scaly structure. Studies of the form of beta-keratin present in foot scales in birds has shown it is more closely related to that in feathers than to the beta-keratin in lizard or crocodile scales. Other recent work has identified the genes involved in switching a scaly bird leg to a feathered one. Changes in the expression of these genes can even result in elongate feathers in artificially selected domestic fowl and pigeons. In the fossil record, a number of dinosaur species closely related to, and including, some early flyers appear to lack scales on top of the feet altogether, and instead had completely feathered feet.

◄ Scanning electron micrograph image of a feral pigeon (*Columba livia*) feather, showing the central rachis, branching barbs and tiny feather barbules, made up of beta-keratin.

◄ Hooklets at the ends of feather barbules allow neighbouring barbs to interlock, forming a watertight aerofoil.

Barn swallow
(*Hirundo rustica*)

Greater vasa parrot
(*Coracopsis vasa*)

Fabulous feathers

What we typically call a feather is a hierarchically branched beta-keratin structure with a central midline rachis from which barbs branch. Angling off these barbs are smaller barbules. These barbules may have even tinier hooklets, which reach those on the adjacent barb to form a 'zip' that locks neighbouring feather barbs tightly together. These feathers are among the most complex integumentary or skin structures known in vertebrate animals. In cross section, rachises and barbs generally show a central hollow, air-filled space, or vacuole. In barbs, feather colour-producing mechanisms lie around this vacuole. Feathers can also have an after-feather, a distinct pinnate feather that branches off the base of the rachis. Many after-feathers are loose and downy in appearance, providing insulation and supporting a stable body contour for a bird during flight.

▲ Colourful beak covering, or rhamphotheca, of an Atlantic puffin (*Fratercula arctica*) during the breeding season.

▲ Serrated beak edges in a female common merganser (*Mergus merganser*), used to grasp fish tightly.

Beak and claw coverings

All birds also possess beta-keratin beak and claw coverings. The skeletal beak is covered by layers of keratin and may further be ridged (in filter-feeding species), spiked, toothed or serrated. This horny keratin covering of the bill is called the rhamphotheca, and in most birds it grows continuously. A strange exception is the puffins, which shed their colourful rhamphotheca after the breeding season. The shapes and colours of beak coverings are related to the distinct ecologies of birds, and they also function in sexual selection. In highly territorial hummingbird species, beaks are used as weapons in combat and may even be serrated.

► Keratinous spikes on the tongue of a king penguin (*Aptenodytes patagonicus*) are useful in capturing slippery prey.

Keratin claw coverings are present on the last bone of the toes, and are typically larger or longer than these bones. Comb-shaped serrations on the foot claws are apparently used in grooming in cormorants, and in nightjars, swifts and frogmouths. The second digit of the foot in cassowaries bears a straight, elongate claw sheath that is used defensively and in aggressive territorial combat. Tiny claws may also be present at the tip of the wing; these are vestigial remnants of the robust hand claws in extinct dinosaurs. Wing claws can develop in avian embryos but in many species are lost during early adulthood.

Bare skin, horns and helmets

Other skin structures in birds include keratin horns and bare skin around the eye, on the neck and on the feet. The coloured crop of diurnal raptors and the inflatable neck skin of sage grouse and frigatebirds are all puffed out by filling the oesophagus with air. The rooster's comb and turkey's wattle are long skin flaps whose red coloration comes primarily from an abundance of tiny blood vessels. The South American horned screamer (*Anhima cornuta*) even has a long keratin horn on its head. Head casques or 'helmets' – found in cassowaries, hornbills and some fowl – are similarly sheathed in keratin. Even the tongue surface of birds is a modified skin structure covered in stiff keratin layers that can be shaped into spikes or ridges, in stark contrast to the soft, fleshy and easily manipulated tongue of mammals. These tongue structures function in keeping a captured prey item in the mouth while it is

▲ The dramatic ornamental keratin spike on the head of a horned screamer (*Anhima cornuta*).

manoeuvred for swallowing or in filter-feeding. Penguins, for example, have long tongue spikes, particularly at the base of the tongue.

Skin on the feet and legs of birds typically extends only to about the equivalent of the ankle. However, in some raptors, including the rough-legged buzzard (*Buteo lagopus*), the foot is covered in feathers. Alternatively, in wading birds or the cursorial ostrich, skin can cover most of the leg. This skin, on the top and bottom of a bird's foot, is typically covered in scales, but in a very few species such as the North American chimney swift (*Chaetura pelagica*), the top foot scales are absent and skin covers this part of the foot. Leg skin can be highly coloured, such as the structural blue colour seen in the Pacific blue-footed booby (*Sula nebouxii*), which functions in courtship.

▲ Male southern cassowary (*Casuarius casuarius*), showing off an array of bizarre keratin structures.

Feather types and distribution

The flight feathers of a bird comprise the remiges on its wings and rectrices of its tail, while the contour and covert feathers give a smooth shape to the body and wings, respectively. This section looks at the various types of feathers, their distribution on a bird's body and how they are deployed. The topography of a bird describes the feathers as they appear in different locations on the body. These feather tracts can be coloured differently or moulted at different times, and are separated by bare patches of skin.

Feather distribution and topography

Birds have variable numbers of feathers covering their bodies depending on the species. Some have only a few thousand (e.g. songbirds) but others, such as the Holarctic tundra swan (*Cygnus columbianus*), have more than 25,000 feathers. Feathers are not usually evenly distributed over the whole bird but are arranged in tracts (pterylae), with patches of bare skin between them, although the latter are absent in ostriches, emus and penguins. The absence of apteria was originally thought to be an ancient feature retained in these taxa, but the independent loss of pterylae is supported at least in penguins. Pterylae are important in moult and different tracts can be moulted independently of one another. Feather density on the body can also vary markedly, with the greatest number of feathers per square centimetre known in penguins, a characteristic thought to be related to streamlining the body in swimming and diving, and for insulation.

Even a casual observer will have noticed that different feathers occur on different parts of the wings and body of birds. These distinct areas have been given standardised names and together form a 'map' to the feathering of birds, commonly called feather topography. Some of the names refer to a general region (e.g. the throat or nape), but others refer to the type of feathers or structures that make up that region (e.g. primaries, secondaries and tertiaries or tertials).

Covert feathers

On the upper and lower surfaces of the wing, shorter feathers called coverts give the wing a cambered cross section – the classic tapered aeroplane-wing shape. On the top of the wing, a more elongate greater covert is paired with each flight feather, attaching near its base. Successively shorter covert feathers extend along the surface of the leading edge of the wing and cover the propatagium, a membrane that stretches between the avian equivalents of the shoulder and wrist. The number of rows of covert feathers is highly variable; songbirds have very short covert wing feathers in relatively few total rows compared with birds like loons (divers), penguins or ducks. A recent study showed that these differences in feathering across birds do not correspond strongly with either wing shape or described flight style, but instead are explained more by phylogeny, or species relationships. In birds that use their wings in diving, coverts may add further structural support in water.

◄ Tundra swan (*Cygnus columbianus*) showing off its feathers. Bird species differ in how many feathers cover their body, but some of the highest feather counts are known from swans.

Bird topography and feather detail

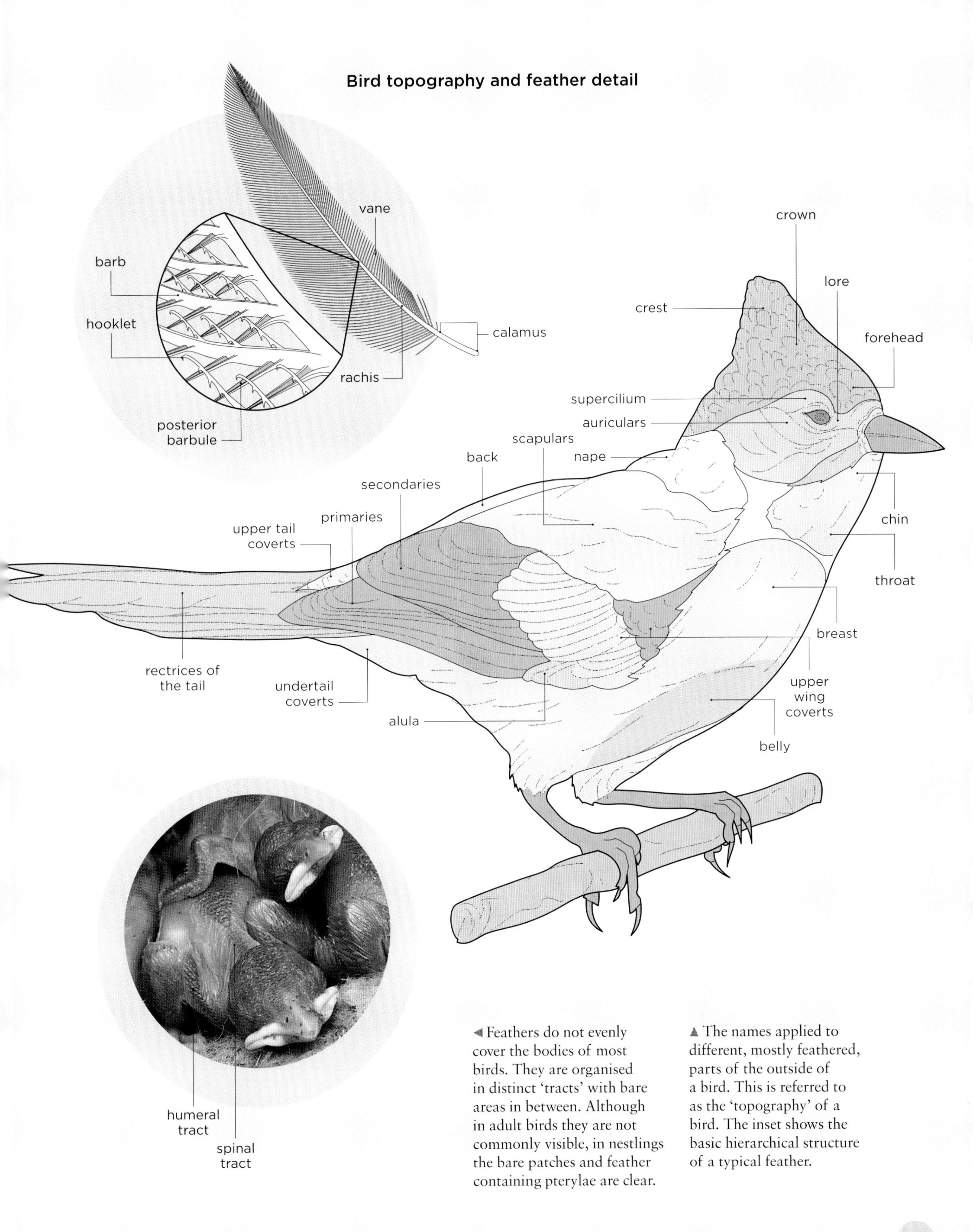

◀ Feathers do not evenly cover the bodies of most birds. They are organised in distinct 'tracts' with bare areas in between. Although in adult birds they are not commonly visible, in nestlings the bare patches and feather containing pterylae are clear.

▲ The names applied to different, mostly feathered, parts of the outside of a bird. This is referred to as the 'topography' of a bird. The inset shows the basic hierarchical structure of a typical feather.

Flight feathers

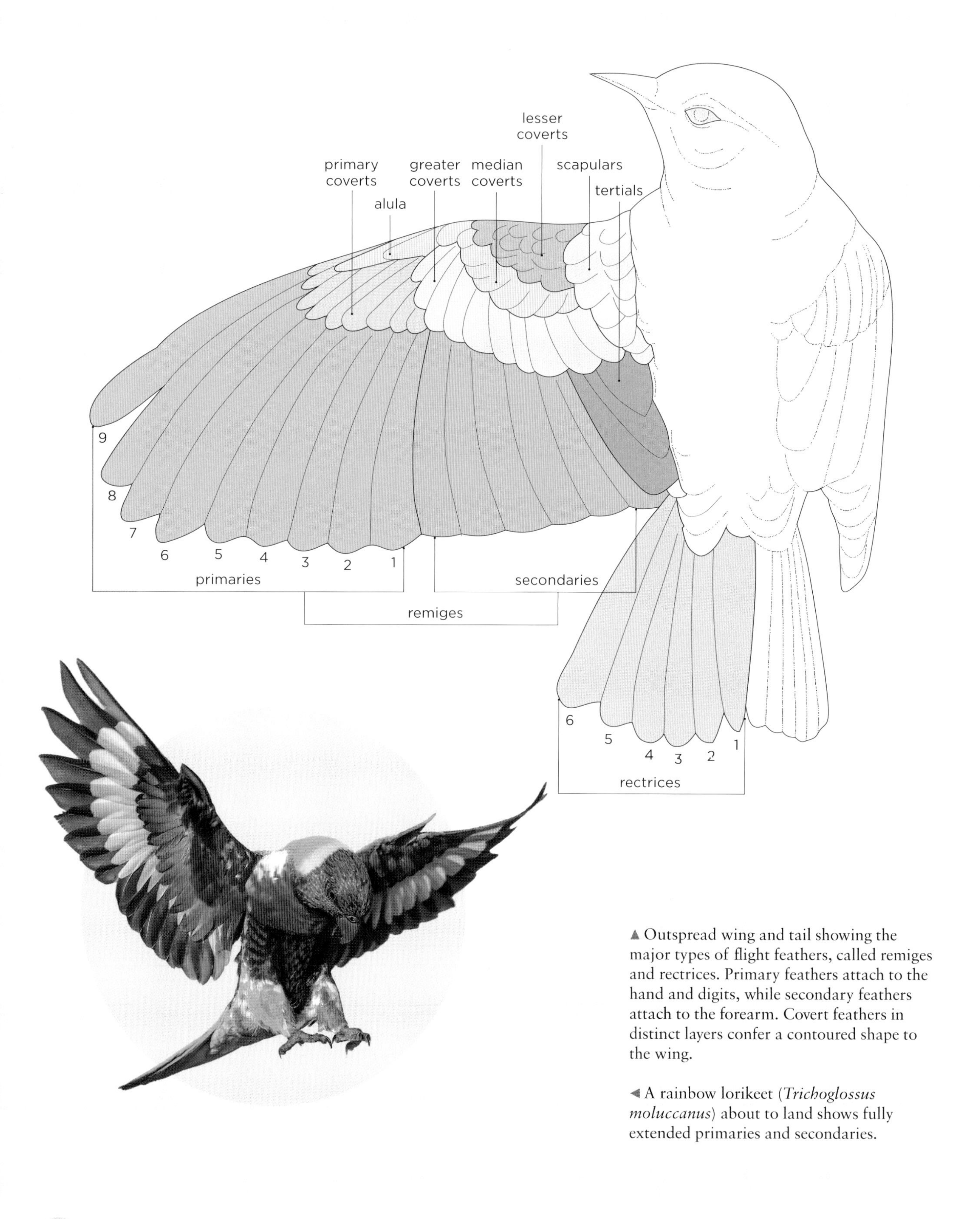

▲ Outspread wing and tail showing the major types of flight feathers, called remiges and rectrices. Primary feathers attach to the hand and digits, while secondary feathers attach to the forearm. Covert feathers in distinct layers confer a contoured shape to the wing.

◄ A rainbow lorikeet (*Trichoglossus moluccanus*) about to land shows fully extended primaries and secondaries.

◄ An African fish eagle (*Haliaeetus vocifer*) in flight shows the distinct separation of the primary feather tips referred to as slotting. Slotting in both gliding and flapping birds is thought to control airflow at the tip of the wing. Each asymmetrically vaned feather tip acts like an independent aerofoil, shaping vortexes of air and possibly improving flight efficiency.

Contour feathers

Contour feathers cover most of the bird's body and give it a streamlined profile. Like flight feathers, they have tiny ligaments attached to their bases to keep them in place and support the body outline, reducing drag. Water is 800 times more dense than air, and thus reducing drag is even more important for deep-diving species. Penguin contour feathers are highly flattened, with modified barbs and barbules that function in further streamlining the body during dives. The outer parts of these feathers have so-called closed or pennaceous portions, in which the hooklets of the barbules zip them together in a tight vane. At the base of these feathers there is usually a plumulaceous section, in which hooklets are absent and the barbs are free. Most of the major groups of birds can be identified by the shapes of tiny nodes on the barbules of this plumulaceous section.

Flight feathers

The elongate flight feathers of a bird's wing are called remiges, with those at the tip of the wing known as primaries or primary remiges. Most birds have 10–12 primaries, although a notable exception is a genetically related group of songbirds known as the nine-primaried oscines. The first primary feather is short in many species that have elliptical wing shapes, and more elongate than the inner primaries in species with high-speed wings. Flight feathers that extend from the ulna, or forearm, are called secondaries. The number of secondaries is more variable, ranging from six in hummingbirds to 30 in the North Pacific black-footed albatross (*Phoebastria nigripes*). Finally, some birds have a few additional elongate feathers at the joint between the forearm and humerus (upper arm bone); these are called tertiaries, tertials or innermost secondaries. Many birds also have short feathers extending from the leading edge of the wing, creating a slot that smoothes the flow of air over the top of the wing. These feathers, called alular feathers, are not part of the primaries but attach to the innermost digit of the hand. Ligaments connect most flight feathers to the bones, allowing wing shape to be constantly modified during flight.

The elongate feathers of the tail are called rectrices. Tail shapes in birds range from forks to broad fans, and may include shorter feathers on the outer part of the tail or feathers that are more elongate or approximately the same length across the whole series. Rectrices can be either asymmetrically or symmetrically vane, meaning that the rachis is either in the centre of the feather or slightly off centre. All of these differences in feathering relate to the functions of the rectrices in the aerodynamics of flight and in sexual signalling.

Specialised feathers

Specialised feathers in most birds include bristle-like forms as well as a remarkable variety of strange shapes for show or signalling. Fine filoplumes and semiplumes have very different sensory or insulation functions, while various tuft-like down feathers present in many newly hatched birds and adults also provide insulation. Powder down feathers are even more specialised, disintegrating to form a powder that helps keep the rest of the feathers in prime condition.

Aside from the standard feather types already discussed (see page 60), birds also have a variety of specialised feathers. These include bristles, which may have limited barbs at their base, and filoplumes, with limited barbs at their tips. Filoplumes are generally distributed among the other feathers of the wing and tail, while bristles tend to be located on the lore (the area around the eye or beak). Tiny, short feathers often ring the eye, and ostriches and some other birds even have modified feather bristles that look like our eyelashes. The central rachis of fluffy semiplumes is much longer than the individual barbs, and these feathers lack interlocking barbules throughout their length. They commonly underlie the body contour feathers, conferring added insulation and contributing to the streamlined body shape in flight.

Soft as down

Tuft-like downy feathers also serve an insulation function, both in growing chicks and adult birds. Simple down feathers (natal down) are commonly present in newly hatched birds, and are modified feathers that generally lack a central rachis and

▲ Body contour feather, with fluffy open barbs (the barbules do not interlock) at the base and a closed pennaceous tip.

► Close-up of a downy feather, showing the non-interlocking barbules that help to trap air in a bird's plumage.

► Downy feathers lack interlocking hooklets and provide both chicks and adults essential insulation.

►► The 'eyelashes' of this common ostrich (*Struthio camelus*) are, in fact, modified feathers.

interlocking hooklets. They may be absent or limited in altricial young such as songbirds or parrots, which hatch naked in a relatively undeveloped state and require parental feeding and extensive care. In other species, adults have down feathers underlying their body contour feathers; these are known as definitive down and are most common in ducks and their relatives.

The feather barbules of the bottom, 'downy' portion of contour feathers in dabbling ducks have expanded nodes that cause the barbs to form a tangled mass, trapping air and conferring improved insulation in water. The modified body contour feathers of great cormorants (*Phalacrocorax carbo*) have a wettable outer portion of the feather/plumage and a highly water-repellent inner portion. This unique structure allows a thin layer of air to be trapped for insulation, while allowing the birds to dive with minimal buoyancy. The lower breast feathers of the southern African double-banded sandgrouse (*Pterocles bicinctus*) are also wettable, but are used for transporting water to nestlings in their arid desert habitat.

POWDER DOWN

Powder down feathers are even more highly modified. These continually growing feathers are mixed in with the body contour feathers, and their barb structure disintegrates into a powder that is distributed over the skin and other feathers by the bird during grooming. The powder formed by the disintegrating feathers is thought to have antimicrobial properties, and to keep feathers waterproof and in prime condition. Most birds have a few of these feathers, but herons and their relatives have far more, especially on the breast.

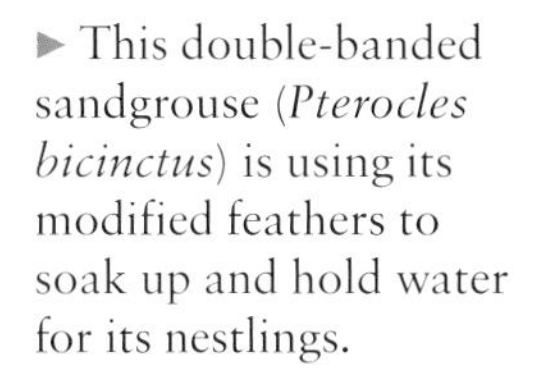

► This double-banded sandgrouse (*Pterocles bicinctus*) is using its modified feathers to soak up and hold water for its nestlings.

►► Powder down dust on a window after a collision.

Feather functions

Across the 10,000 or so species of living birds, feathers show extreme modifications for a wide variety of different functions. The most important are in generating lift and reducing drag during flight, and in signalling and visual communication. Tail feathers function in manoeuvrability and landing, but long tail feathers are generally aerodynamically costly and may instead help to attract choosy mates (see page 69). Contour feathers create a streamlined aerodynamic or hydrodynamic shape, reducing drag in air and water. Small colour patches in plumage generally camouflage a bird, while large contrasting patches will make it stand out.

Feathers and movement

Bird wings are dynamic aerofoils, changing their shape during the flight stroke. In this action, feathers move against one another on the upstroke when the wing is extended. In soaring and in parts of the flight stroke when the wing is fully extended, the whole wing surface is generating lift. In most species, thrust is generated during the downstroke, but the wings in penguins, puffins and hummingbirds create thrust on both the upstroke and downstroke. The primary feathers are responsible for generating the thrust,

▲ Air spaces between feather structures help water to bead up on the feather surface, keeping the bird dry.

▼ Close-up of the modified sound-dampening structures in the barbules at the edges of the wing feathers of a western barn owl (*Tyto alba*).

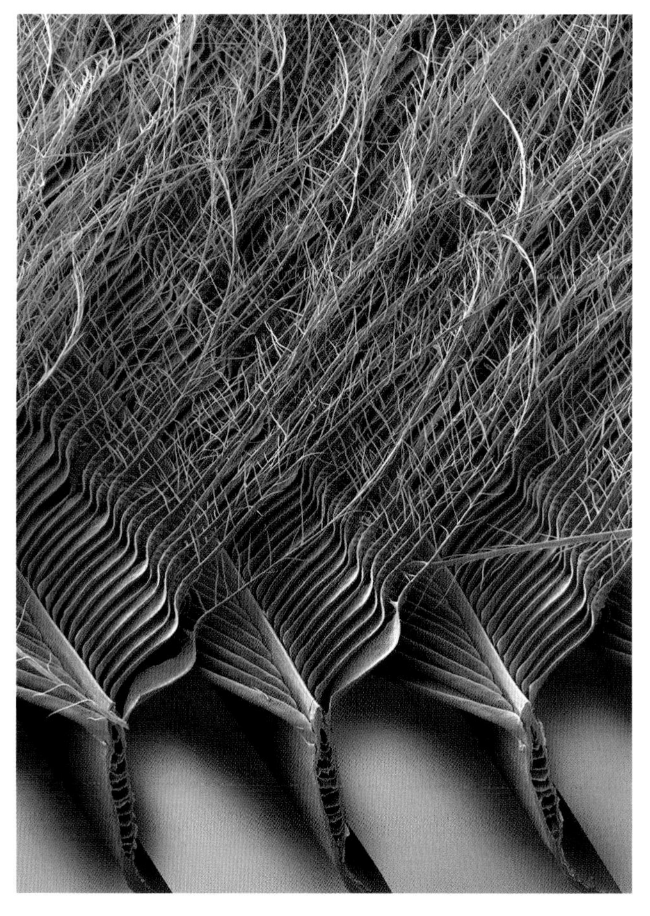

▲ The vanes on the sides of the primary feathers that face the wind in this mallard (*Anas platyrhynchos*) are reduced to create an aerodynamic shape like the wing of an aircraft.

moving the bird through the air during the flight stroke, while the secondary feathers form a surface that primarily generates lift, keeping the bird airborne. The overall organisation of flight and covert feathers on the wing creates its cambered shape.

In order to maintain both thrust and lift, birds need to minimise the volume of air passing through their wings and the formation of turbulent vortexes, which create drag. In efficient flight, air passes smoothly over and under the aerofoil surface, but not through it. The tips of primary feathers in many birds, especially species that soar, are tapered, and separate and bend upwards during flight. These 'slotted feather' wing tips are thought to act as a set of mini aerofoils or aeroplane wings, smoothing air passing over them and thus reducing aerodynamic drag. In waterfowl and, to a lesser degree, some owls and gulls, L-shaped projections from barbs close to the base of the flight feathers confer a shiny, waxy texture to the underside of these feathers, called the tegmen, which is thought to further reduce the passage of air through the wing in flight.

The wing primaries and secondaries, as well as some tail feathers, may have asymmetrical vanes. In these feathers, the rachis does not lie in the centre of the feather but off to one side, closer to the leading edge (i.e. the side facing the direction in which the bird is moving in flight), creating an aerodynamic shape like the camber of an aircraft wing. This feather asymmetry has been lost in some, but not all, secondarily flightless birds; it is still seen in some flightless rails. Further feather modifications affect other ways of moving. For example, the tail feathers of climbing birds (e.g. woodpeckers) and diving birds have a thicker rachis that functions, respectively, in propping the bird on vertical surfaces and as a stiff rudder. Barn owls (*Tyto alba*) have sound-dampening flight feathers that are important in an ecological context for sneaking up on unsuspecting prey. The modified barbs on the edges of these feathers create a toothed shape, altering air flow to minimise turbulence and dampen sound.

THE COLOUR SPECTRUM

Colour is incredibly variable across a bird's body, and research in this area has tested the relative importance of ecological factors and visual communication in determining where and how species produce different colours on their bodies. The light environment hypothesis predicts that colour variation among species is the result of differences in the direction and/or strength of selection among habitats with different light properties. For example, the understorey of a rainforest is rich in long-wavelength hues (oranges, reds), while the canopy is rich in ultraviolet (UV) and blue hues. If naturally selected ecological traits are most important in terms of colour, you may therefore expect camouflaged brown colours in birds that live in the canopy and green plumage colours in birds of the understorey. On the other hand, if social factors and sexual selection are more important, you would expect to see conspicuous UV/blue plumage colours in birds of the canopy, and yellow and red colours in species inhabiting the understorey. A comparative study published in 2004 by Doris Gomez and Marc Théry found support for the latter hypothesis: that social factors have been of key importance in the evolution of avian colour patterns.

◄ The greater bird-of-paradise (*Paradisaea apoda*) sports bright iridescent throat feathers and colour patterns, which it uses in its elaborate display behaviour.

Feathers for show

Many bird feathers are modified as ornaments, such as the flag-like primary feathers of the West African standard-winged nightjar (*Caprimulgus longipennis*), used in the species' showy display. Bright hummingbird feathers have highly modified feather barbules that, when interlocked, give the feather surface a grooved appearance that reflects sunlight directionally during flight displays. The primary feathers of cassowaries lack barbs and barbules, and are just shiny, stick-like oversized rachises that may be sexually selected. However, perhaps the most dramatic examples of sexually selected modified feathers are the tail-feather fans of Indian peacocks (*Pavo cristatus*) and the elaborate plumage the birds-of-paradise use in their captivating performances.

Feathers also function in sound-making used alongside showy displays. For example, both the wing bones and flight feathers of the Central and South American manakins are modified for this purpose, creating snapping and buzzing sounds. Similarly, the tail feathers of some hummingbirds and of Wilson's snipe (*Gallinago delicata*) produce loud, high-pitched sounds during the males' courtship dives. In many other birds, feathers are deployed in visual displays in concert with vocal production. In doves and pigeons, wing feathers also produce sounds, but for a different kind of acoustic function: they make a distinct 'wing whistle' during take-off that some researchers have proposed serves as an alarm call.

Colour patterns

Complex plumage patterns can be important for attracting a mate. Indeed, Richard Prum has argued that many traits in birds are simply 'beautiful' to the opposite sex and are not necessarily communicating information directly about the quality of the prospective mate and its territory. Other authors have shown that the way feathers develop affects the kinds of colour patterns that can be produced. This idea was explored by Trevor Price and colleagues at the University of Chicago, who noticed that the size of the head in a developing embryo was much larger than other body parts, giving more opportunity for novel, intricate feather patterns to form here as the bird develops.

Mechanisms of feather colouring

Birds colour their feathers using a combination of two general mechanisms: light absorption involving pigment-based colouring; and light scattering by structured tissue (structural coloration), including both iridescent and non-iridescent forms. Both kinds of coloration are also found in the skin, scales and eyes of many birds. Different groups of birds are characterised by the addition of new pigments and new ways of generating structural colours, and they may also combine coloration mechanisms and even apply 'cosmetic' colours.

Colourful pigments and structures

Birds can produce colours using pigments or by structural methods. The most common pigments are melanins and carotenoids. Melanin pigments generate a range of hues, from reddish brown in the breast feathers of American robins (*Turdus migratorius*), to yellow in the neck feathers of Southeast Asian comb-crested jacanas (*Irediparra gallinacea*) and black in North American ring-billed gulls (*Larus delawarensis*). Carotenoid pigments are responsible for the bright orange legs of mallards (*Anas platyrhynchos*), the yellow contour feathers of American goldfinches (*Spinus tristis*) and the red beaks of Australian zebra finches (*Taeniopygia guttata*).

In addition to pigment-based colours, birds can produce colours by structuring components of their integument in a way that selectively reflects some wavelengths of light. These 'structural colours' are responsible for some of the brightest colours in nature, including the stunning tail feathers of the Indian peafowl (*Pavo cristatus*), the turquoise throat feathers of the superb bird-of-paradise (*Lophorina superba*) from the rainforests of New Guinea, and the metallic, purplish-red throat colours of Anna's hummingbird (*Calypte anna*) from the western USA.

▲ Melanin pigments create the bright colour patterns in the Eurasian hoopoe (*Upupa epops*).

► Structural purple colour and carotenoid red colour combine in the same feathers in the breast plumage of a southern double-collared sunbird (*Cinnyris chalybeus*).

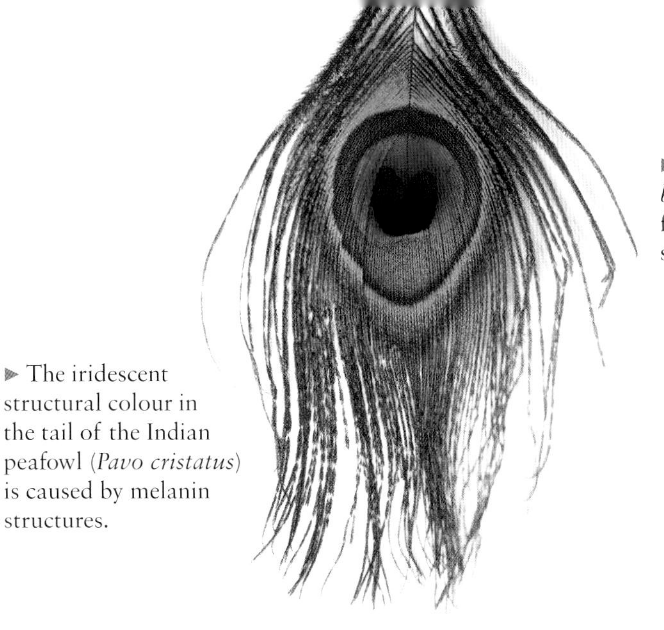

▶ The iridescent structural colour in the tail of the Indian peafowl (*Pavo cristatus*) is caused by melanin structures.

▶ Bearded vultures (*Gypaetus barbatus*) stain their neck feathers rusty orange using soil rich in iron oxide.

Colour combinations

Combining different colour-producing mechanisms gives birds even more possible colours. For example, melanin pigments can enhance the richness or saturation of blue structural colours, as in the back feathers of eastern bluebirds (*Sialia sialis*). The drab green colours of New World warblers are caused by a combination of one form of melanin, eumelanin, and yellow carotenoids, while the silver colour of the American anhinga (*Anhinga anhinga*) is caused by a combination of structurally coloured white barbs and melanin-pigmented barbules. Some species even show off structural and pigment-based colour patterns in the same feather, as in the shield-like breast feathers of the colourful southern double-collared sunbird (*Cinnyris chalybeus*) of South Africa.

Another way birds add colour to their feathers is through external cosmetic coloration. Cosmetic colours are diverse, are found in 13 bird families across the bird tree and are produced through varied mechanisms. For example, the rusty neck coloration of bearded vultures (*Gypaetus barbatus*) is produced from soils across their African and Asian range that are rich in iron oxide and stain the feathers. Secretions from the preen gland (see page 81) and skin of red knots (*Calidris canutus*) add extra shine to their feathers, while the rock ptarmigan (*Lagopus muta*) deliberately dirties its feathers to blend in with its arctic and subarctic habitats.

Colour-changing feathers in Anna's hummingbird

▲ Male Anna's hummingbird (*Calypte anna*) throat feathers have structural coloration, which appears dull black (left) or bright purple (right) depending on the direction of the sun.

Pigment-based colours

Pigments can be produced either from dietary proteins or repurposed pigments found in nature. Melanin-based pigments are synthesised by birds and confer browns, blacks, greys, and some reds and golds. In contrast, carotenoid pigments must be obtained from plant parts (e.g. seeds and berries) or invertebrates rich in carotenoids, and give birds their vivid orange, red and yellow colours. Pigments serve both display and non-display roles in avian lives, and can be found in all parts of a bird, from the feathers to the eyes and internal organs.

Melanin pigments

Melanin is the most common pigment in nature, and in humans it is the main determinant of the colour of our hair, eyes and skin. This mechanism of coloration arose deep in vertebrate evolution (we know this because we see melanin-based colours in bony fish, mammals and reptiles), but it has been uniquely elaborated in birds. There are two chemically distinct forms of melanin pigments in birds: eumelanin and phaeomelanin. Eumelanin pigments are generally dark brown or black, while phaeomelanin pigments are yellow, rufous or brown. Both forms of melanin are endogenous pigments synthesised by birds internally and biochemically from the amino acid tyrosine.

In all vertebrates, eumelanin and phaeomelanin pigments are packed into cellular organelles known as melanosomes. Black feather colours are generally associated with feathers rich in elongate melanosomes, which are predominantly composed of eumelanin pigments with a phaeomelanin core. On the other hand, brown and rufous colours are generally associated with feathers containing nearly spherical melanosomes full of phaeomelanin. This means the shape of the melanosomes and the chemistry of the melanins inside are linked. Using this relationship between melanosome shape and feather colour in living birds, scientists have been able to predict the colour patterns of extinct dinosaurs by studying the shapes of their fossilised melanosomes. For example, the Jurassic troodontid dinosaur *Anchiornis huxleyi* likely had white and black wing bars and a reddish-brown crest, possibly used in mating displays.

Compared to black and reddish-brown colours, the grey colours of feathers are generated in more heterogeneous ways. Melanosomes in the grey feathers of the North American grey catbird (*Dumetella carolinensis*), for example, are thought to contain mainly the eumelanin pigment responsible for black coloration, but are highly variable in shape and intermediate in length between melanosomes associated with brown and black colours. This mechanism of creating grey in birds is very different from that in mammals. Humans lose the ability to produce melanin as we age, resulting in grey or white hair. In other mammals with grey fur, the colour is not due to age, but rather very thin bands of alternating white and black colours within individual hairs. The end effect of this patterning is an overall grey appearance – for example, as seen in grey squirrels.

◀ The grey colours in this grey catbird (*Dumetella carolinensis*) are associated with large round melanin granules, known as melanosomes, inside the feathers.

▶ Different melanin colours are associated with different melanosome shapes, as seen in the plumage of this Eurasian hoopoe (*Upupa epops*).

phaeomelanin-containing
melanosomes
eumelanin-containing
melanosomes

Carotenoid pigments

Carotenoids, such as carotene, are organic fat-soluble pigments that add colour to ripe tomatoes and autumn leaves. Unlike melanin pigments, which can be produced by birds using available amino acids, carotenoids must be obtained from a plant- or invertebrate-rich diet. Carotenoids are available and at least seasonally abundant in the diet of most bird species, and the majority are further altered (through metabolisation) after being ingested by a bird. This process changes the chemical structure, and therefore the colour, of the pigment.

Research on the carotenoid-rich throat feathers of North American house finches (*Haemorhous mexicanus*) has taught scientists a great deal about how sexual selection works in birds. Since these pigments are known to function as both colourants and antioxidants, male house finches are able to use their bright red throat colours as a signal to potential mates that they are in prime condition and would make a good parent.

Feathers aren't the only place carotenoids can be found. These pigments also circulate in the blood and are what make the yolks of birds' eggs a bright yellow. Carotenoids are deposited in the photoreceptors in the retina and filter incoming wavelengths of light to influence what birds see, and they colour the red and yellow beaks and skin of many species. Bare skin on the chest can be coloured in raptorial birds like caracaras but shows as yellow or red only when the crop pouch of the oesophagus is distended and full of food. Similarly, when the oesophagus of the magnificent frigatebird (*Fregata magnificens*) is inflated, a large area of bright red chest skin becomes visible.

Uncommon pigments

Carotenoids aren't the only pigment used by birds to produce yellow, orange and red colours. There is still some mystery as to why carotenoid-based colours are lacking in some groups of birds and different, newly evolved pigments are used to create essentially the same colours. Pterin pigments confer yellow colour to the irises of some birds, including the American great horned owl (*Bubo virginianus*). An additional, unique uncharacterised pigment creates the bright yellow crown feathers of crested penguins (*Eudyptes* spp.). While green parrots, like other green birds, create their colours through a combination of structural blue colour and a yellow pigment, all parrots use a group of unique pigments called psittacofulvins to produce a variety of yellow, orange and red 'carotenoid-like' colours. Another group of pigments, metalloporphyrins, can also produce red hues. They include blood-derived haemoglobin pigments that cause the red eye coloration in the Neotropical phainopepla (*Phainopepla nitens*), and unique turacoverdin pigments that produce the bright red and green feather colours of African turacos (*Tauraco* spp.). Turacoverdins are the only known example in birds of a pigment that creates a specifically green colour.

Carotenoid colours in house finches

◀ Different forms of carotenoids cause variability in sexually selected red (left) and yellow (right) throat colours of house finches (*Haemorhous mexicanus*). Most carotenoids consumed by a bird result in yellow coloration and the conversion to red carotenoids requires a special enzyme.

▲ Lesser flamingos (*Phoeniconaias minor*) feed on brine shrimp, inside which are microscopic carotenoid-producing algae. Special enzymes allow the birds to break down these carotenoids into pink and orange forms, giving them their distinctive hue.

Structural colours

Structural colours have expanded the colour palette of birds. There are no known blue pigments in nature that birds can add to their skin, so instead they rely on nanometre-scale structures within their integument to selectively reflect certain wavelengths of light. These structures can be highly ordered to produce iridescent colours that change dramatically with viewing angle, or less ordered to produce non-iridescent colours that are less sensitive to viewing angle. Because these colours are produced by structures and not pigments, they are susceptible to damage and wear.

Iridescent structural colours

Iridescent colours in feathers change depending on the viewing angle and can be associated with feathers containing millions of hyperelongate melanin-containing melanosomes (e.g. the turquoise flight feathers of the Brazilian teal, *Amazonetta brasiliensis*) or have rice-shaped bubble-filled melanosomes (e.g. the violet throat feathers of the American Costa's hummingbird, *Calypte costae*). In contrast, the reddish, greenish or bluish tones of the shiny colours created by either of these mechanisms are generated by tiny, nanometre-scale changes in the spacing and size of layers of these differently shaped melanosomes. Differences among bright, iridescent colours can also stem from fine-scale changes either in the organisation and thickness of the arrays of long, skinny melanosomes, or through the addition of air bubbles, as seen in the hummingbird example above. Recent fossil discoveries tell us that dinosaurs were showing off their iridescent feathers as far back as the Jurassic – for example, the duck-sized dinosaur *Caihong juji* probably sported hummingbird-like iridescence in its throat and neck feathers. The four-winged paravian theropod dinosaurs in the genus *Microraptor* likely displayed glossy, iridescent feathers across their body, similar to those of the extant boat-tailed grackle (*Quiscalus major*) of the southern United States.

Melanin structures

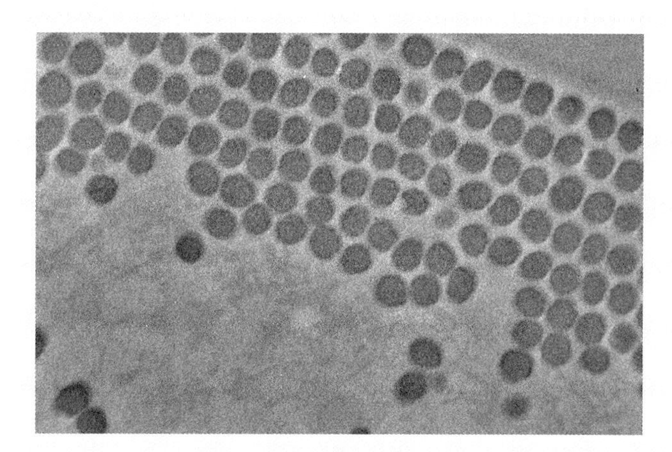

▲ Colour-producing stacks of melanosomes in the iridescent turquoise barbules of a Brazilian teal (*Amazonetta brasiliensis*). These crystalline stacks form by self-assembly during feather development.

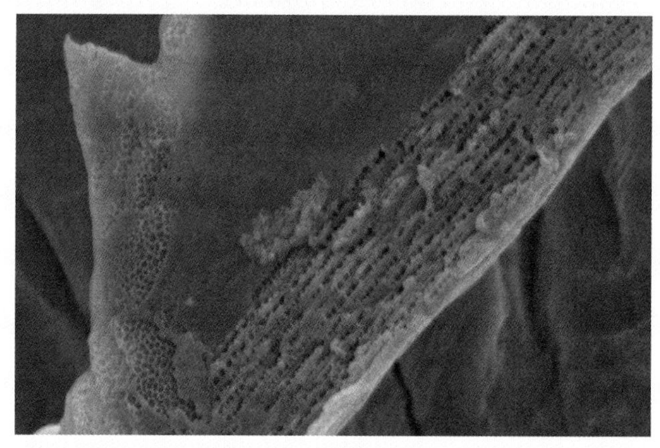

▲ Interior of a feather barbule from the magenta throat feather of a male Anna's hummingbird (*Calypte anna*), showing the bubble-filled pancake-shaped melanosomes that cause the iridescence.

Because iridescent colours depend on the fine-scale arrangement of melanin and beta-keratin inside feathers, they can be very sensitive to environmental changes. For example, the blue-green back feathers of the American tree swallow (*Tachycineta bicolor*) can change to yellow in a matter of seconds when the ambient humidity changes, making them nature's weathervanes. Another way to test this out is by smashing the tail feather of an Indian peafowl (*Pavo cristatus*) with a hammer: when the layered structure in the feather is destroyed, the hue changes from blue-green to black, this colour coming from melanin pigments in the feather.

Non-iridescent structural colours

Melanin structures are not the only way that birds can produce structural colours. Spongy structures of keratin and interspersed air bubbles can cause non-iridescent structural colours by selectively reflecting only certain wavelengths of light. In some cases, a layer of melanosomes beneath this spongy layer serves to absorb extra wavelengths of light and increase the saturation or vividness of the colour (e.g. in the contour feathers of the common kingfisher, *Alcedo atthis*). In other cases, such as the bright white crown feathers of the South American opal-crowned manakin (*Lepidothrix iris*), structural colour is produced without the assistance of melanosomes.

Another common form of structural colour in birds, thin-film iridescent colour, is caused primarily by the interaction of light rays reflecting from the top and bottom of a thin layer of keratin (i.e. the keratin cortex) at the surface of feather barbules. Here, although a row of melanosomes defines the bottom edge of the cortex, variation in the thickness of the cortex itself, not the size of the melanosomes, is the primary determinant of different hues. An example of this type of structural colour is seen in the chestnut-winged starling (*Onychognathus fulgidus*) in Africa.

▲ Structural blue colour in the bill of a helmet vanga (*Euryceros prevostii*), likely caused by light scattering from a spongy layer of beak tissue overlying a layer of melanin pigment.

▲ Close-up of a single tree swallow (*Tachycineta bicolor*) feather, showing the colour change from dry blue barbules (bottom) to waterlogged yellow barbules (top).

Glossy colours

A further way that structures interact with pigments is in producing shiny feathers. For example, the shiny black feathers of the American northern raven (*Corvus corax*) are the combined effect of light absorption by melanin pigments and light scattering by a thin layer of keratin on the surface of tiny feather barbules. Without these structures, the colour would be matt black, as seen in the red-winged blackbird (*Agelaius phoeniceus*). And it is not only black feathers that shine. Several species of *Campephilus* woodpecker have glossy red crests that are produced by a combination of red carotenoid pigments and smooth, flattened feather barbules that act like the reflectors on a car's rear light. Research into glossy colours in birds is fairly new and we are just beginning to understand how the gloss is produced and why it might have evolved.

◄▼ Modified flat feather barbs create the gloss on a pileated woodpecker (*Dryocopus pileatus*) crest (left) helping it stand out. Matt black colours in red-winged blackbirds (*Agelaius phoeniceus*; below) are less sensitive to changes in light angles.

▲ The glossy black colours in this northern raven (*Corvus corax*) are caused by reflections from organised melanin layers in its feather barbules.

Glossy and matt feathers

▲ White is a form of structural coloration, caused by air bubbles in the beta-keratin. Rock ptarmigan (*Lagopus muta*) moult into white winter plumage to camouflage themselves against snow.

Bright whites

At first glance, it is tempting to assume that all white coloration is simply the default state of feathers in the absence of all other colour mechanisms. However, white colour can also be a form of structural coloration based on a microscopic arrangement of the feather tissue. Examples include the white bills of the Eurasian coot (*Fulica atra*) and the Australian magpie (*Gymnorhina tibicen*), and the often white contour feathers of the belly in many birds. Another good example is the rock ptarmigan, which moults into a brilliant white plumage in winter to blend in with the surrounding snow.

White feathers can contain millions of tiny air bubbles or pockets embedded in the beta-keratin. Since the air bubbles are of various sizes, they scatter all wavelengths of light randomly, much in the way light scatters from air between ice crystals in snow. Most white colours are produced through random scattering in this way. This mechanism is different from structural blue coloration, in which air bubbles and beta-keratin fibres are uniformly sized to reflect only single (e.g. blue) wavelengths of light, although in some cases structural white feathers have evolved from structural blue feathers. We now know that these changes can occur both through loss of the melanosomes that are closer to the centre of the feathers and serve to enhance the saturation of otherwise desaturated (white) colours, as well as modifications to the crystal-like feather structure. Hopefully, future work will shed light on the function of white coloration in the context of signalling.

Feather care and moult

Birds spend a considerable amount of their day caring for their feathers, cleaning, preening and oiling them. Evidence suggests that species with gaudy plumages spend more time doing so than species with drab plumages. Preen oils from a gland at the base of the tail help maintain feather quality, as do behaviours such as dust bathing and water bathing. Unlike most other skin structures, old feathers are shed and new feathers routinely regrown in what is known as moult cycles. Birds may moult once a year to replace their feathers just before the breeding season, or several times per year.

Feather maintenance

Feathers are replaced in moults one to several times a year, but in between these cycles birds have to maintain their quality. Ectoparasites such as tiny feather mites have been an issue for feathered species for a long time – they are known deep in the fossil record from a 50 million-year-old feather preserved in amber. Depending on the environment, dirt may also accumulate on feathers, making their colours duller, and general wear and tear affects their overall quality. Dust and water baths help keep feathers clean in all birds and are also thought to keep the number of ectoparasites to a minimum. Certain pigments, especially melanin pigments, also resist wear and tearing.

Preening – where birds run their bill through their feathers – involves rezipping the barbule hooklets of flight and contour feathers if they have been separated during flight or wear, as well as removing dirt, grime and other environmental contaminants from all body feathers. Because of the relative importance of bright colours in signalling to other members of the species, birds with brighter plumages tend to spend more of their day preening than those with more muted colour patterns.

In most birds, preening also involves spreading preen oil on feathers to help maintain their condition.

▼ Feather mite on a barb. Ectoparasites such as this have been an issue for feathered species for a long time – they are known from a 50 million-year-old feather preserved in amber.

▼ Trumpeter swans (*Cygnus buccinator*) spend a significant amount of time preening to keep their feathers in good condition and water repellent.

FEATHERS THAT CLEAN THEMSELVES

Birds have to work hard to keep their feathers clean (so that they can attract a mate) and orderly (so that they can fly). But some species, like the widely distributed mallard duck (*Anas platyrhynchos*), have come up with a unique strategy. Flight feathers are made up of tiny structures called barbules that interlock to form a kind of mesh. Although the keratin protein that makes up a bird's feathers sticks to water, water droplets can float on channels of air between the barbules, in the same way that a person can lie on a bed of nails. So when the bird tilts its wings, water drops roll off the feathers, carrying soil and dust with them – like water off a duck's back.

This is produced by a gland at the base of the tail, called the preen gland, oil gland or uropygial gland. Not all birds have a preen gland, however, as is the case in ostriches, the emu, cassowaries, bustards and frogmouths.

Preen glands are generally large in some aquatic species such as petrels and pelicans, where waterproofing of feathers is particularly important. However, although it has been widely assumed that preen oils make feathers more water repellent, studies in which scientists removed the oils from feathers showed no immediate effects on buoyancy or water repellence. Instead, the effects of preen oils on feathers may be more long term: mallards without access to their preen gland oils had poor-quality and waterlogged feathers after several months. Preen oils might therefore be more important for maintaining the orderly zipper-like arrangement of feather barbs and barbules that is critical for keeping out water. Other, less studied functions of preen oils may be in staving off infection by bacteria that reside on eggshells, communicating by scent with other birds (through airborne molecules found in preen oils) or changing feather coloration.

◄ Water droplets float in the air channels in the barbules of mallard (*Anas platyrhynchos*) feathers, rolling off when the bird tilts.

▼ Birds preen by running their bill through their feathers, rezipping the barbule hooklets in their flight and contour feathers.

Replacing feathers in moult

Regardless of maintenance, feathers wear out over time, so birds need to replace them. New feathers emerge after a moult or replacement cycle from the same follicle in the skin, and the old feathers are either pushed out, or pulled off, during preening. Birds moult to change their plumage throughout their life, from a juvenile plumage indicating that they are not competing for mates, to the full adult plumage involved in mating or differing ecological contexts.

Precocial birds (those that hatch in a developed state) have feathers at hatching, while in altricial birds the chicks are born naked or with very sparse down and a complete plumage largely develops well after hatching. The plumage of chicks is later replaced by a juvenile plumage, which is defined as a covering of 'true' feathers with two-dimensional vanes and a pennaceous (closed) structure, rather than the plumulaceous or 'downy' feather structure of nestlings. One exception is the large-bodied ratites, including ostriches, the emu and kiwi, which have two-dimensional feathers as newly hatched chicks despite lacking barbule hooklets.

Most bird species indicate their status as juveniles or adults through their plumage. Birds may moult only once a year just before the breeding season, as in the polygynous red-winged blackbird, or several times a year, as in mallards, where males have both a nuptial/breeding plumage and a female-like 'eclipse' plumage pattern. There are about 35 North American passerine species in which some males retain a dull 'female-like' plumage in their first breeding season, before later moulting into a 'basic' (or adult) plumage. One particularly interesting example of this phenomenon, known as delayed plumage maturation, is the tree swallow. Here, it is the female rather than the male that shows a delay in moulting into the adult plumage. It is worth noting here that, although we have a good understanding of the social and ecological factors important in the evolution of delayed plumage maturation in terms of colour patterning, we know very little about how feather shape changes through different moult stages.

◀ King penguin (*Aptenodytes patagonicus*) in its first catastrophic moult. Penguins moult all their feathers at nearly the same time over two or three weeks.

▼ Male common blackbird (*Turdus merula*) in fresh adult plumage (left) relative to its condition during moult (right).

ANTING

The behaviour known as anting might also contribute to maintaining feather quality. In this, some birds – including ravens and crows – will cover themselves with ants during the time of year when they are moulting their feathers. Two behaviours have been described: active anting, in which the bird places individual ants on its body using its beak; and passive anting, where the bird extends its wings and crouches down over the soil or grass to acquire ants. Scientists have suggested that the birds may be applying scents to their body to combat parasite load or that they are obtaining a benefit from the medicinal/soothing effects of the chemicals released by the ants (e.g. formic acid). However, the exact explanations for this behaviour remain unknown.

▶ Anting is common in carrion crows (*Corvus corone*) and other corvids. The birds may be using the ants' scent to deter parasites, but the exact purpose of this behaviour is unknown.

Moulting strategies

There are striking differences in the timing of feather replacement through the year (e.g. seasonal variation) in different birds, and four basic strategies have been described. In the simple basic strategy that is thought to be ancestral to crown birds, a species replaces all of its feathers just once per year. Examples include ratites (ostriches, emu) and some large, long-lived seabirds (petrels, penguins). In these species, the moult is sometimes described as 'catastrophic' – the feathers on the body are largely lost all at once, and individuals are highly vulnerable to predators and weather conditions during this period.

In the second, 'complex basic' moult strategy, juveniles have an additional plumage after the downy nestling stage but before the sexually mature adult, or basic, plumage is formed. Examples include many Neotropical birds. The third strategy is the simple alternate strategy, in which an additional (pre-alternate) moult is added to all cycles. This is the least common strategy among birds, with an example being the common loon or great northern diver (*Gavia immer*). The fourth strategy is the complex alternate strategy, in which species have at least two distinct plumages in the first annual cycle, followed by two plumages in all subsequent cycles. Examples include many dabbling ducks, shorebirds and several passerines (i.e. those with alternate plumages in adults, such as the American goldfinch, *Spinus tristis*). In all but the first simple basic strategy, moult is sequential – that is, feathers are not lost and replaced all at once, but in stages from different feather tracts on the body. This strategy is thought to minimise vulnerability during the moult period.

Birds may also replace different kinds of feathers at different times of the year. Nearly all birds replace their primary flight feathers sequentially, so that they are still able to fly during the moult. Dabbling ducks and geese, however, are unique in this respect in that they replace their primary feathers all at once. This means that they can be flightless for up to 40 days and potentially very vulnerable to predators. They reduce this risk by moulting in wetland habitats, taking to the middle of lakes and remaining concealed from predators. In addition to differences in timing, birds vary in how fast they moult their feathers. For example, most passerines moult their feathers over the course of five to 12 weeks, but some hawks and owls can take up to two years to replace their feathers!

Some of the moult stages in a European herring gull (*Larus argentatus*)

▲ First-year European herring gull plumage. Complex juvenile plumages in this species, which reaches sexual maturity only after four years, signal age cohort and non-breeding status.

▲ Third-year European herring gull after a partial moult, showing some of the light grey feathers on the back that characterise the adult plumage. Gulls have one partial and one complete moult per year.

◄ Gentoo penguin (*Pygoscelis papua*) in moult. Penguin body contour feathers are among the most densely spaced in birds, and when shed during the moulting period they create a mass of fluff!

▲ The European herring gull adult breeding plumage is typically in place from April to September. This bird is in prime condition and finally, after around four years, may have its first breeding event.

▲ This is the adult non-breeding plumage of the European herring gull, typical from September to April. However, gulls can be hard to distinguish, especially from hybrids of two gull species or adults of other species.

CHAPTER 3

Muscles, Skeleton and Nervous System

It is hard to imagine what it would be like to be a bird. The ability to stay perched on a narrow branch while sleeping seems impossible to those of us who toss and turn in bed, and generating enough sustained force with our upper limbs to overcome the pull of gravity and take flight is beyond even our most talented athletes. Just as remarkable are the capacities of bird nervous systems. Birds see dimensions of colours invisible to us and have perfect pitch, and many can detect food odours from great distances. Although the term 'bird brain' is colloquially used as an insult, many birds, perhaps all, have complex cognitive abilities, supporting such skills as problem solving, innovation and tool use. Vocal learning is much more common in birds than in mammals, and is present in all songbirds, hummingbirds, and parrots. Many birds have complex spatial memory and the ability to navigate over vast distances. Bird brains, and the neuromuscular systems they control, are remarkable indeed.

◀ Short-eared owl (*Asio flammeus*).

Bird skeletons

The design of birds' bodies is similar to the basic vertebrate body plan. However, bird bones and muscles show many adaptations for flight, even in species that are flightless. The forelimbs (wings) are specialised for flight, and the hindlimbs (legs) for perching, walking or running. To further aid flight, many bones are fused into rigid structures and are hollow to increase strength. Although bird bones are lightweight relative to their strength and stiffness, contrary to popular belief bird skeletons are not lightweight relative to the overall body mass.

Fused bones

Many bones in the bird skeleton are fused, including several bones of the spine (vertebrae), bones of the pelvis and the clavicle (collarbone). The fusion of these bones reduces overall mobility of the body trunk of birds, compared to other vertebrate animals. This relatively rigid box-like structure of the bird body provides secure attachment points for the muscles that power flight. The clavicles of birds are fused into the furcula, and the fused thoracic vertebrae support the ribs and the large keel that projects from the sternum. The pelvic bones are also fused to the backbone and form the synsacrum, which aids in absorbing the shock of landing and take-off.

Extending from the rigid body frame are the cervical vertebrae (neck) and the caudal vertebrae (tailbone). Birds have 12–25 neck-bones, but the number of these is not related to overall neck length. Birds' long and flexible necks compensate for the rigidity of the body frame and allow many to turn their head far enough to look directly behind themselves. At the other end of the body, birds have four to nine caudal vertebrae and a structure of fused bones called the pygostyle, which provides an attachment point for the tail feathers. Woodpeckers use their tail feathers to brace themselves against tree trunks, and their pygostyle has a special disc-shaped protrusion for muscle attachment for this purpose.

Hollow bones

Some bird bones contain marrow, as with most other terrestrial vertebrates, but many instead contain air-filled spaces. This so-called pneumatisation of bones increases their rigidity relative to their weight, just as a hollow pipe is stiffer than a solid cylinder of the same diameter. The hollow spaces within bird bones are typically reinforced with internal struts that increase strength, and some are also continuous with the airways of the lungs and air sacs. Flightless birds retain pneumatisation of their bones, but the size of the air spaces is smaller than in flying birds and their bones are thus more dense. Even the skulls of birds are pneumatised, and the cranium (braincase) consists of inner and outer layers surrounding an air space, with the struts that span this air space visible through a bird's scalp. Because the process of pneumatising the skull can take many months, a skilled bird recorder can identify whether a bird is a juvenile or adult by gently parting the feathers on the head and observing the skull through the scalp.

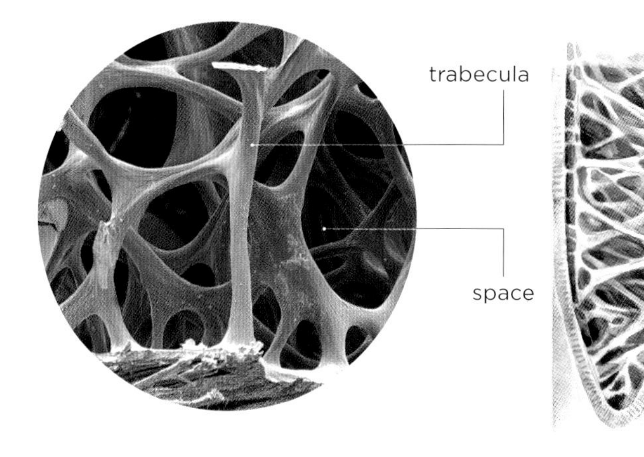

▶ Most bird bones are pneumatised, with hollow air spaces inside. Struts, or trabeculae, crossing the internal air spaces, provide rigidity and strength.

Avian skeleton and variation

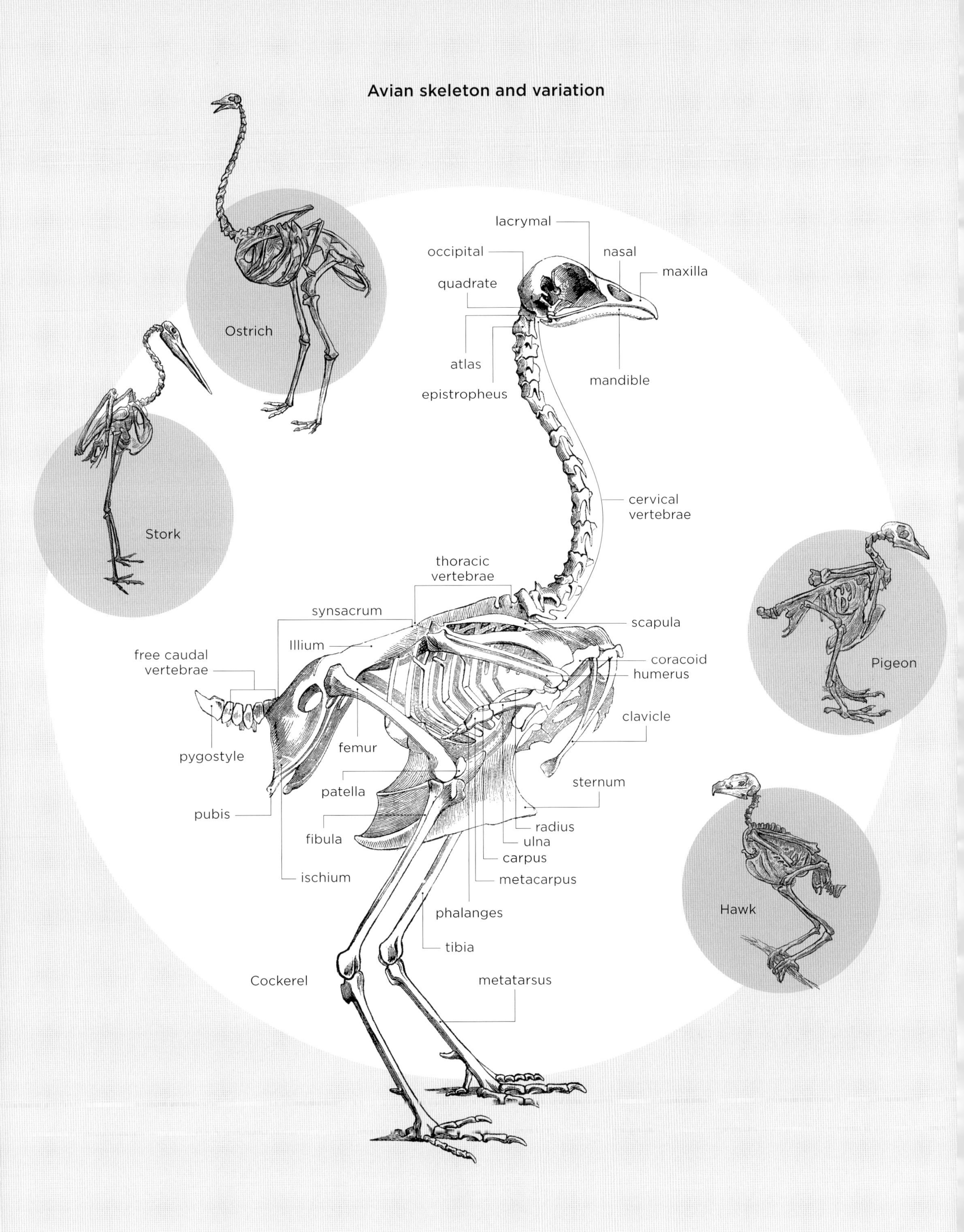

Bird skulls

The most striking feature of a bird's skull is the huge eye sockets. In fact, in some birds the eyes are so large that the inter-orbital septum, a thin layer of bone between the eye sockets, is missing. Birds also have large brains for their body size, so their skulls have a correspondingly large cranium. The remainder of the skull bones are relatively small fused frontal bones that provide an attachment point for the beak.

Skull features

The surface of the bird cranium is smooth and lacks the sutures seen in mammals and other vertebrates. Within each eye socket (orbital cavity) a ring of small bony plates called the sclerotic ring helps to support the lens in the bird's very large eyeballs. Some bird eyes are so large that they can barely move within their cavity, requiring the animals to move their whole head to change their gaze direction. Combined, the eye sockets and cranium form the largest portion of the skull.

Neck and jaw bones

Compared to the rigid fused bones of the body frame, bird necks and skulls are relatively flexible. The neck of most species is long and agile, allowing the birds to look in a variety of directions, forage, and preen their body feathers. The bones of the beak are flexibly joined to the skull, allowing a variety of movements. Bird beaks lack teeth but they do have a bony core that is covered with an integument (see page 58), which may be hard and protective, as in songbirds, or soft and flexible, as in waterfowl.

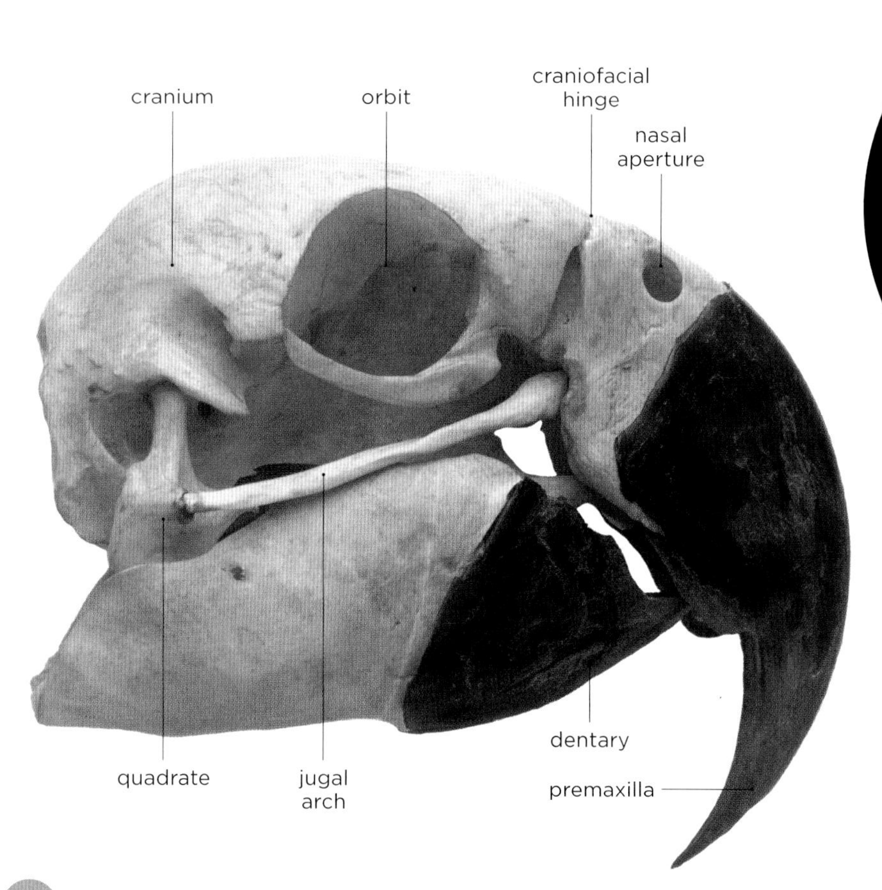

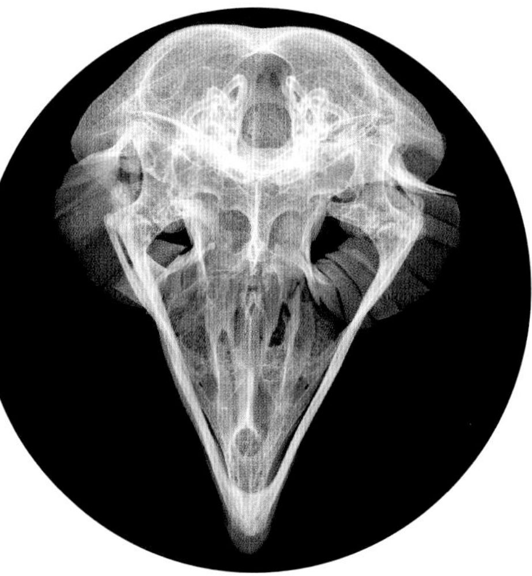

▲ Coloured X-ray of an eagle owl's skull (*Bubo* sp.) illustrating bony sclerotic rings (in brown) that support the very large eyes.

◄ A parrot skull, illustrating the craniofacial hinge that allows the upper jaw to move independently of the skull itself, a feature unique to birds. The cranium is smooth, lacking the suture joints visible in mammalian skulls.

THE HYOID APPARATUS

The bones that support the tongues of birds are collectively called the hyoid apparatus. Hyoid structure varies across species in relation to diet. In seed-eating birds, the tongue is relatively short and the hyoid bones relatively small. Woodpeckers have remarkable hyoids to support their ability to protrude their tongues great lengths into crevices while foraging for insects. The base of the hyoid apparatus forks into two branches that wrap around the back and over the top of the cranium, almost touching the upper bill. When woodpeckers protrude their tongue, muscle contractions pull the slender hyoid bones down and around the skull to extend the tongue. When these muscles are relaxed, the hyoid bones retract in their muscular sheaths and again wrap around the skull.

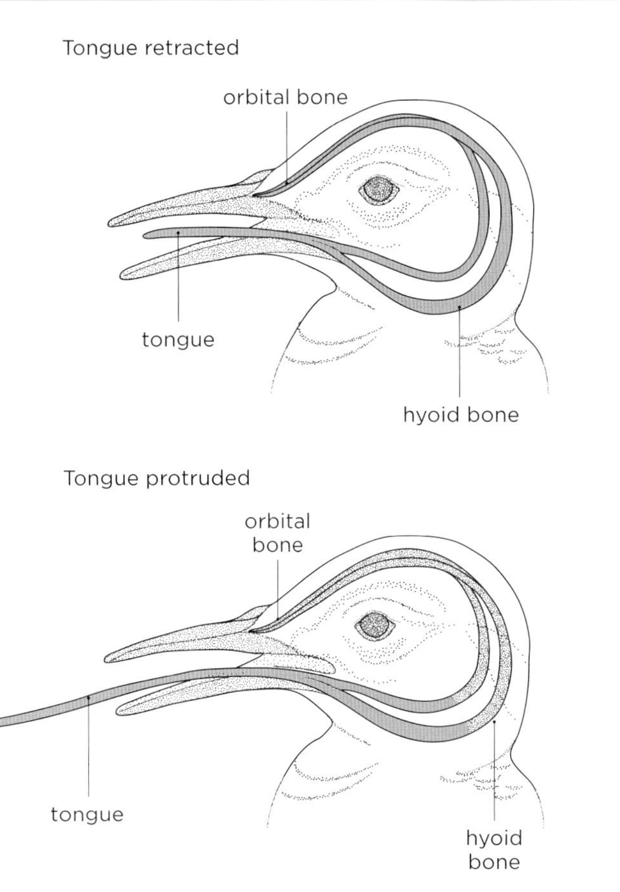

The upper beak, the maxilla, and lower beak, the mandible, are attached to the skull via movable rectangular quadrate bones on each side of the skull. The movements of the quadrate bones permit beak movements that vary across species. In some birds, the joint of the upper maxilla and the frontal bones of the skull forms a hinge that allows the upper beak to move independently, an action known as rhynchokinesis. This is unique to birds among living vertebrate animals – in all others, including humans, the upper jaw is fused to the skull and only the lower jaw can be moved.

◀ The eyes of many birds are so large that they have limited mobility in the orbits of the skull. To move their gaze, most birds need to move their entire head.

Bird muscles

Bird bodies are designed by natural selection for flight, and the structure of the major muscles of the body provide a striking example of this. The most conspicuous muscles of birds are the large flight (pectoral) muscles on their chest. Limb muscles are usually smaller, with their mass concentrated closer to the bird's centre of gravity – bird toes, for example, have no muscles and are controlled through a pulley system of tendons connected to muscles higher on the leg. Having most muscle mass close to the centre of the body contributes to a bird's aerodynamic fusiform shape.

Muscle fibres

Anyone who eats poultry will know that different muscles have different colours and textures. This is a result of the relative amounts of different muscle fibre types they contain. Fast-twitch fibres contract quickly but also tire more quickly, while slow-twitch fibres contain more myoglobin, which helps transport oxygen, and are also more closely associated with fat for slower, but sustained contraction. Birds that generally walk and use short, rapid flights, like poultry, have greater amounts of fast-twitch fibres in their flight muscles (breast meat), and greater amounts of slow-twitch fibres and fat in the leg muscles.

Flight muscles

The largest muscles of birds are the flight muscles, making up to a third of total body weight in active flyers like hummingbirds, but only about a tenth of body weight in occasional flyers like rails. The large pectoralis muscles power the downstroke of the wings. Interestingly, the upstroke of the wings is not produced by back muscles like the latissimus dorsi of mammals, but by the supracoracoideus, a muscle lying beneath the pectoralis and attached to the sternum and keel. The supracoracoideus tendon passes through an opening in the shoulder and connects to the upper surface of the humerus (wing bone). Thus, like a pulley system, the wing strokes upward instead of downward when the supracoracoideus muscle contracts. Having both the pectoralis and supracoracoideus attached to the sternum and keel allows these large flight muscles to have a low centre of gravity, improving stability during flight (see drawing opposite).

Besides the flight muscles, other skeletal muscles provide movement of the head, neck and beak, as well as the lower limbs. Skeletal muscles connecting to the pygostyle control movement of the tail feathers. Birds, like mammals, also have other forms of muscle, including smooth muscles in the gastrointestinal tract and blood vessels, and the cardiac muscle of the heart.

Muscles involved in flight

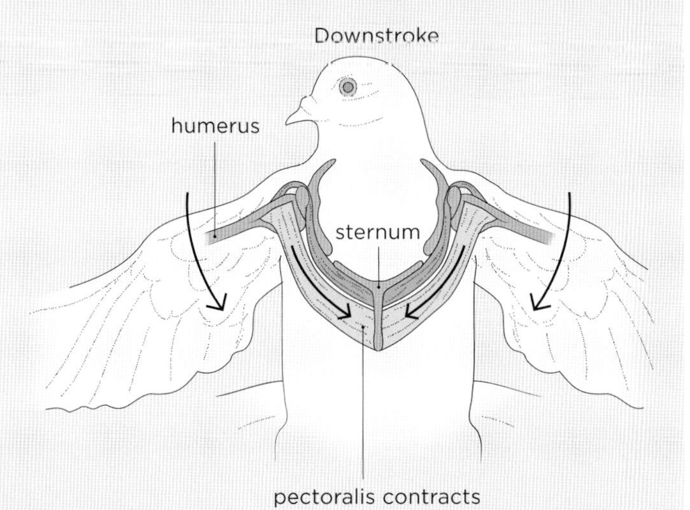

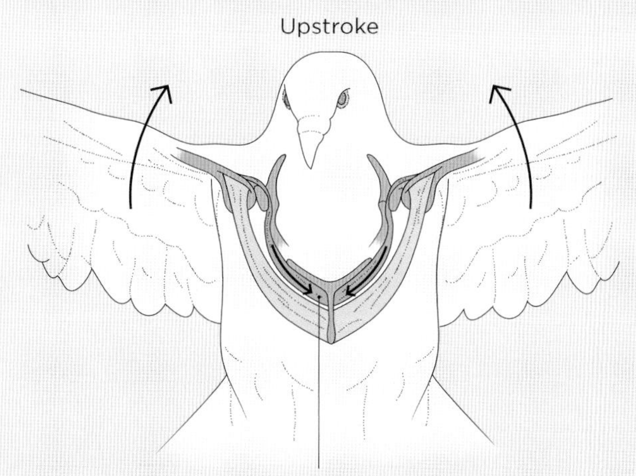

▲ The opposing muscle pairs that power the upstroke (supracoracoideus) and downstroke (pectoralis) both attach to the sternum or keel, helping to create a low centre of gravity.

◄ Flight is powered by downstrokes that alternate with recovery upstrokes. The shape of the wing can be altered during each stroke to induce power on the downstroke and reduce drag on the upstroke, as illustrated in this time-lapse image of a North American ferruginous hawk (*Buteo regalis*).

THE SYRINX

Another specialised muscular organ found in birds is the syrinx. Birds do not produce sounds with their larynx, as do mammals. Instead, they produce sounds with the syrinx, located at the point where the trachea divides into two bronchi before entering the lungs. The precise location and structure of the syrinx varies across bird taxa, and its complexity is related to the variety of vocalisations produced – it is most specialised in oscine songbirds. In this group, the syrinx is formed of cartilage rings at the junction of the trachea and bronchi, where six pairs of muscles control soft tissue that can interrupt the flow of air from each bronchus as the bird exhales, acting like the laryngeal vocal cords in mammals. Songbirds can control each side of the syrinx independently to produce two sounds simultaneously or in sequence, allowing for the production of very complex songs.

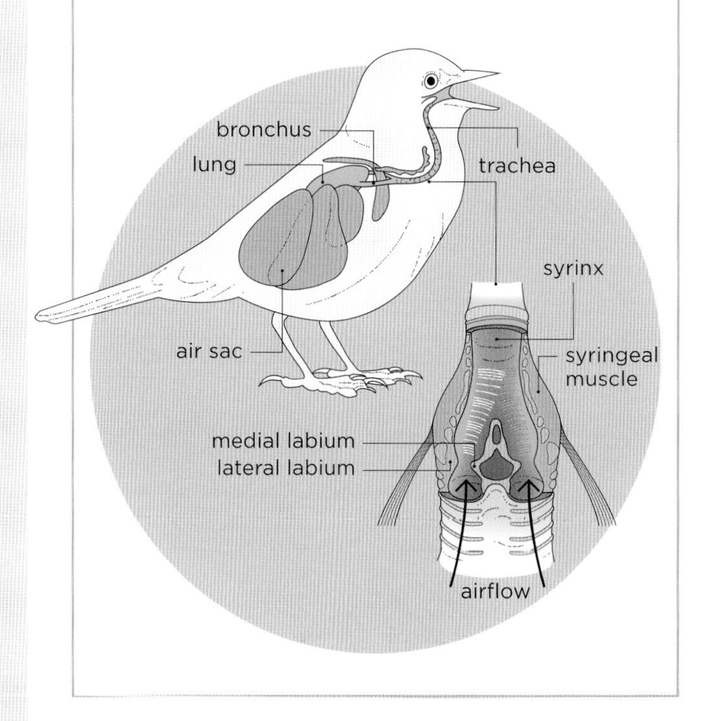

Bird forelimbs: wings

Bird wings are homologous to the forelimbs of other vertebrates, with reduction and fusion of some of the bones. Despite external appearances to the contrary, a bird wing and the foreleg of a cat both share the same skeletal design. Bat wings are mostly modified hands, but bird wings are made of arm bones and fused hand and finger bones, which support primary flight feathers. Some of these bones have quill knobs at the locations where the flight feathers attach.

Wing bones

The wing bones are the humerus, radius and ulna, and the carpal bones and digits. The humerus, closest to the body, has a large hole near the shoulder that connects air sacs to the hollow space within the bone. The radius and ulna are the bones of the forearm. Furthest from the body, the carpal bones and digits are much reduced and fused compared with those of most terrestrial vertebrates, forming the carpometacarpus and three short fingers (phalanges).

Powering the wings

The flight muscles of birds attach to the keel and pectoral girdle (shoulder bones) to power and control wing movements. The pectoral girdle consists of the sternum, with its large protruding keel, and the coracoids, scapulae and clavicles. In most species the clavicles are fused to form the furculum (wishbone), which acts as a spring to help power wing strokes, recovering energy spent during the downstroke to assist in the upstroke. At each shoulder the scapula, coracoid and clavicle join and form a small canal that allows the tendon from the supracoracoideus to loop over the shoulder like a pulley system, powering the upstroke of the wing (see page 95).

▲ A bateleur (*Terathopius ecaudatus*) in the Kalahari uses large pectoral muscles to power a wing downstroke during take-off.

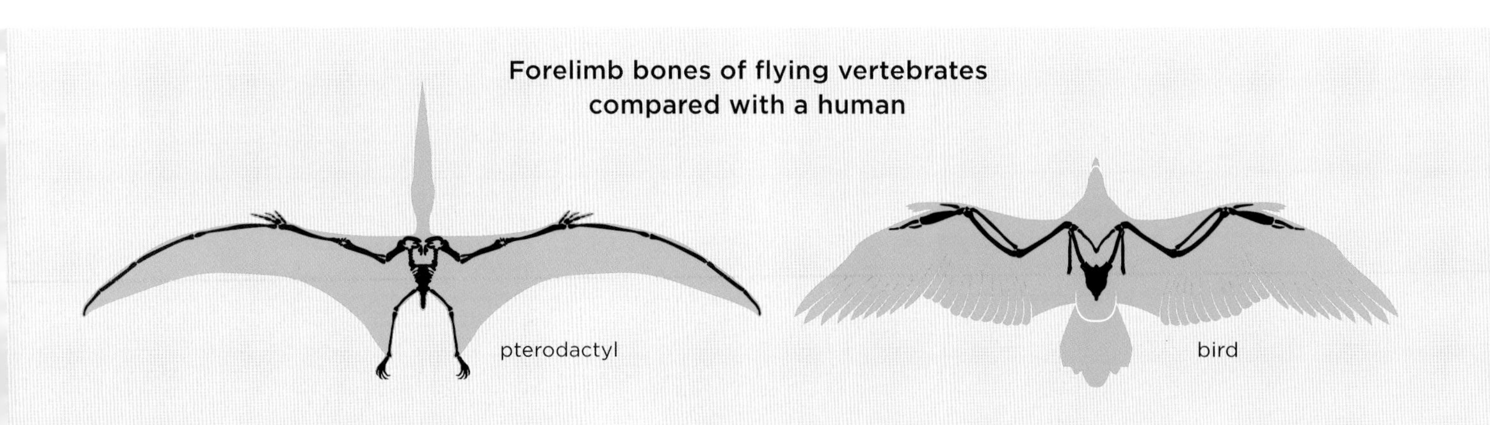

Wing claws

Some birds have claws that extend from the phalanges. These were common, and likely more functional for climbing, in ancient extinct groups of birds. In most living species they are thought to be vestigial structures, but in the Amazon young hoatzins (*Opisthocomus hoazin*) use their claws to climb back to their nest if they are flushed from it by a predator. Australian southern cassowaries (*Casuarius casuarius*) have a long claw extending from the second digit of the wing, the function of which is unclear but which may be used in preening and/or fighting. Distinct from claws, some birds such as Australian masked lapwings (*Vanellus miles*) and South American southern screamers (*Chauna torquata*) have bony spurs on their wings, typically used as weapons during aggressive interactions.

▶ The wings of New Zealand southern brown kiwi (*Apteryx australis*) are so reduced and vestigial that they are not visible beneath their hair-like feathers.

▲ Masked lapwings (*Vanellus miles*), found in northern Australia, New Guinea and New Zealand, have carpal spurs on the wing that they use to aggressively defend territories during the breeding season.

Flightless wings

The wings of flightless birds have the same general design as those of flying birds but may be reduced in size and strength. Those of New Zealand southern brown kiwi (*Apteryx australis*) are the most extreme example – they are tiny and generally not visible under the bird's plumage, consisting of only a few small bones with hair-like feathers. Penguin wing bones are shorter and flatter than those of flying birds, providing a stout flipper-like structure that can withstand the greater forces generated during underwater flapping.

Bird hindlimbs: legs and feet

Birds inherited bipedal toe-walking from their theropod dinosaur ancestors. This form of locomotion freed their forelimbs to eventually evolve into wings. Bird legs are supported by the pelvis, the right and left sides of which are elongated and fused to form the synsacrum. Unlike mammals, the majority of birds do not have a symphysis, the joint of the two pubis bones that closes the pelvic girdle into a circular structure. Instead, the pelvis usually remains an open semicircle to facilitate egg-laying.

Bones of the leg

Bird legs are formed by the femur, tibiotarsus and tarsometatarsus. The femur is relatively short and is commonly hidden from view by body feathers. The visible, and typically unfeathered, portion of the leg is composed of the tibiotarsus and tarsometatarsus. The latter is sometimes called the tarsus by ornithologists, but is actually formed mostly by fused metatarsal bones, homologous to human foot bones. Bird feet are composed of a variable number of toes whose bones are pedal phalanges. Most birds have four toes, but some have three and ostriches have only two. The anatomical arrangement of the toes is variable among different groups of birds (see box).

Toes and claws

At the end of each toe are claws that are specialised in different species. They may be hooks for climbing trees (as in creepers and nuthatches) or long, curved talons (as in raptors). Some ground-foraging songbirds have elongated claws on their backward-pointing toe (hallux), which may act like a snowshoe, helping prevent the birds from sinking into soft earth. In herons, the claw of the third toe has a serrated edge, like a comb, used in preening feathers. In addition to claws, some birds have bony spurs extending from the tarsometatarsus that can be used as weapons, as commonly seen in peafowl, pheasants and other Galliformes.

Most of a bird's legs has little musculature or blood supply, with the muscles surrounding the femur and upper tibiotarsus. Movement of the toes and lower leg is achieved by long tendons running from these muscles down the length of the leg to the toes.

◄ The toes of African jacana (*Actophilornis africanus*) are long, distributing their body weight so that they can walk atop lily pads and other floating vegetation.

▼ Many fowl, including the Indian peafowl (*Pavo cristatus*), have bony spurs protruding from the tarsometatarsus, which they use as weapons during fighting.

◄ Common ostrich (*Struthio camelus*) toes are much reduced, with only two large toes that are well adapted for running at high speed.

When birds perch, the flexor tendons running down the back of the tarsometatarsus lock the toes closed, allowing the animals to remain in place with minimal energy and even while asleep. Releasing this grip is accomplished by straightening the legs, which relaxes the flexor tendons.

TOES AND TAXONOMY

The configuration of toes differs among groups of birds and has long been used as a classification tool by taxonomists. The arrangement of the toes is thought to reflect adaptations based on how different birds use their feet. By convention, toes are numbered from the inside (homologous to our big toe). Most perching birds, doves and hawks have anisodactyl feet, with the hallux projecting backward and the other three forward. Next most common are zygodactyl feet, found in parrots, woodpeckers, cuckoos and other birds, in which both the first and fourth toes project backward. Other toe arrangements are more rare – for example, ostriches have only two toes, perhaps associated with running locomotion.

Bird toes and taxonomy

anisodactyl

zygodactyl

heterodactyl

syndactyl

pamprodactyl

Bird brains

Birds perform a wide range of complex behaviours, including flight and long-distance navigation, making and using tools, remembering locations of food, and learning vocalisations. All of these are a product of the nervous system. Although the term 'bird brain' has historically been used as an insult, we now know that bird cognition is as complex as that of most mammals, and that the brains responsible for this cognition are also at least as complex. Indeed, the number of neurons in bird brains is comparable to that of primates.

Bird brains versus mammal brains

The anatomical structure of bird brains is distinct from that of primates and other mammals, comprising two smooth hemispheres with a lower surface area than the wrinkled and convoluted cerebral cortex of humans and other primates. Historically, it was thought that birds lacked a cortex, and that subcortical brain regions evolved to control 'instinctive' behaviours. Indeed, many textbooks still promulgate this error. However, studies of gene expression in different brain regions have now demonstrated that most of the structures in the bird brain once thought to be formed from subcortical tissue are, in fact, homologous to the cortex, and both are formed from a neural layer called the pallium.

Bird brains also differ from those of mammals in the distribution of white matter (myelinated neural tracts) and grey matter (neuron cell bodies). The mammalian cortex is multilayered and regionally differentiated by function (visual cortex, auditory cortex, etc.), and has an underlying layer of white matter that connects these regions. In contrast, the avian pallium is organised into functional regions (brain regions or nuclei) in a non-layered manner, with white matter tracts connecting these regions.

Brain structure

The anatomy of bird brains is dominated by the large cerebrum, which partially overlaps the two large optic tecta (optic lobes). These support birds' complex visual perception, as do the large optic nerves, which convey neural signals from the retinas of the eyes to the brain. One exception is the extinct Hawaiian Kaua'i mole duck (*Talpanas lippa*), whose

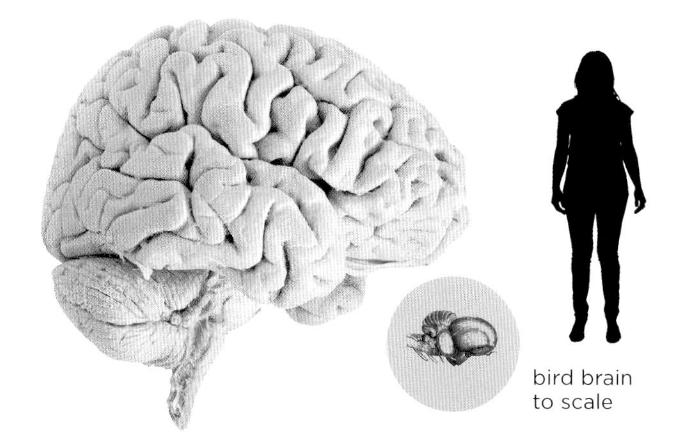

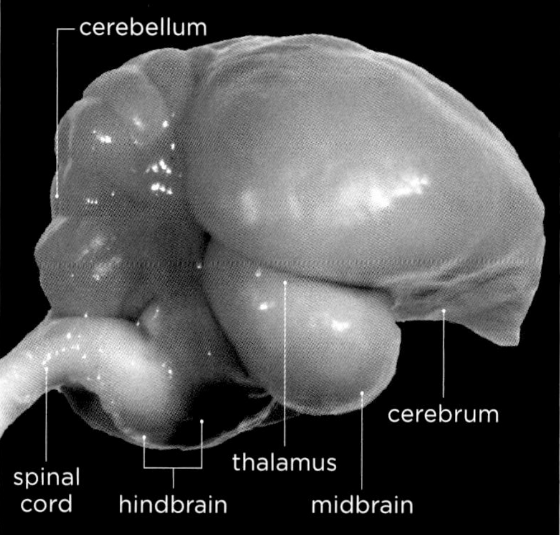

▲ Human brains (top) are dominated by a large convoluted, folded cortex that is organised in layers. In contrast, bird brains (above) are dominated by large cerebral hemispheres that are homologous to the cortex, but are smooth and organised differently into functional regions. The classical view was that bird brains are more primitive than those of mammals, but modern analyses indicate that the overall components of bird and mammal brains are similar.

Organisation of the bird brain

Classical view

complex cognitive behaviour

instinctive behaviour

bird brain to scale

Modern view

three major forebrain subdivisions responsible for complex behaviour

pallial

striatal

pallidal

bird brain to scale

skull anatomy implies it had very small eyes and optic nerves. This cave-dwelling duck was likely blind, or near blind, and foraged primarily using smell and touch. The olfactory bulbs of most birds are relatively small compared to those of many mammals, but are enlarged in some birds that use smell extensively in foraging, such as New Zealand North Island brown kiwi (*Apteryx mantelli*) and vultures.

The rear part of a bird's brain is dominated by a large cerebellum. This area of the brain is important in sensorimotor coordination and timing of movement, among other functions, and its large size in birds may be an adaptation for the complex neuromuscular coordination required for flight. The midbrain and hindbrain of birds are important for sensory integration and regulation of physiological processes, and are similar to equivalent areas in other vertebrates. The brain connects to the peripheral nervous system through the spinal cord and 12 cranial nerves, similar to the arrangement in other vertebrates.

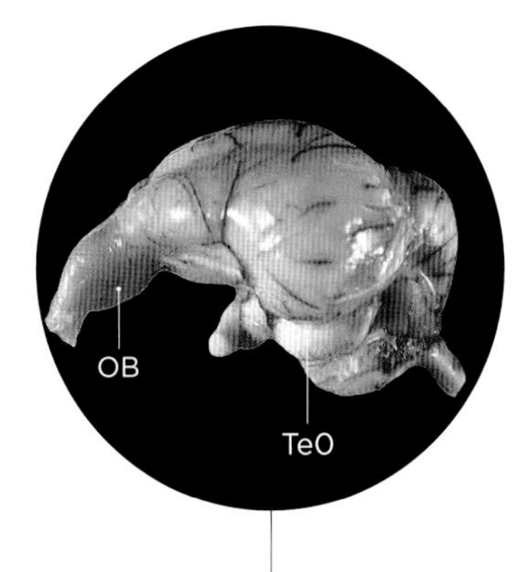

► Birds that rely on a keen sense of smell to find food, such as turkey vultures (*Cathartes aura*) that range over North and South America, tend to have enlarged olfactory bulbs (OB) relative to the size of other brain regions such as the optic tectum (TeO).

The song-control system

Studies of songbird brains have twice revolutionised the field of neuroscience, providing evidence of large sex differences and extensive seasonal changes. These breakthroughs began with the discovery of brain regions that support imitative vocal learning in the group. Similar neural circuits have since been discovered in parrots and hummingbirds, which also have imitative vocal learning. These circuits, collectively known as the song-control system, have a similar design to the brain circuits that are important for learning speech in humans.

Sex differences in brain structure

Shortly after the discovery of the song-control system in birds, researchers noted that there are huge sex differences in these brain regions. Prior to this time, it was assumed that any variation between male and female brains involved only subtle differences in neurochemistry or the connections between nerve cells. However, in songbirds the sex differences in the brain are remarkable. In Australian zebra finches (*Taeniopygia guttata*), males sing and females never sing. Parts of the song-control system, such as the region called the vocal contral region (HVC), are up to ten times larger in males than females, and some connections are entirely absent in females. Sex differences in HVC size have now been discovered in many other species. Although the extent of the sex difference is most extreme in zebra finches, the distinction is found even in species where both males and females sing duet songs, or in species where females sing more than males. It is clear from this that the neural circuits for singing are remarkably different in male and female oscine brains.

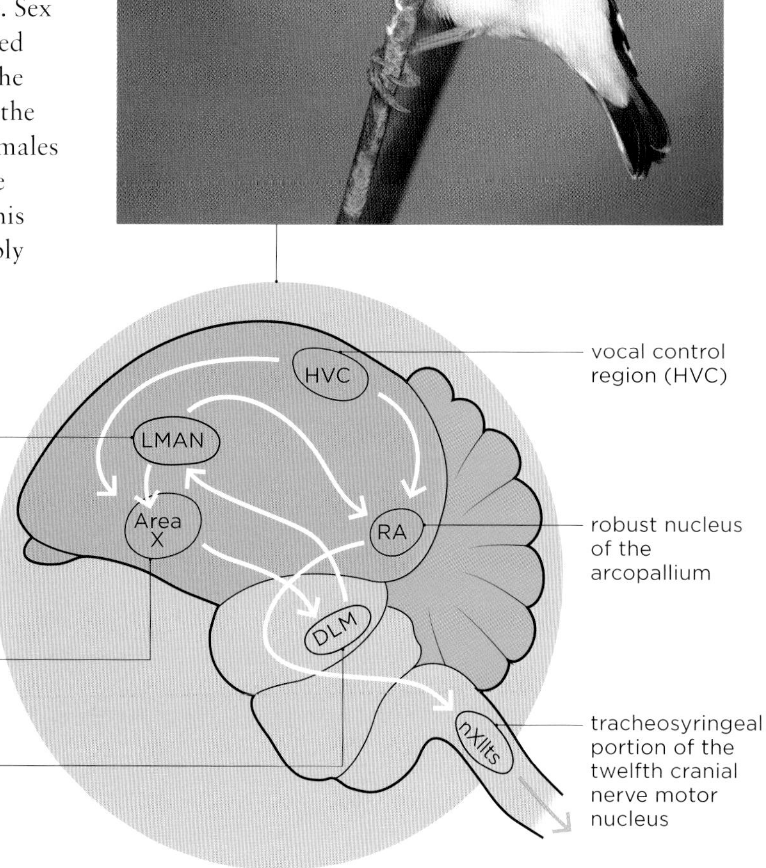

► Domesticated strains of Australian zebra finches (*Taeniopygia guttata*) are the most extensively studied songbird in neuroscience. Studies of this, and other, species has revealed the functioning of neural circuits required for imitative vocal learning.

GYNANDROMORPHIC BRAINS

▲ This gynandromorphic northern cardinal (*Cardinalis cardinalis*) is male on the left (red) side of its body and female on the right.

Following the discovery of sex differences in HVC size, scientists were keen to determine whether this depends on sex hormones or on genetic differences between males and females. In some rare cases, hatched birds have one-half of their body male and the other half female, a state known as gynandromorphy. The fortuitous discovery of such a gynandromorphic zebra finch (*Taeniopygia guttata*) in a laboratory breeding colony allowed researchers to examine the nature of neural sex differences. Both sides of the body of gynandromorphs share the same blood supply and hormones, but they are genetically distinct, male and female. In this case, the HVC was large on the male side and small on the female side. Thus, although hormones can modulate HVC development, the sex differences observed in zebra finches are largely independent of hormones and instead reflect sex differences in genes.

Seasonal brain changes

Another revolutionary discovery derived from avian studies is that the brain regions making up the song-control system are highly plastic. For example, the size of the HVC can change seasonally, which involves the generation of new neurons (neurogenesis) in adults. In fact, the number of neurons in the HVC of North American song sparrows (*Melospiza melodia*) can almost double from winter to spring. This greater number of neurons in the breeding season is thought to increase the complex sensorimotor control required for singing, and may be involved in seasonal song learning in some species. Historically, it was thought that adult terrestrial vertebrates could not grow new brain cells in adulthood. Some studies suggested that adult neurogenesis did occur in rodents, but it was the work on songbirds that verified that this process routinely occurs in adult vertebrates.

▼ Seasonally breeding songbirds, such as North American song sparrows (*Melospiza melodia*), sing most frequently during their spring breeding season. The brain regions controlling song also change seasonally, and are almost twice as large during the breeding season than in autumn or winter.

Somatosensory system

The somatosensory system reacts to sensations on the surface and inside of the body, including touch, heat, cold, pressure and pain. How neural somatosensory input is organised is less understood in birds than in mammals, and it varies across species. Broadly speaking, somatosensory signals to the brain are organised into two groups: those from the body (wings, legs, claws, body surface), and those from the beak and tongue. Birds use somatosensory information to control their internal physiology, and to respond behaviourally to their environment.

Topographic organisation of the brain

In many animals the organisation of the brain provides a 'map' of the organisation of the body. For example, input from your fingers is processed in areas of your sensory cortex that are adjacent to areas that process input from your thumb. In addition, the amount of cortex that processes sensation from your fingers is much larger than that from, say, your elbow, reflecting your fingers' greater sensitivity. Such topographic organisation is found in parts of the avian somatosensory system, although much of it remains a mystery. For example, barn owls (*Tyto alba*) have a unique bulge at the front of the brain – the somatosensory Wulst claw area – that processes information from their toes and talons, indicating the importance of these in the life of the species.

Sensory receptors

Bird skin contains a variety of receptors that detect mechanical force and noxious stimuli. Herbst corpuscles detect acceleration and deceleration of vibrations in the skin, providing sensory information about objects that come into contact with the body. These receptors are distributed across the skin and occur in association with muscles, joints and blood vessels. Herbst corpuscles found in clusters at the base of feather follicles respond to the vibration of feathers, suggesting that birds may be very sensitive to the flow of air over their bodies during flight. Much in the same way that many animals use hair and whiskers to sense the world around them, birds may use their feathers to extend their sense of touch.

In addition to Herbst corpuscles, bird skin contains nociceptors, which respond to noxious or painful sensations, and Grandry corpuscles, which detect lateral movement of the skin, providing the bird further sensation of touch. Birds are also able to detect and respond to temperature changes – for example, by changing their behaviour or the timing of breeding. However, how they do this and the location and nature of the thermal receptors remain a mystery.

Bill sensations

Many birds have concentrated receptors at the tip of the bill, called bill-tip organs. Parrots have clusters of such organs along the inside upper bill, providing them with sensory feedback as they explore objects with their beak. Ducks and geese have touch papillae clustered around the rim of the bill that may provide detailed sense of touch – mallards (*Anas platyrhynchos*), for example, can recognise objects based on touch alone using their bills. Many birds use long bills to probe sand or soil as they forage, including kiwi, ibises and shorebirds. In these species the bill-tip organs are organised into clusters of pits in the bone around the tip of the bill. The pits contain Herbst and Grandry corpuscles covered by soft skin, providing the birds a sense of remote touch and allowing them to detect hard-shelled molluscs from a distance by vibrations transmitted through the sand or soil.

► Many shorebirds, such as this curlew sandpiper (*Calidris ferruginea*), have exquisitely sensitive bills. Bill-tip organs are sensitive enough to detect hard-shelled molluscs in silt or sand remotely, without actually making contact with them.

Chemical sense: smell and taste

Chemoreception, the senses of taste and smell, depends on chemicals binding to specialised receptors in the mouth or the olfactory epithelium. For many years bird smell and taste were poorly understood, but we now know that birds have rich gustatory and olfactory senses, which they use when foraging and communicating with one another.

Sense of smell

Historically, it was assumed that most birds had no sense of smell, due in part to misconceptions held by the pioneering ornithologist John James Audubon. However, we now know that birds have an excellent sense of smell, which they use in both foraging and communication. Olfaction occurs as air is inhaled through the nares, or nostrils. In most birds these are at the top and base of the bill, but in kiwi the nares are at the tip of the bill (presumably to aid in foraging by smell) and in some diving birds they open inside the bill (presumably to reduce water entering the nares during a dive). Air entering the nares passes through nasal chambers lined with the olfactory epithelium. Here, chemicals within the air dissolve and bind to neural receptors, which then transmit information to the olfactory bulb in the brain.

Olfaction can be used during foraging to detect the nature and location of food. For example, American turkey vultures (*Cathartes aura*) can detect carrion from great distances even if it is hidden from view. Similarly, Antarctic prions (*Pachyptila desolata*) fly in zigzag patterns over the open ocean until they detect odour plumes of dimethyl sulfide emitted by phytoplankton that are being grazed by predators. These odour plumes indicate areas of high productivity and food abundance. In a similar example, Old World great tits (*Parus major*) are attracted to odours emitted by tree leaves as these are being consumed by caterpillars.

Odour is also important for communication in many birds. The waxy preen oils on bird feathers contain volatile chemical compounds that may provide a signal to other birds. American dark-eyed juncos (*Junco hyemalis*) can recognise their own species by odour, and in Europe spotless starlings (*Sturnus unicolor*) can tell males from females by

▼ Birds of the order Procellariiformes – the tube-nosed seabirds – have a sensitive sense of smell that they use to find food as they roam over the open ocean. This northern giant petrel (*Macronectes halli*) has found carrion, but it also feeds on krill, squid or fish.

▼ Antarctic prions (*Pachyptila desolata*) use their sense of smell to detect odour plumes of dimethyl sulfide, a gas emitted when krill eat phytoplankton. These odour plumes provide a cue to local food sources in an otherwise featureless seascape.

their smell. Several petrel species can recognise the odour of individual chicks, and a number of songbird species appear to use body odour in choosing a mate. Far from being anosmic, birds use their sense of smell in a rich variety of ways.

Sense of taste

Taste involves the detection of chemical compounds dissolved in saliva. The sensory structures that detect these chemicals are clustered in taste buds throughout the mouth. In mallards, taste buds are distributed at the base of the tongue, the floor of the bill and near the bill tip, suggesting that the birds can taste food held at the front of the bill prior to swallowing it. Although birds do not chew their food with teeth, the number of taste buds in their mouths, relative to food size, is comparable to that of mammals, suggesting that they have similar sensitivity to taste cues. Birds are also capable of detecting similar taste categories to humans, including sweet, umami, bitter, calcium, salt, sour and fat. Particular tastes are preferred by some species but not others. For example, hummingbirds are attracted to sweet tastes, but chickens are not. Birds do not, however, appear to be sensitive to capsaicin, the primary chemical that makes chilli peppers taste 'hot'. Capsaicin triggers pain receptors in mammals and contributes to our perception of flavour, but birds do not appear to respond to it and can consume chilli species that are unpalatable to mammals.

▼ Taste and touch receptors are found in various locations in bird bills, mouths and tongues. Although they appear to be hard, insensitive body parts, bills, like that of this Yellow-legged Gull (*Larus cachinnans*), provide birds with rich tactile and chemosensory information.

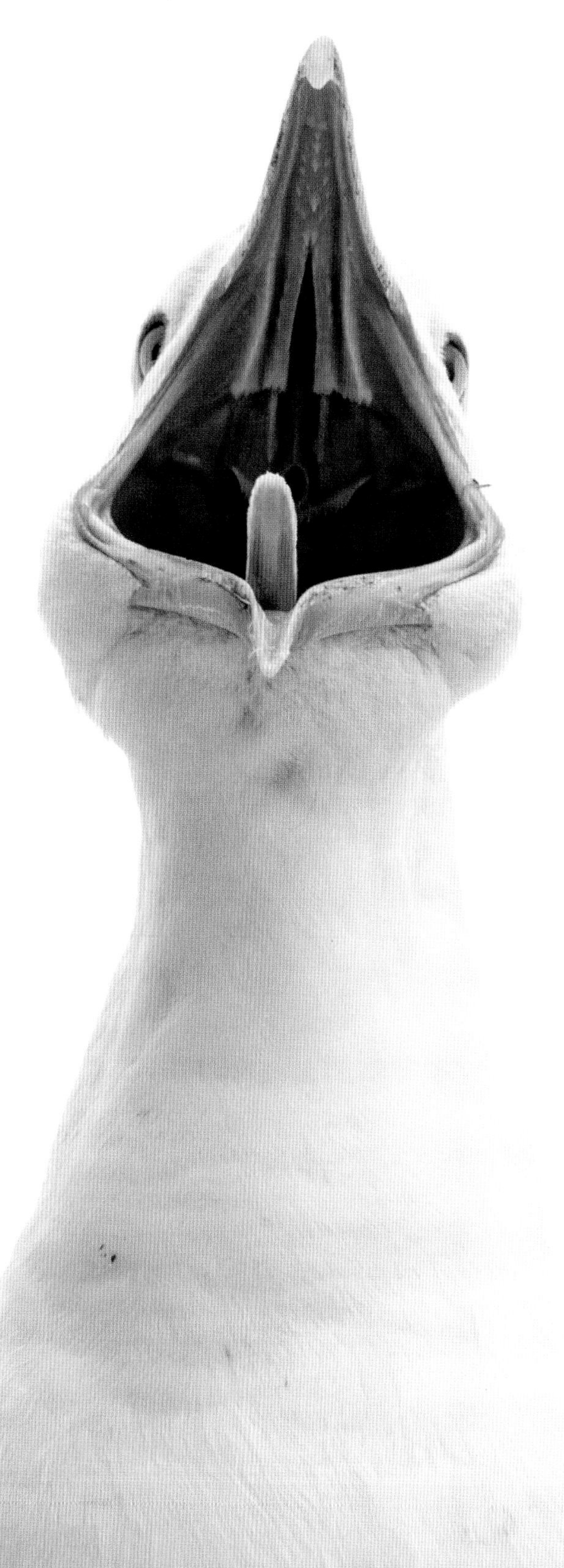

▼ Mallards (*Anas platyrhynchos*) and other dabbling ducks appear to have taste receptors located in several places, including near the bill tip. This suggests that they are likely able to taste potential food while dabbling their bill, prior to consuming it.

Ears and hearing

Birds lack the external pinnae common in most mammals that we typically identify as 'ears', with the opening of the ear canal on either side of the head behind the beak protected by auricular feathers. Sound is transmitted by pressure waves that are received at the eardrum and transduced to neural signals in the cochlea. From this relatively simple physical stimulus, birds perceive a world rich in sounds: soft and loud, near and far, melodic and dissonant. Birds use sounds to find prey, and to avoid becoming prey. They also produce remarkably complex vocalisations, which require equally complex auditory perception.

Ear structure

The pinnae of mammals help us localise sound sources by reflecting and filtering sound waves. In some owls, a ruff of feathers around the face serves the same function by reflecting sound like a parabolic dish. This has been most extensively studied in barn owls but is also present in other species. Note that the 'ears' of long-eared owls (*Asio otus*) or great horned owls (*Bubo virginianus*) are actually feather tufts that function as visual signals rather than in hearing. At the base of the ear canal is the eardrum (tympanum), beyond which are the middle and inner ears, with similar structures to those of mammals. The middle ear is an air-filled space.

In birds, a single ear bone, the columella, transmits vibrations from the eardrum to the inner ear. The avian middle ear also contains a mechanoreceptive sense organ, the paratympanic organ, which responds to changes in air pressure. This organ may function as an altimeter and/or a barometer, allowing birds to detect changes in air pressure related to flight or the approach of stormy weather. Beyond the middle ear, the inner ear contains the cochlea, the structure for sensing sounds, as well as vestibular organs for sensing balance and motion. Three semi-circular canals respond to angular and rotational acceleration of the bird's head. Beneath these are otoliths, which detect gravity and linear acceleration.

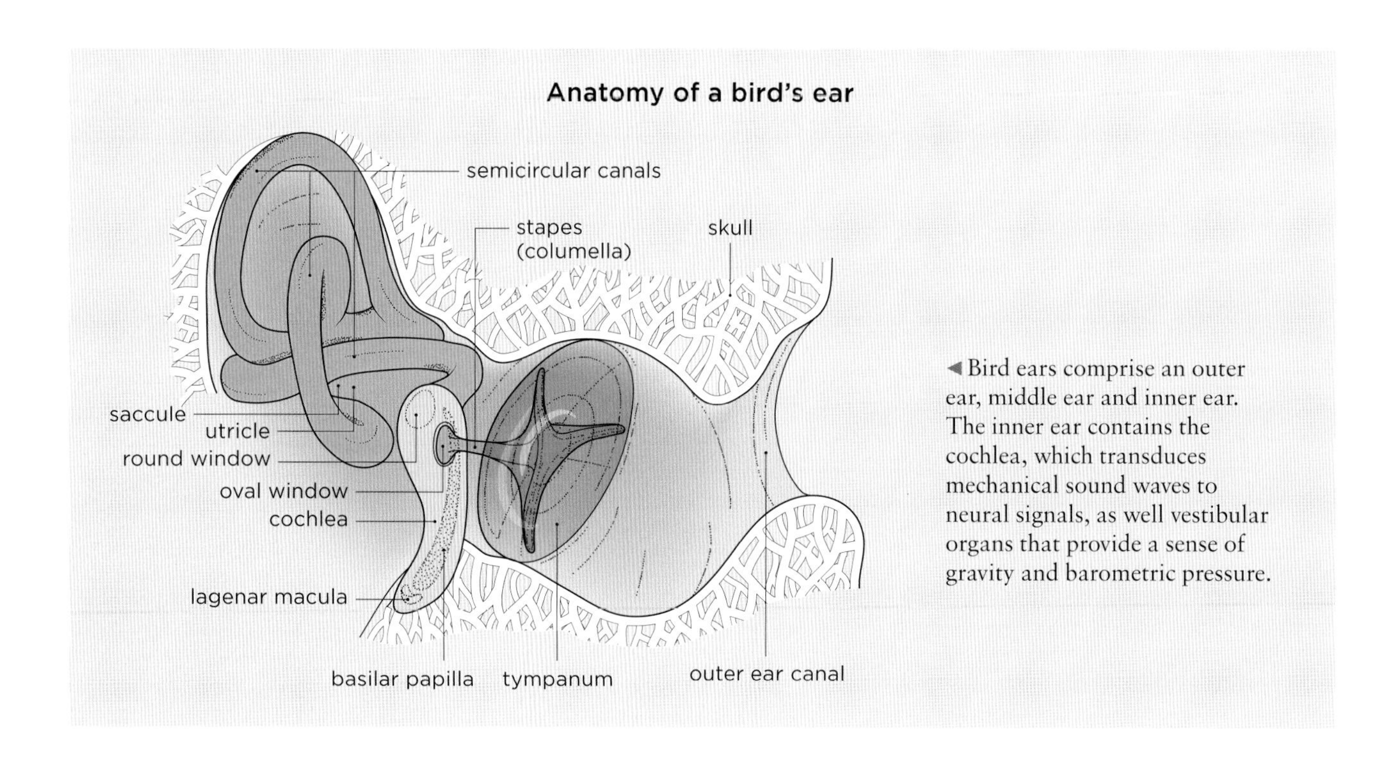

◄ Bird ears comprise an outer ear, middle ear and inner ear. The inner ear contains the cochlea, which transduces mechanical sound waves to neural signals, as well vestibular organs that provide a sense of gravity and barometric pressure.

Sound localisation

Locating the source of a sound is an important part of decoding a complex auditory scene and requires determining its distance and direction. Perceiving the direction of a sound depends on comparing incoming sound signals between the two ears. Sound coming from a bird's right will arrive at the right ear slightly sooner than at the left ear and will also be louder at the right ear. The brain decodes these inter-ear time and loudness differences to determine the direction of the sound source. Of course, birds with small heads have a shorter distance between their ears, and are correspondingly less accurate at determining sound direction. Owls are far superior at sound localisation than songbirds, and can determine sound direction to within 3–4°. These nocturnal raptors have remarkable neural circuits in which the neurons that convey information from cochlear nuclei to the brainstem have very long axons. This results in delays in transmission, aiding the perception of very subtle differences in sound timing between the ears.

Perception of the distance to a sound source depends on overall loudness (amplitude) as well as sound quality. As sounds travel through the environment, they are predictably degraded as they deflect off surfaces such as leaves. Songbirds are able to discriminate the proximity of a rival singer based on the degradation of familiar songs in addition to their loudness.

The cochlea

The avian cochlea differs from the snail-like coiled cochlea of mammals, being straight and slightly flattened in cross section. Within the cochlear duct, the organ of Corti is attached to the basilar membrane and converts vibrations into neural signals. When the fluid within the cochlear ducts moves due to vibrations transmitted through the middle ear, hair cells at specific locations are stimulated, depending on the frequency of the sound. Birds are most sensitive to sounds in the range 1–4 kHz, although this extends in some species down to about 0.3 kHz and up to 8–10 kHz. Although several birds are also known to be sensitive to very low frequency infrasound (sounds below 20 Hz), it is likely that such sounds are detected by structures other than the ear. Most birds appear to have low sensitivity to high-frequency sounds that are commonly heard by mammals.

▲ On birds with featherless heads, including this African hooded vulture (*Necrosyrtes monachus*), the opening of the ear canal is clearly visible behind and below the eyes. In most birds, the ear canal is hidden behind feathers.

► Although birds lack the ear pinnae of mammals to help detect the direction of sound, some species – including barn owls (*Tyto alba*) – have a facial ruff that acts like a parabolic reflector to aid in locating a sound source.

◄ The so-called ear tufts of northern hemisphere long-eared owls (*Asio otus*) and other owl species are not ears at all, but are likely used as visual signals to other owls.

▼ The ability to detect a beat in rhythmic sounds may be rare in non-human animals, but at least one domesticated male sulphur-crested cockatoo (*Cacatua galerita*) has been shown to dance in time to the beat of music.

► Songbirds perceive pitch differently to humans. Most songbirds that have been studied, including European common starlings (*Sturnus vulgaris*), seem to have perfect pitch and recognise sounds by absolute pitch rather than melodic changes between sounds.

Pitch and rhythm perception

Birdsong is so named because it often sounds musical to our ears. But do birds hear melody and pitch in the same way that we do? Behavioural studies demonstrate that birds have different perception of their auditory worlds. Human recognition of melodies is dominated by relative pitch perception: we can identify the same melody even if it transposed to a higher key (pitch range). Birds appear to favour absolute pitch perception, and categorise sounds based on their frequency rather than the changes in pitch between sounds. Although songbirds like European common starlings (*Sturnus vulgaris*) can be trained to respond to pitch changes, their default response is to behave like someone with perfect pitch and recognise the identity of individual notes rather than the melodic changes between them. Less is known about how birds perceive rhythm of sounds. Some evidence suggests that parrots can respond to the rhythmic beat of sound sequences, but many other birds are apparently unable to do so.

Eyes and vision

Sight is the primary sensory modality for most birds, and as such many have remarkable visual capabilities. Although the bird eye is similar in basic structure to the mammalian eye, it has some reptilian characteristics and other features that are unique to birds. Bird eyes are remarkably large, and the photoreceptive cells in the retina are densely packed, often with multiple regions of high visual acuity. The visual world of birds is impossible for us to imagine, with more dimensions of colour, and they also process images at a faster rate than humans.

Eye structure

A bird eye has a bony sclerotic ring that divides it into anterior and posterior chambers. This bony ring, shared with fish and reptiles, is thought to aid in adjusting the shape of the lens to focus light on the retina. The lens is shaped differently in different groups of birds, with that of diurnal species having a flatter anterior surface compared to nocturnal and diving birds. Birds can also adjust the shape of the outer cornea through contraction of specialised muscles, called Crampton's muscles. The focusing capacity of the cornea disappears when birds are diving underwater as there is no difference in refraction between the cornea and water. Thus, birds that use their eyes above and below water, like cormorants, must alter how they focus their eyes in each setting (see box). The shape of the eye also varies among different bird species. In some, the anterior chamber protrudes forward from the sclerotic ring, and in others the eye is more round. The relative positions of the cornea, lens and retina balance a trade-off between the size and brightness of the image projected on the retina.

The iris acts as a diaphragm to control how much light can pass through the pupil and lens to the posterior eye chamber. The innermost retinal layer on the back of the posterior chamber contains photoreceptive cells. Some species have a layer called the tapetum lucidum under the retina, which reflects

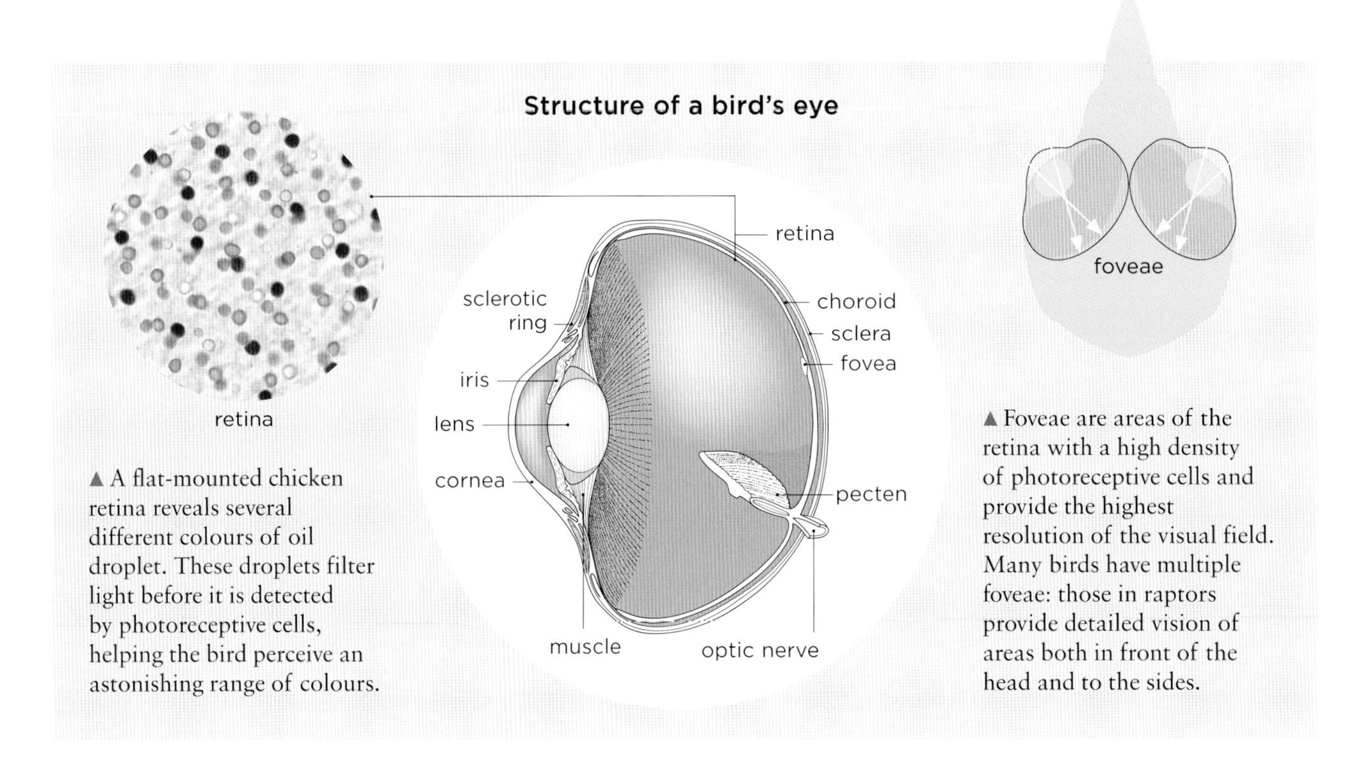

▲ A flat-mounted chicken retina reveals several different colours of oil droplet. These droplets filter light before it is detected by photoreceptive cells, helping the bird perceive an astonishing range of colours.

▲ Foveae are areas of the retina with a high density of photoreceptive cells and provide the highest resolution of the visual field. Many birds have multiple foveae: those in raptors provide detailed vision of areas both in front of the head and to the sides.

▲ Frontal view of a shoebill (*Balaeniceps rex*), illustrating the overlapping binocular visual field of a predator.

light back through the retina to improve vision in low light. The posterior chamber of the eye also contains a reptilian structure called the pecten. This is highly vascularised and, with the aid of an eye-specific protein called globin E, which acts like haemoglobin in the blood, allows oxygen and nutrients to diffuse through the fluid of the eye to the retina. With this boosted supply of nutrients and oxygen to the retina, a greater density of photoreceptors can be packed tightly together in the retina, increasing visual acuity.

Foveae

Foveae are regions of the retina with the highest densities of photoreceptors and thus the highest visual acuity. In mammals, the shape of the fovea varies across species, but in birds both the shape and the number of fovea are variable. For example, birds that forage on the wing, including raptors, typically have two foveae – one centrally directed and the other directed toward the binocular visual field. Birds also have a greater variety of photoreceptors than mammals, with up to five types of visual pigments in multiple types of photoreceptive cells (e.g. rods, cones, double cones). In addition, up to six types of coloured oil droplets in the anterior end of photoreceptive cells can filter light before it reaches the photopigments, where the light is detected. The range of wavelengths detected by bird eyes extends well beyond mammalian vision and into the ultraviolet range.

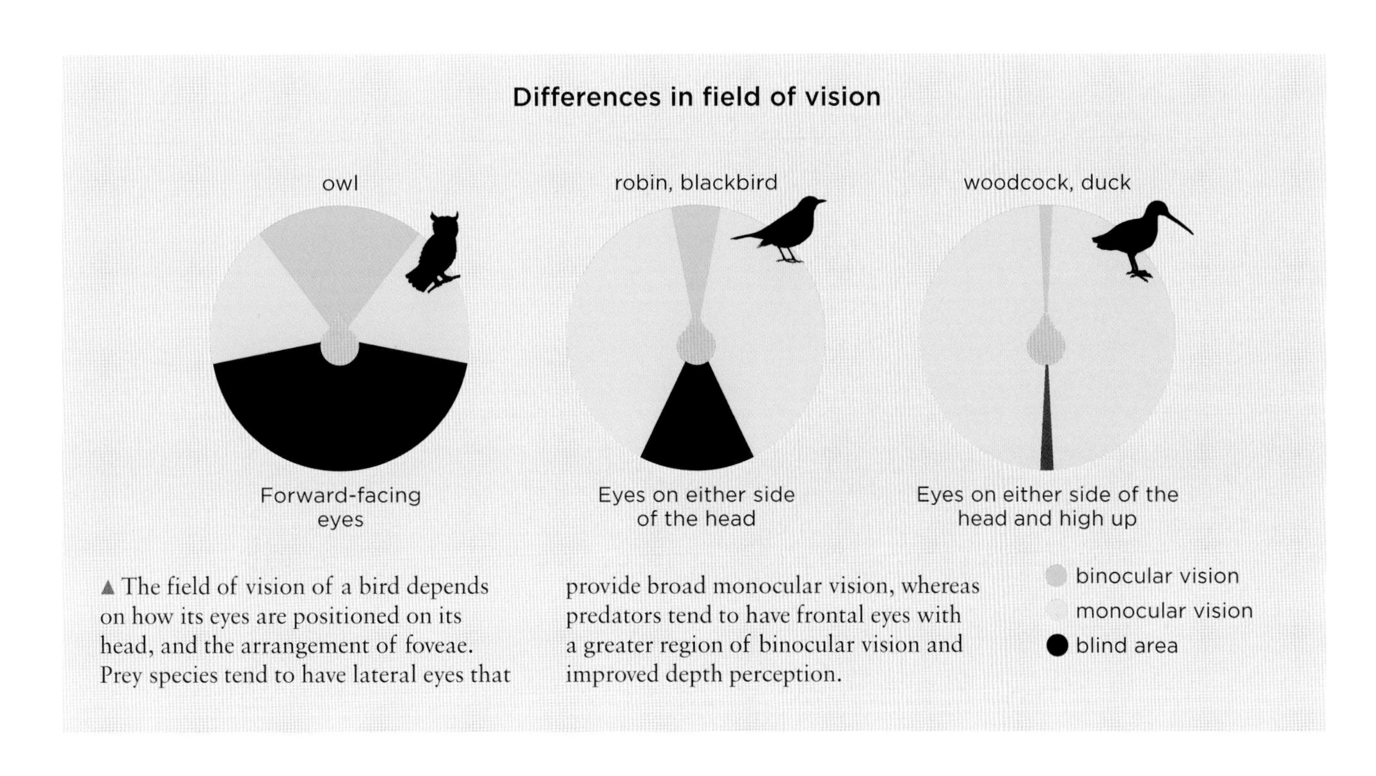

▲ The field of vision of a bird depends on how its eyes are positioned on its head, and the arrangement of foveae. Prey species tend to have lateral eyes that provide broad monocular vision, whereas predators tend to have frontal eyes with a greater region of binocular vision and improved depth perception.

Northern harrier
(*Circus hudsonius*)

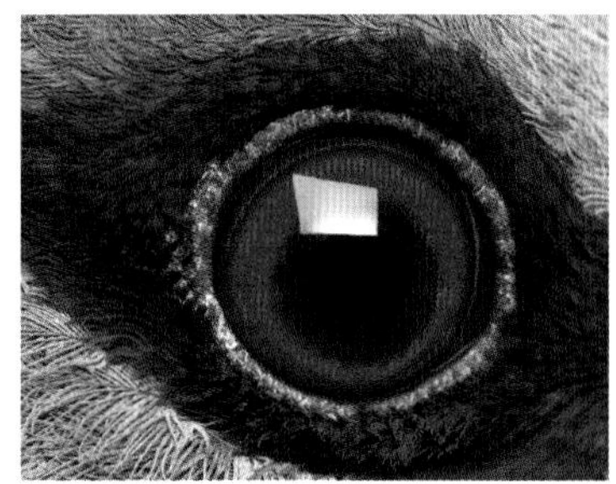

Victoria crowned pigeon
(*Goura victoria*)

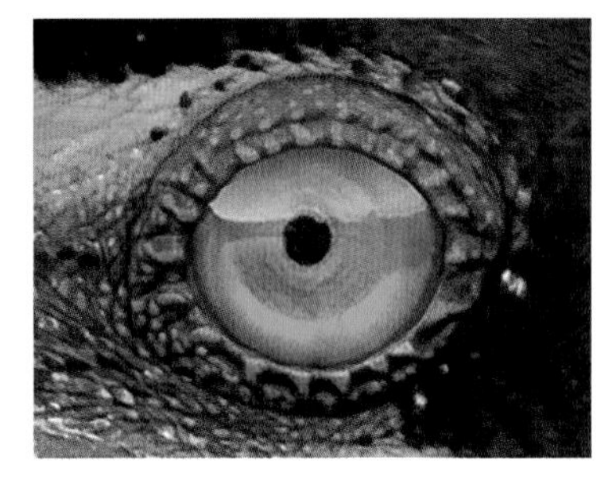

Double-crested cormorant
(*Phalacrocorax auritus*)

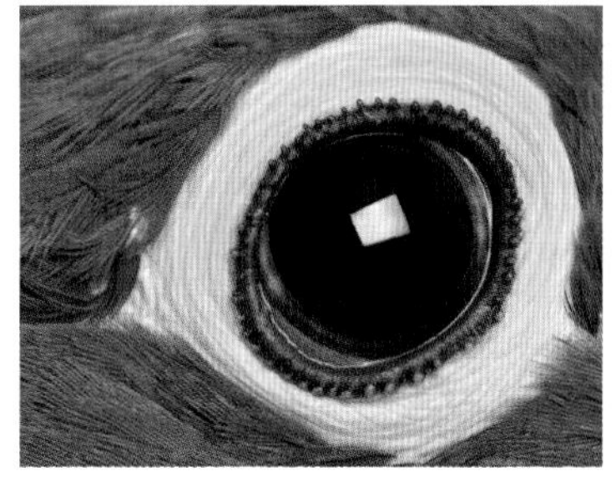

Hyacinth macaw
(*Anodorhynchus hyacinthinus*)

▲ Bird eyes exhibit incredible diversity in shape, size and colour.

Bird vision

Birds' remarkable eyes send the visual information they receive to the brain, where large regions are dedicated to visual processing and perception. Visual perception includes different components, such as acuity, motion and colour. Visual acuity is the ability to resolve fine spatial details, and varies with illumination levels. Bird visual acuity varies across species, with remarkably high acuity in diurnal raptors (hawks and eagles) under bright light conditions, and relatively low acuity in nocturnal owls. High visual acuity appears most important for detecting small objects, such as prey, from great distances. In bright daylight, eagles have about twice the acuity of young humans and about a hundred times the acuity of barn owls. However, the acuity of eagles and humans falls quickly in dim light, but does not decline as steeply in owls. This means that on a dark night barn owls have better acuity than humans. Contrast sensitivity is another measure of visual perception, and refers to the ability to detect varying increments of light and dark on the greyscale. Compared to mammals, birds have low contrast sensitivity – in other words, for a given spatial resolution, the contrast between light and dark regions needs to be greater for birds to detect them.

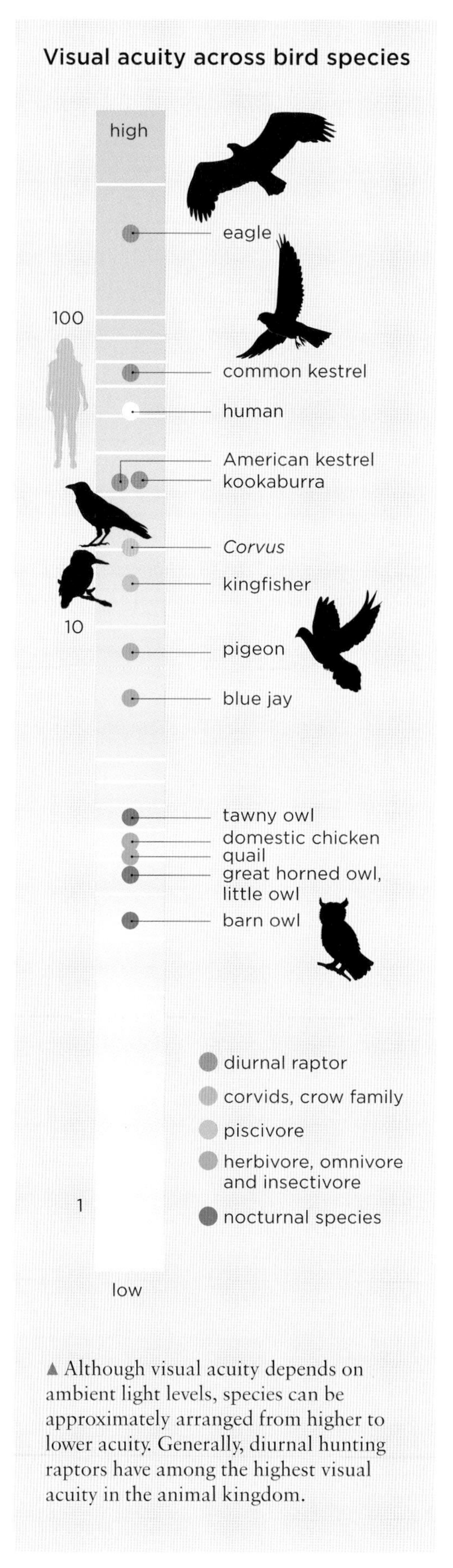

▲ Although visual acuity depends on ambient light levels, species can be approximately arranged from higher to lower acuity. Generally, diurnal hunting raptors have among the highest visual acuity in the animal kingdom.

Flicker and colour

Another form of visual perception is temporal resolution, or how well images are resolved in time. The flicker-fusion threshold is the point at which the rate a light flickers on and off is perceived as continuous illumination. For example, we perceive movies at the cinema as smooth motion rather than individual frames because the rate of presentation exceeds our flicker-fusion threshold of about 60 Hz. Birds have a higher flicker-fusion threshold, at 100 Hz or more, which may aid their ability to resolve visual information quickly in time as they fly through a cluttered environment.

Light does not have colour. Rather, visual systems create a perception of colour based on combinations of different wavelengths of light. Because birds have multiple photoreceptors in their eyes that can detect an expanded range of wavelengths, they can perceive colours outside our visual range that we can only imagine. This means that objects that appear identical to us can look starkly different to birds. So while the crest feathers of male and female chickadees and titmice often appear identical to humans, birds have no trouble telling them apart because to them they appear quite distinct.

▲ Male Eurasian blue tits (*Cyanistes caeruleus*) have a UV-reflectant crown patch. When sunscreen was applied to male birds, hindering UV reflectance, the birds' mates produced fewer male offspring, supporting the notion that UV reflectance plays a role in male attractiveness.

Great white egret spearing a fish

▲ Herons and egrets that spear fish from the water need to correct for the visual displacement of underwater objects. When viewing their prey from above water, the actual location will differ from the viewed location, depending on the viewing angle. Great egrets (*Ardea alba*) are distributed worldwide, and are clearly adept at this task.

SEEING INTO WATER

As already mentioned, some birds need to correct for refraction at the air–water surface when they are hunting for fish. Western reef herons (*Egretta gularis*) adjust their head position such that the actual depth of a fish is highly correlated to its perceived depth before they strike. Plunge-diving predators, such as the Afro-Asian pied kingfisher (*Ceryle rudis*), also need to cope with refraction in order to successfully capture prey. The diversity of bird eyes and visual systems allows different species to cope with a variety of foraging challenges under bright and dim light, and in both air and water.

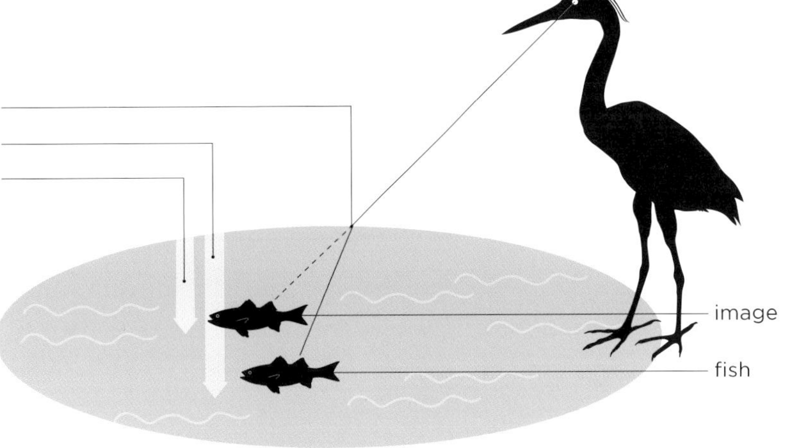

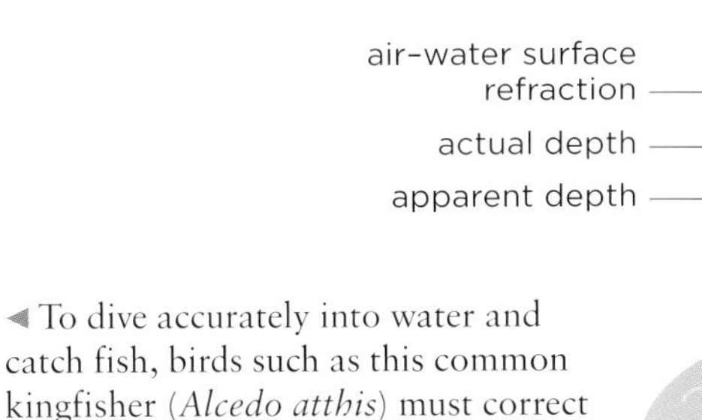

◀ To dive accurately into water and catch fish, birds such as this common kingfisher (*Alcedo atthis*) must correct for refraction of light at the air–water interface, and must also be able to focus their vision in both air and water.

▼▼ An osprey (*Pandion haliaetus*), with talons spread, comes in to catch a fish lurking just below the water's surface.

Magnetoreception

Although it seems alien to us, many animals are able to sense the Earth's magnetic field and use it during orientation and navigation. Magnetic compasses interact with, and may be redundant alongside, other compasses such as a sun compass or star compass. Some songbirds appear to calibrate their magnetic compass using sky cues at sunset, and depend on seeing polarised light at this time.

◄ Many birds, including homing or messenger pigeons (*Columba livia domestica*), have been shown to use the Earth's magnetic field for orientation and navigation.

Earth's magnetic field

Across the globe, the Earth's magnetic field varies in strength, angle of inclination and polarity. Its strength ranges from about 60,000 nanoTesla (nT) at the poles, to 30,000 nT at the equator. The direction of magnetic force is parallel to the surface of the Earth at the equator, but vertical to the Earth at the magnetic poles, thus this angle of inclination ranges from 0° to 90°. Finally, the polarity of the Earth's magnetic field creates north and south poles. Of these, birds primarily use magnetic inclination for their magnetic compass sense.

A compass in the beak?

There are two hypotheses as to how birds sense the Earth's magnetic field, both of which may be correct. The first potential mechanism is that magnetic iron compounds within cells could provide magnetic sensing. Structures containing magnetic iron-based minerals have been reported in birds and other animals, but whether these minerals form a compass or are part of iron metabolism is still unclear. There is evidence that iron-mineral structures are located in the upper beak of birds, and that magnetic information may be conducted to the brain via the trigeminal nerve. However, this potential mechanism does not appear to be required for compass use and instead may detect the strength of the magnetic field. There may also be an iron-based sensor in the lagena, which is part of the vestibular system of the ear, and neural recordings suggest that this structure may detect magnetic fields.

A compass in the eye?

A second possible mechanism is that photoreceptive molecules in the retina allow magnetoreception. Light-sensitive molecules, potentially cryptochromes, may react differently to light depending on the magnetic field, and thus may provide the biochemical basis by which cells in the retina could allow visual detection of magnetic fields. A brain region located in the visual Wulst, called cluster N, has been shown to be activated at night when birds are using their eyes for magnetic migratory orientation. This region is not active during the day, and appears to have reduced activity in non-migratory species and during the non-migratory season. It is thus possible that the magnetic compass depends on biochemical processes in the eye and neural processing in visual regions of the brain.

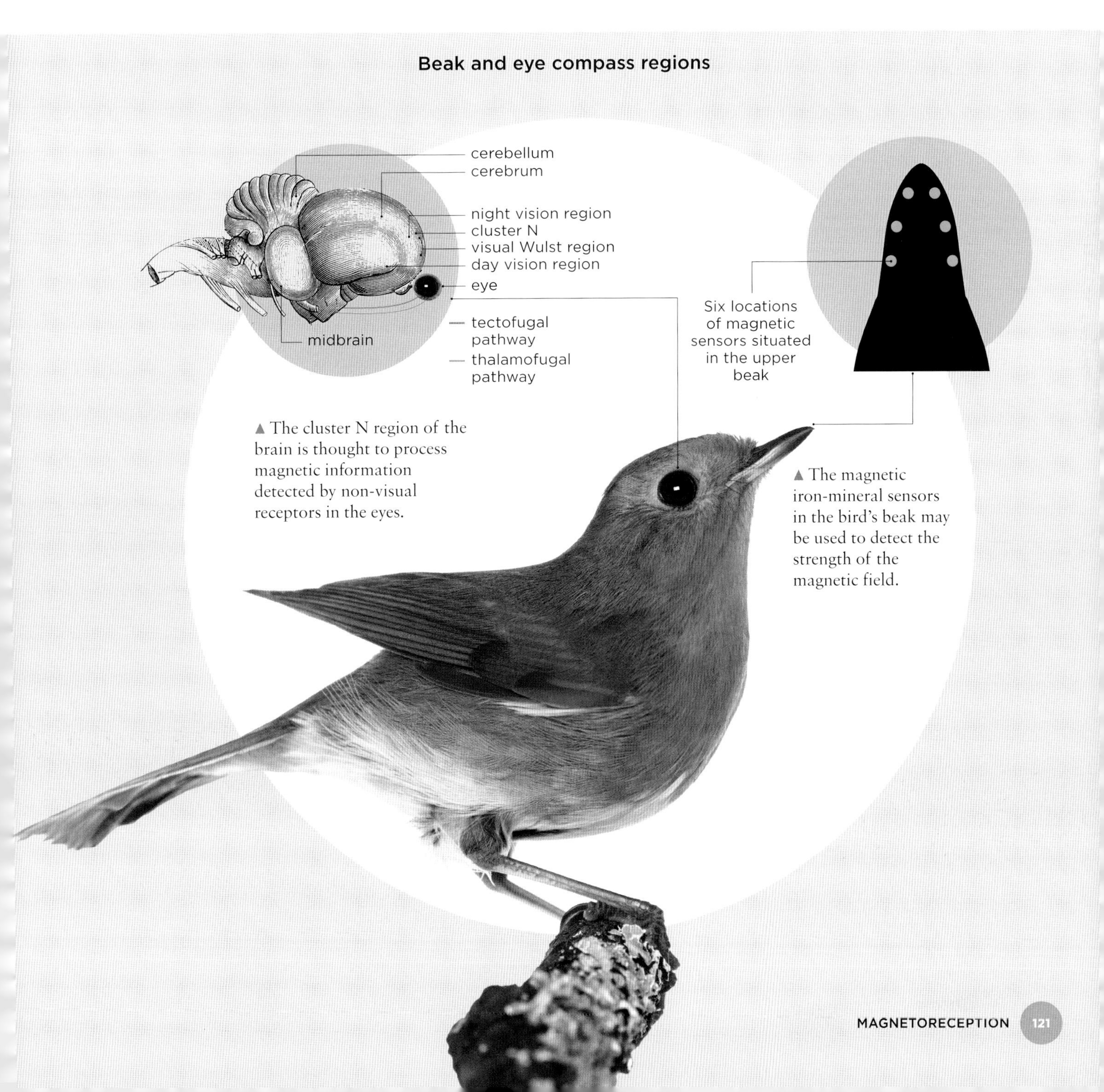

▲ The cluster N region of the brain is thought to process magnetic information detected by non-visual receptors in the eyes.

▲ The magnetic iron-mineral sensors in the bird's beak may be used to detect the strength of the magnetic field.

Hippocampus and spatial memory

The hippocampus is a region of the brain important for memory, especially spatial memory. The avian hippocampus is homologous to the mammalian hippocampus, but rather than being located in the medial temporal lobes of the brain, it is at the top of each cerebral hemisphere. Despite its very different appearance, the avian hippocampus has a very similar function to that of mammals, including being critical for the recollection of spatial locations.

Memory for food locations

The importance of the hippocampus for spatial memory has been highlighted by studies of food storing in birds. Many species of chickadees and titmice scatter-hoard, hiding scores of individual food items at various locations. They then retrieve these hidden food items days or weeks later using their spatial memory. Birds that store seeds, such as North American black-capped chickadees (*Poecile atricapillus*) and Eurasian willow tits (*Poecile montanus*), have a larger hippocampus relative to their overall brain size compared to non-storing species like great tits. Food-storing nuthatches and corvid species also tend to have a larger hippocampus than non-storing species within these groups. For example, Eurasian jays (*Garrulus glandarius*) rely on stored acorns and have a larger hippocampus for their body size than do Eurasian jackdaws (*Coloeus monedula*), which rarely store food.

Adaptations of the hippocampus

The hippocampus also varies within species. For example, black-capped chickadees living in the harsher climate of Alaska have a larger hippocampus than those living in more temperate regions. Presumably, the Alaskan birds depend more on stored food during winter, and natural selection has driven the evolution of a larger hippocampus with more neurons.

Hippocampus in a chickadee

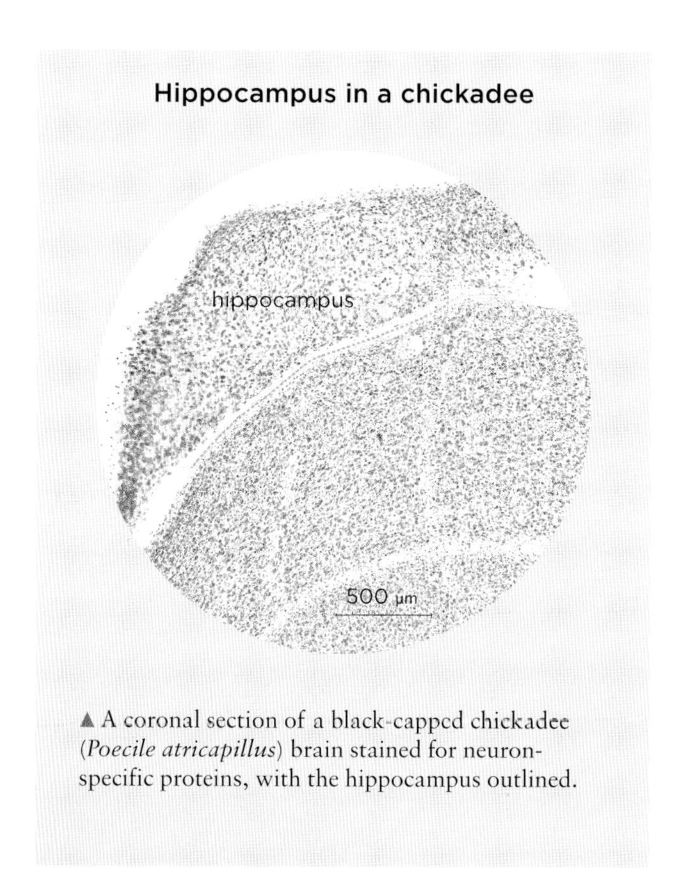

▲ A coronal section of a black-capped chickadee (*Poecile atricapillus*) brain stained for neuron-specific proteins, with the hippocampus outlined.

► Black-capped chickadees (*Poecile atricapillus*) scatter-hoard food and then retrieve these food items using spatial memory. Food-storing species of titmice have larger hippocampi than closely related species that do not store food.

◄ Female brown-headed cowbirds (*Molothrus ater*) from North America are obligate brood parasites. They search for nests of other species in which to lay their eggs, and have a larger hippocampus than the males.

Alaskan chickadees also perform better on spatial memory tests, and common-rearing experiments support the idea that these population differences are inherited. At a finer scale, North American mountain chickadees (*Poecile gambeli*) that live at higher elevations, where winters are harsher and food storage potentially more important, also have a larger hippocampus and better spatial memory than birds living at lower elevations.

Another case of hippocampus size reflecting spatial memory in birds comes from brood parasites. Females of the North American brown-headed cowbird (*Molothrus ater*) are obligate brood parasites, and they search for, and remember, the locations of many nests of other species in which to lay their eggs. Researchers have found that the males, which do not search for nests, have a smaller hippocampus than the females, and that in other cowbird species that are not brood parasites this sex difference in hippocampus size is absent.

Seasonal changes

Food storing in black-capped chickadees is a seasonal behaviour, with most hoarding occurring in autumn or winter. Some evidence indicates that black-capped chickadees have a larger hippocampus with increased rates of neurogenesis when they are storing most actively. Seasonal variation in hippocampal neurogenesis has been documented in several studies, but the exact timing of maximum neurogenesis varies between these. This suggests that changes in the brain may reflect year-to-year variations in the timing of peak food storing (and food retrieving), rather than being strictly tied to the seasons.

◄ This common grackle (*Quiscalus quiscula*) nest has been parasitised with three cowbird eggs.

Physiology

By adaptively modifying their physiology, birds have evolved to live in nearly every habitat on Earth, from the poles to the tropics. Consider a small songbird on a hot summer's day. How does it stay cool at high noon when the air temperature is well above its body temperature? Can the bird get all the hydration it needs from the food it eats or must it drink water? How does it modify its activity (feeding, breeding, territorial fighting, singing and so on) through the day to reduce overheating, yet at the same time manage to reproduce successfully and avoid predators? Answers to such questions fall within the purview of avian physiology, the topic of this chapter. Here, we address how birds regulate their internal environment across vastly different external environments, from the Antarctic ice caps to the hottest deserts, from tropical rainforests to temperate taiga, and across the oceans.

◀ Black grouse (*Lyrurus tetrix*).

The importance of balance

Birds maintain relatively constant physiological conditions within their body (known as homeostasis) across vastly different external environments, and it is this ability that enables them to be so widespread. They regulate their physiological state for water and salt content, temperature, pH, nutrient and oxygen levels, and waste and carbon dioxide levels. Just as the thermostat in a house detects changes in room temperature relative to a set point and adds either hot or cold air as needed, so a bird's homeostatic system relies on negative-feedback pathways to adjust the animal's physiology when unbalanced.

A useful way to consider how birds regulate their internal environment in the face of vastly different external environments is by constructing 'budgets' for relevant physiological conditions, such as energy, protein, water and salt. For example, a 90 kg ostrich living in the hot African savannah loses about 8 litres of water each day, mainly through urine and faeces production (64 per cent), respiration (29 per cent) and via its skin (7 per cent). To balance these losses, the ostrich must drink about 7.5 litres of water each day – the other 0.5 litres that is required is produced from the metabolism of food. An ostrich that is unable to find this much drinking water cannot maintain water homeostasis and would not survive for long.

Osmoregulation

For most birds (and other vertebrates), maintaining an adequate water and salt balance within the body is a major challenge, because dehydration is a constant danger when living in terrestrial or marine systems.

Life on Earth requires water, and the bulk of a bird's body – normally 60–70 per cent – is water. Water is sometimes called the universal solvent and is needed for most chemical reactions and for maintaining many bodily functions. For example, if the water content of blood is inadequate, blood pressure falls and this compromises the delivery of key nutrients to vital organs. Salt balance must be considered along with water balance, because all organisms – including

▼ Budgerigars (*Melopsittacus undulatus*; left) are native to Australia, where hot, dry conditions are common. They must drink to stay appropriately cool, and their distribution is therefore constrained by the location of free water. Other birds, such as this rock dove (*Columba livia*; right), stay cool by ruffling their feathers to increase heat loss from their body and decrease heat gain from solar radiation.

▲► The common ostrich (*Struthio camelus*) inhabits some of the hottest, driest places in Africa and so must find ways to conserve its body water, despite inevitably losing some through respiration, excretion (faecal and urine loss) and perspiration (cutaneous water loss). The birds achieve this by regularly ingesting water, minimising the inevitable losses and using their osmoregulatory system to maintain water and salt balance.

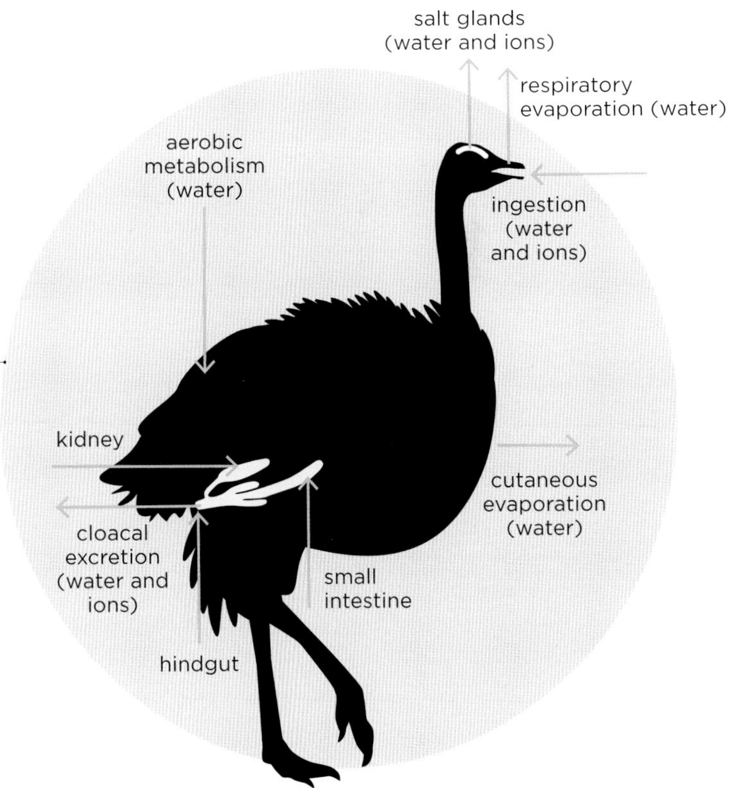

birds – move water across cell membranes passively by the process known as osmosis (hence the term 'osmoregulation' for the entire physiological system that is involved).

Osmosis only occurs though when water, a highly potent solvent, moves across semipermeable membranes (such as the cell membranes within a bird's body) from a less concentrated solution to a more concentrated one. These solutes, or components dissolved in the solution, include important chemicals such as sodium and chloride (the constituents of common table salt), as well as potassium and calcium. Osmosis is clearly demonstrated when limp vegetables such as carrots are placed in water to make them crisp again – water moves into the carrots, because when they are limp they contain a relatively higher concentration of nutrients and salts.

In general, birds maintain their water and salt balance by drinking more and losing less through their skin and during respiration. They primarily achieve water and salt balance with the key organs of the osmoregulatory system: kidneys, intestines and salt glands.

Kidneys and urine production

Birds and mammals are the only vertebrates that can produce urine that is more concentrated than their body fluids (critical for maintaining water balance) and they achieve this thanks to unique adaptations to their kidneys. In brief, the paired kidneys initially filter the blood to produce a very dilute urine. The urine then flows through a convoluted plumbing system within the kidney that includes tubes (called nephrons) arranged in a series of adjacent loops. As the urine moves through the nephrons, water is reabsorbed and the concentration of salts in the urine increases, so the longer the loops, the more concentrated the urine. Mammals have much longer, specialised nephron loops than birds and can thus concentrate their urine to a greater extent – up to 25 times that of body fluids, compared to a more modest three times for birds. Birds and mammals can further increase urine concentration by increasing the rate of filtration of the kidneys and, through the action of hormones, modifying the permeability of the nephrons.

Birds have evolved two other relatively unique ways to reduce their water loss compared to mammals. The first is related to processing the urine after it leaves the kidneys. Unlike mammals, birds

Urinogenital system of a dove

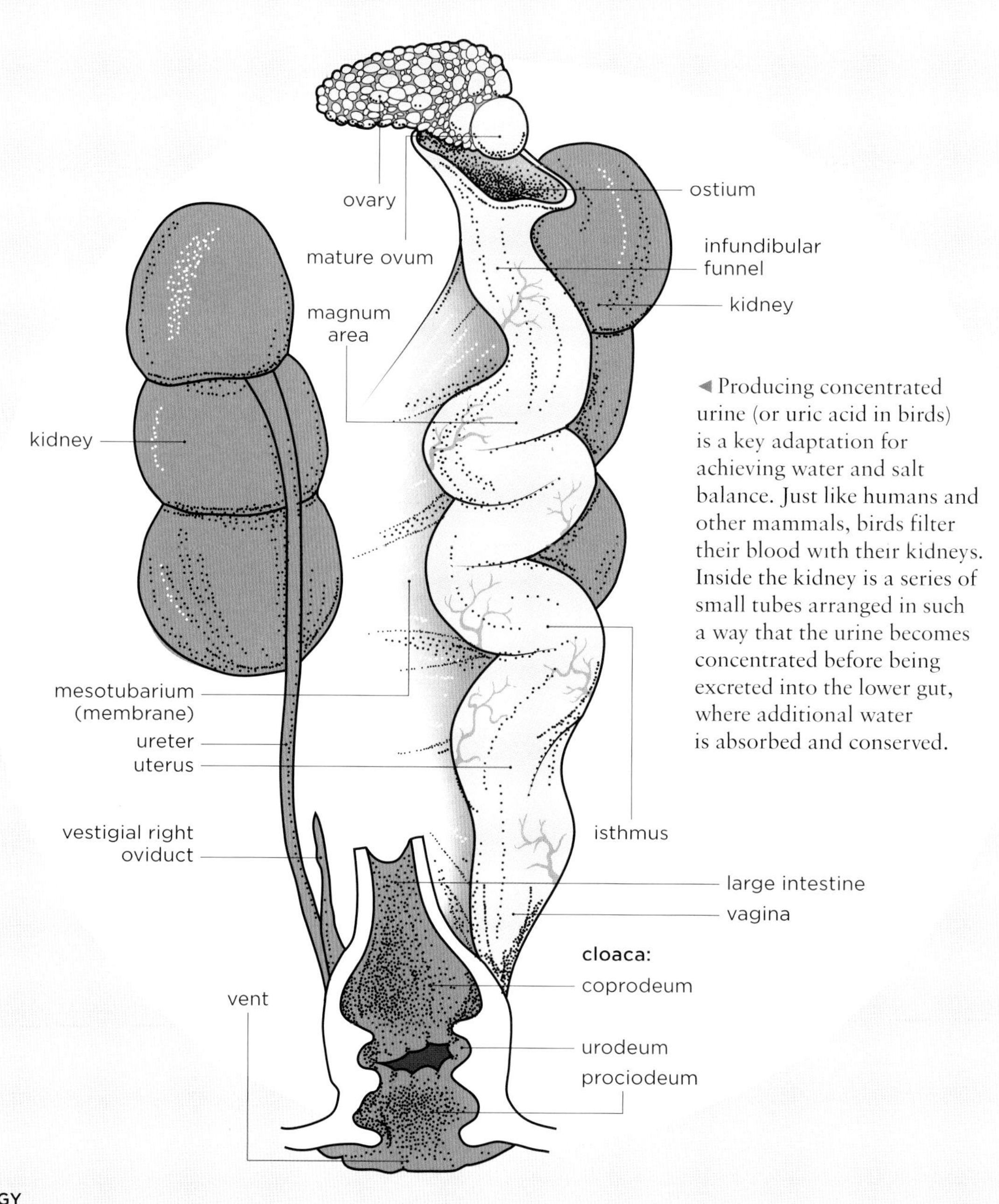

◄ Producing concentrated urine (or uric acid in birds) is a key adaptation for achieving water and salt balance. Just like humans and other mammals, birds filter their blood with their kidneys. Inside the kidney is a series of small tubes arranged in such a way that the urine becomes concentrated before being excreted into the lower gut, where additional water is absorbed and conserved.

have no urinary bladder and the product from the kidneys is released directly into the lower part of the gut, where important reabsorption of water and some salts occurs. The second adaptation relates to another important function of the kidneys – the excretion of nitrogen waste products generated by protein metabolism. Unlike fish, most amphibians and mammals, which excrete protein waste as ammonia and/or urea (otherwise known as urine), birds and reptiles excrete this mainly as uric acid, a white pasty compound familiar to anyone whose car has been splattered by gull or tern droppings. A benefit of producing uric acid is that little water is required (0.001 litres per gram of nitrogen excreted) compared to the water required for ammonia (0.5 litres) or urea (0.05 litres) excretion. However, the energy cost of producing uric acid (as well as urea) as the nitrogenous waste product (around 18 kJ per gram of protein) is relatively high compared to the negligible energy cost of producing ammonia. For birds, the evolutionary balance between water conservation and energy conservation appears to have fallen towards water conservation.

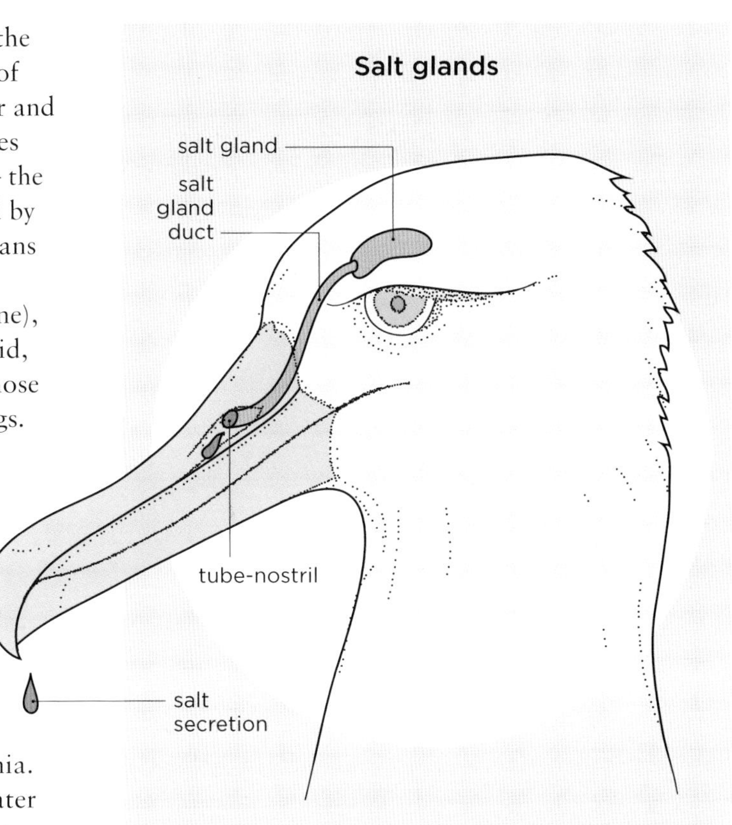

Salt glands

The final piece to the puzzle of how birds achieve water and salt balance in challenging environments such as deserts and oceans involves a useful pair of glands, the salt glands, usually located just above the orbit of the eye. These enable birds to, for example, drink only seawater because they are able to excrete concentrated salt solutions. The activity of the salt glands depends on a bird's recent exposure to salt, as the glands will grow with use and shrink with disuse. The glands rid the body of excess salts primarily through active transport of sodium chloride across cell membranes. The highly concentrated salt-gland solution exits through the bird's nostrils on top of the beak. If you closely observe marine birds such as penguins, gulls or pelicans, you will often see them shaking their head from side to side – this action jiggles the accumulated salt-gland solution off the tip of the bill, making room for more of the salty secretion.

▲ Salt glands in birds are arranged above the eyes (only one is shown above) and enable them to get rid of excess salts when, for example, they drink only seawater and eat mostly salty prey such as fish or squid. The concentrated salt solution is excreted through the nostrils, which for this southern giant petrel (*Macronectes giganteus*) are within its quite large 'tube nose' at the top of its bill.

Respiratory and circulatory systems

The high metabolic rates and extensive daily movement of birds requires an extraordinary capacity to deliver oxygen and energy to their muscles and organs, as well as rapid removal of waste products. The circulatory and respiratory systems of birds accomplish these integrated tasks. The respiratory system transports oxygen from the atmosphere to the circulatory system and moves carbon dioxide in the opposite direction. The circulatory system delivers oxygen from the lungs to the tissues of the body and also delivers digested nutrients from the gut. These systems have several specialisations to meet the high demands of flight.

Flow-through respiratory system

The avian and mammalian respiratory systems are quite different in basic design. Bird lungs are paired to an extensive network of air sacs that extend throughout the chest and abdominal cavity, and even into hollow bones. The air sacs and lungs work together such that air flows in only one direction through the lungs, requiring two complete inhalation/exhalation cycles to move a volume of air through the respiratory system. This unidirectional routing is important, because it means that the air moving through a bird's lungs, with continuous one-way flow, is constantly refreshed and has a high oxygen content. That is, fresh oxygenated air flows through the lungs both when a bird inhales and when it exhales.

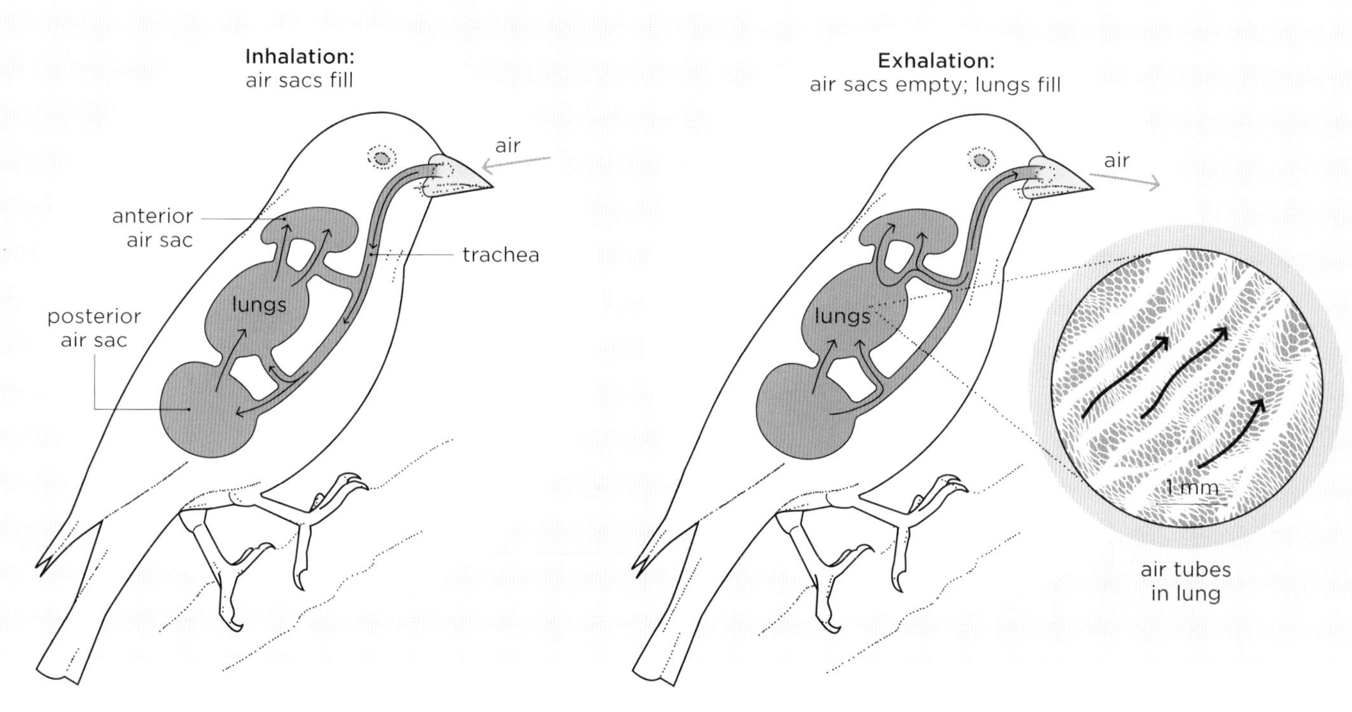

In contrast, airflow in the mammalian lung is bidirectional or tidal (air moves in and out of the blind-ending alveoli, or air capillaries, with every inhalation and exhalation via the trachea and branched bronchi and bronchioles). This mixes the incoming new air with oxygen-depleted air that has already been in the lungs. Birds avoid this mixing of incoming and old air, providing a more efficient rate of gas exchange (oxygen in, carbon dioxide out) than mammals. In addition, just as alveoli greatly increase the surface area of mammalian lungs, the bird lung is subdivided into a large number of thin-walled, tubular structures called parabronchi, which are covered by a rich supply of capillaries to maximise the rate of gas exchange between the lung and the circulatory system.

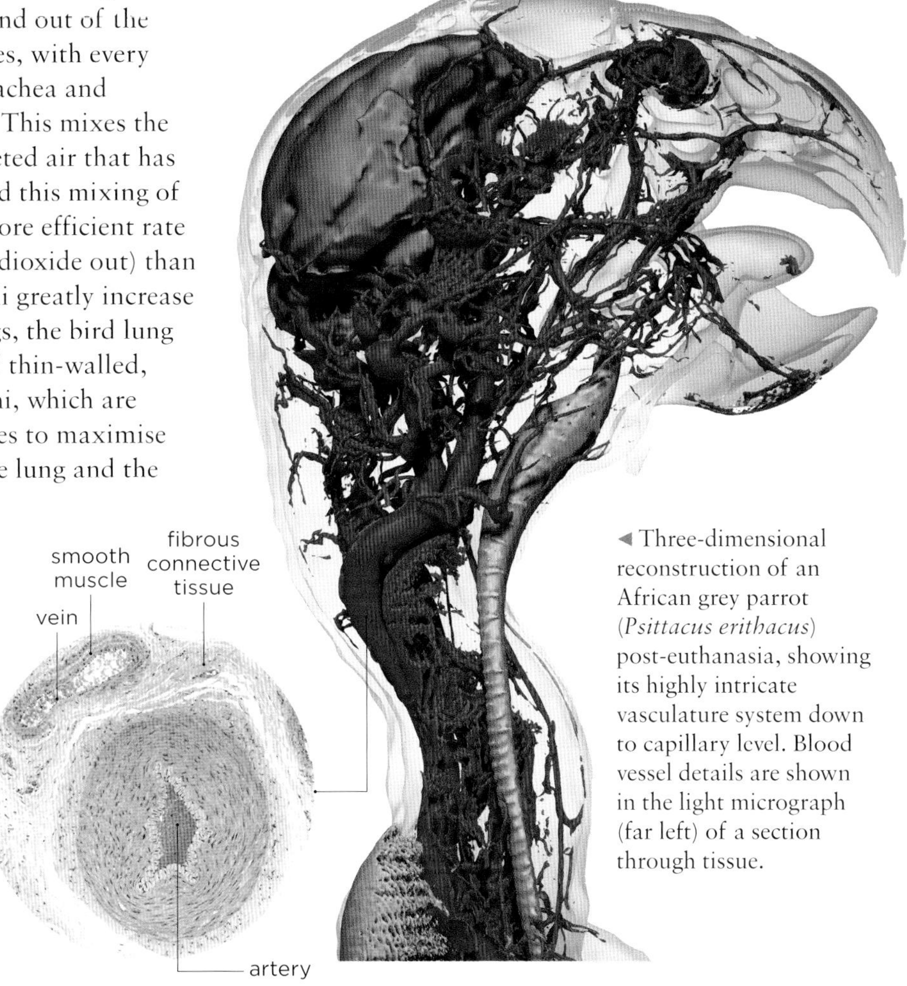

◀ Three-dimensional reconstruction of an African grey parrot (*Psittacus erithacus*) post-euthanasia, showing its highly intricate vasculature system down to capillary level. Blood vessel details are shown in the light micrograph (far left) of a section through tissue.

◀ Birds' lungs and air sacs work together to allow fresh oxygenated air to flow through the lungs both when birds breathe in and when they breathe out. This provides a very efficient transfer of oxygen to, and carbon dioxide from, the blood.

▼ High-efficiency respiration helps Asian bar-headed geese (*Anser indicus*) migrate over the Himalayas.

Circulatory system

Much like that of mammals, the avian circulatory system consists of two major circuits. A pulmonary circuit conducts blood at relatively low pressure through the lungs, where gas exchange occurs: red blood cells become oxygenated and carbon dioxide is exhaled. The other circuit, powered by the left half of the heart, circulates blood to the rest of the body. Arteries conduct blood away from the heart, and veins conduct blood towards it. Arteries have thicker, more muscular walls than veins. The large arteries leaving the heart divide and subdivide into smaller and smaller branches, until they become capillaries, the smallest vessels of the circulatory system. It is in the capillaries that blood exchanges gases (e.g. oxygen), nutrients, waste products and chemical messengers like hormones with the surrounding tissues. After passing through the capillaries, blood is collected in venules; these merge into larger and larger veins before returning to the heart.

Pumping blood: the four-chambered heart

Birds have a four-chambered heart, similar to that of mammals. However, birds and mammals evolved this innovation independently from the more ancestral three-chambered heart still found in amphibians and many reptiles. The heart has two muscular ventricles that act as pumping chambers, connected to two thinner-walled receiving chambers called atria. The two sides of the heart pump blood independently through a lung (pulmonary) circuit and a body (systemic) circuit. In animals with three-chambered hearts, blood from the two circuits can mix in the single ventricle. In birds, the pulmonary circuit can be maintained at a lower blood pressure than the body circuit to protect the delicate blood vessels in the lungs, because the ventricles can induce different force. Blood enters the heart at the right atrium via the veins of the body, and then flows into the right ventricle. Contractions of the right ventricle pump blood through the lungs, where oxygen and carbon dioxide are exchanged, and then on to the left atrium. From the left atrium, blood flows to the left ventricle, which is larger and more muscular than the right, where it is then pumped out to the blood vessels of the body. One-way valves between the atria and ventricles prevent back-flow in this circuit.

DIFFERENT PLUMBING

Although birds and mammals both have four-chambered hearts, they evolved these dual circuits independently. Ancestral vertebrates had dual aortic arches carrying blood away from the heart, and these are present in embryonic stages of mammals and birds. Reptiles retain a double arch into adulthood, but mammals and birds, with their fully separated, dual-circuit, four-chambered hearts, retain only a leftward or rightward arch, respectively.

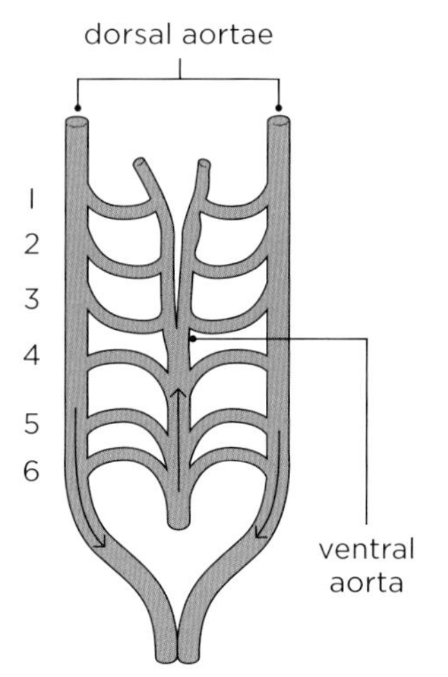

ancestral vertebrate

▲ In ancestral vertebrates multiple aortic arches provided blood to the gills. These are seen in early embryology of many modern vertebrates.

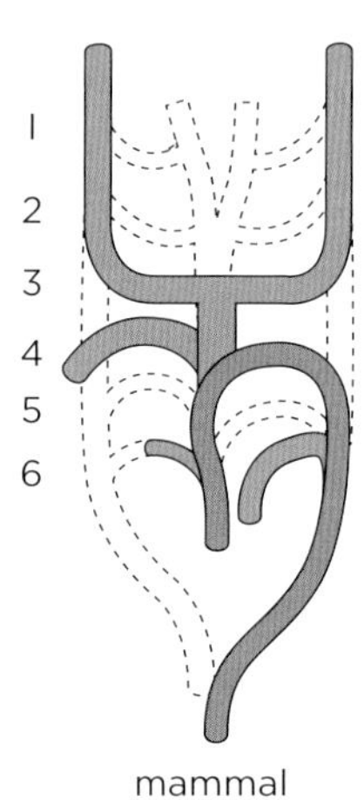

mammal

▲ In mammals, only three aortic arches develop into blood vessels, and a single systemic aorta develops from the fourth arch on the right side.

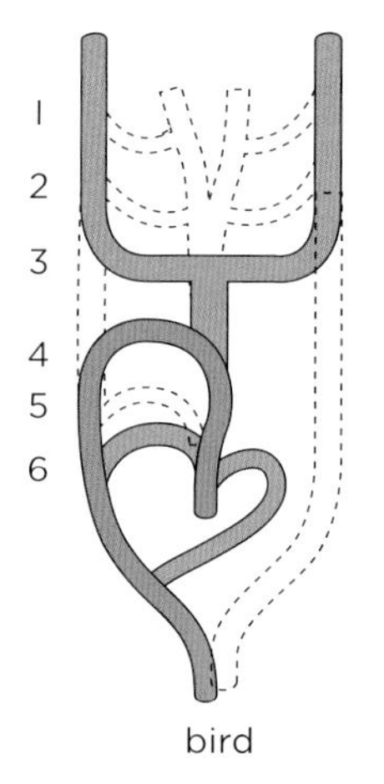

bird

▲ In birds, only three aortic arches develop, but the single systemic aorta develops from the fourth arch on the left side.

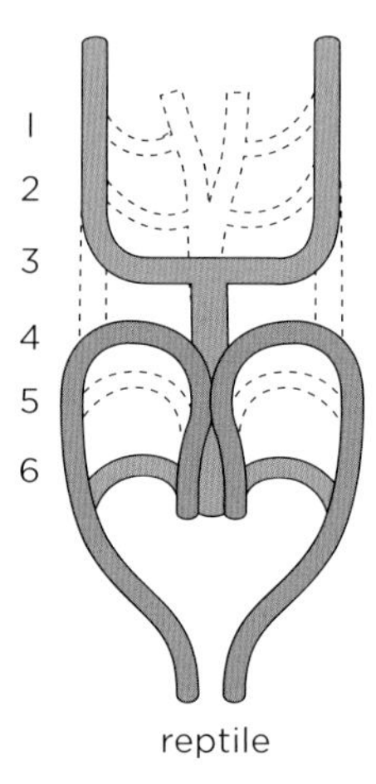

reptile

▲ Many reptile species also retain three of the aortic arches, but these develop into two systemic aorta.

Avian heart morphology and comparisons with other groups

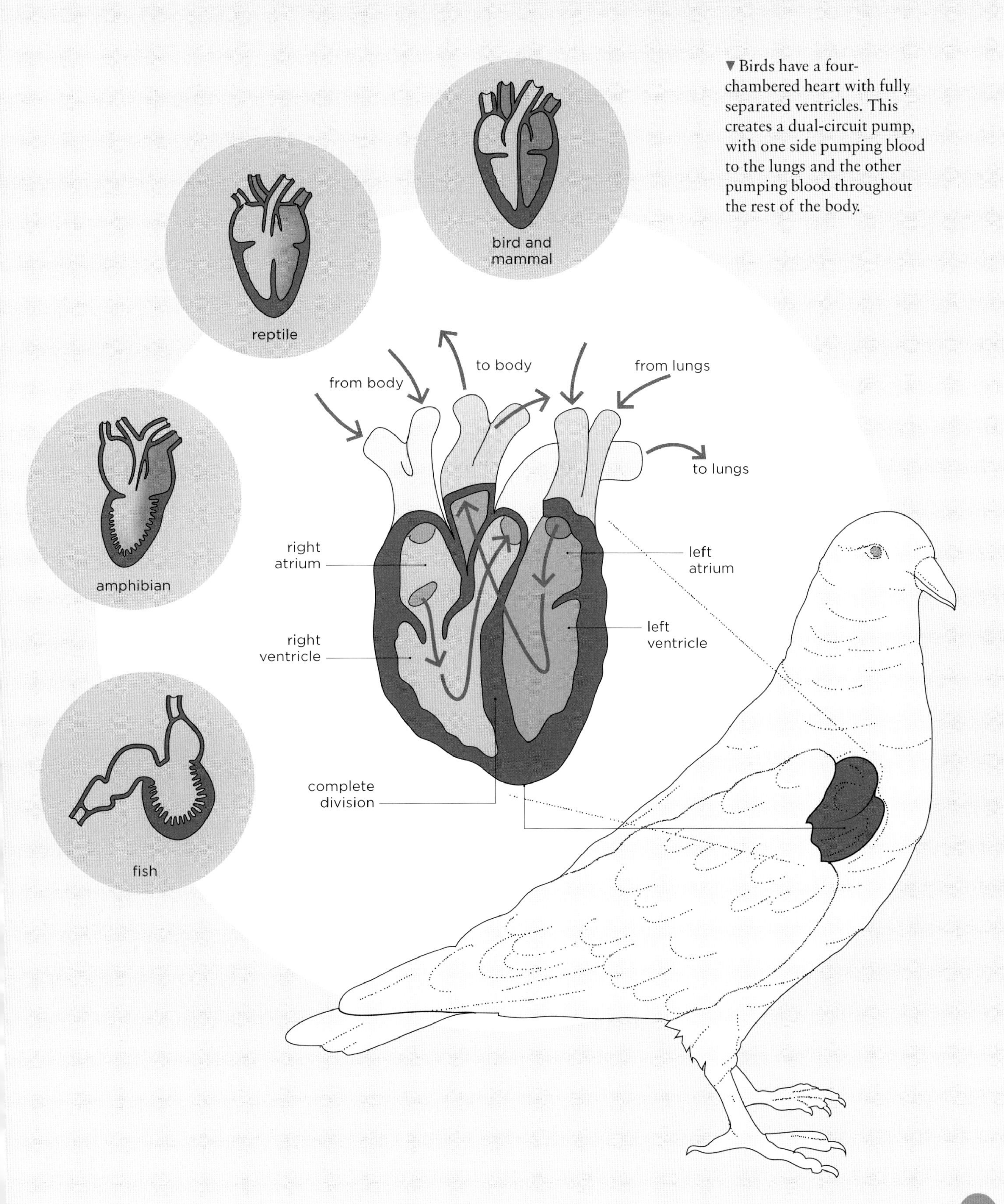

▼ Birds have a four-chambered heart with fully separated ventricles. This creates a dual-circuit pump, with one side pumping blood to the lungs and the other pumping blood throughout the rest of the body.

Heart size and heart rate

Compared to mammals, birds have larger hearts relative to their body size, and they tend to have lower heart rates. This may seem counterintuitive given the high metabolic demands of flight, but an avian heart still manages to pump more blood per minute than a mammalian heart thanks to very efficient filling of the ventricles and high blood pressure.

In general, heart rates are higher for smaller animals, and this is true in birds. Turkeys can have a resting heart rate of about 90 beats per minute (bpm), while in smaller songbirds the resting heart rate is 600–700 bpm. Of course, heart rates change dramatically depending on the activity of a bird. A hummingbird's heart rate is more than 1,200 bpm during active flight, vultures have a heart rate of about 300 bpm at take-off but only about 100 bpm when soaring, and emperor penguins (*Aptenodytes forsteri*) slow their heart rate from about 73 bpm at rest to about 57 bpm during a dive.

Heart size is also plastic. Prior to prolonged migration, many birds increase their heart size to increase cardiac output. The heart, along with many organs, decreases in size during prolonged migratory flights to meet the bird's protein requirements, and then needs to be rebuilt during stopover periods.

Bird blood

Blood transports gases, electrolytes, nutrients, hormones, immune factors and other elements throughout the body. It is mostly composed of water, with its liquid component being known as plasma. Within the plasma are a number of cell types, including erythrocytes (red blood cells), leucocytes (white blood cells) and thrombocytes (similar to mammalian platelets).

Avian red blood cells are larger than those of mammals and have a nucleus (unlike mammalian red blood cells). It is thought that the small, dimpled blood cells of mammals arose early in the group's evolution, to increase surface area to volume ratio in response to a period of low atmospheric oxygen during the Triassic period. In contrast, birds retained larger, nucleated red blood cells similar to those of reptiles. Avian red blood cells are relatively short-lived, with a lifespan of about 40 days, and need to be continually replaced. They contain several forms of haemoglobin, the four-part protein or respiratory pigment that transports oxygen.

The structure of haemoglobin varies across species, and in birds that fly at high altitudes (e.g. the central Asian bar-headed goose, *Anser indicus*) or that have extended deep dives (e.g. the emperor penguin), haemoglobin has a higher affinity for oxygen. The proportion of blood that is composed of red blood cells (its haematocrit) varies across species and also with altitude. Large birds such

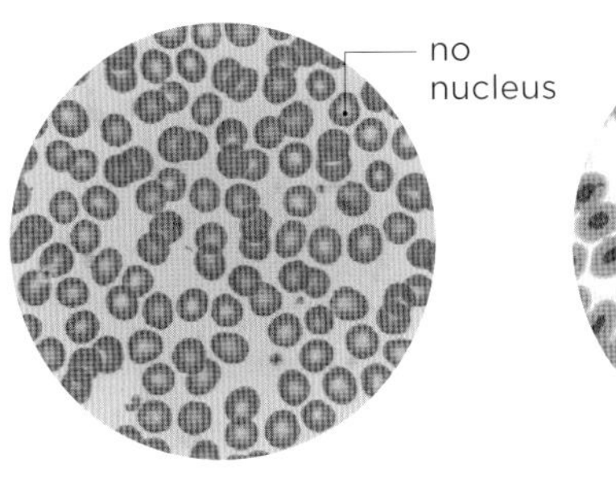

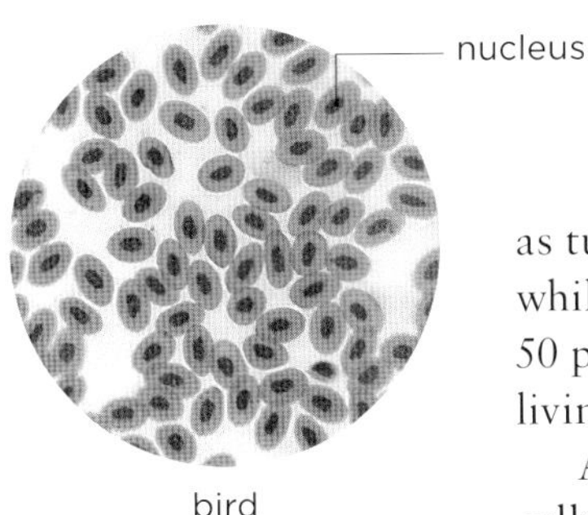

◄ Emperor penguins (*Aptenodytes forsteri*) make extended dives to catch prey, reaching depths of 500 m and lasting up to 20 minutes. During these dives they show bradycardia (slowed heart rate) and reduced blood flow to conserve oxygen.

▲ Mammalian red blood cells are small and dimpled, lacking a cell nucleus. In contrast, birds have retained the large nucleated red blood cells similar to those in reptiles.

as turkeys have a haematocrit of about 40 per cent, while in smaller songbirds the proportion is about 50 per cent – or even 60 per cent in those species living at high altitudes.

Avian blood contains many types of white blood cells, all of which function in immunity to defend against pathogens and parasites. In addition, it contains thrombocytes, which act like platelets in mammalian blood to aid clotting and also have a role in immune function. Following injury or damage to blood vessels, thrombocytes are activated to clot blood and reduce bleeding. This clotting process is slower than in mammals, but as thrombocytes do not form large aggregations in arterial blood like mammalian platelets, it is unlikely that birds suffer from cardiovascular diseases such as strokes.

KEEPING COOL, KEEPING WARM

How do waterfowl keep their feet warm in cold water or while standing on ice? The short answer is that they don't. As in other vertebrates, birds use countercurrent heat exchange, so that extremities like the legs and feet are kept at a cooler temperature than the rest of the body. In the legs, veins returning to the body run closely parallel to the arteries. As warm blood flows through the arteries, heat is transferred and therefore warms the venous blood, which in the process becomes cooler itself. In this way, the feet are kept at a low temperature, decreasing the thermal gradient with the environment and conserving body heat.

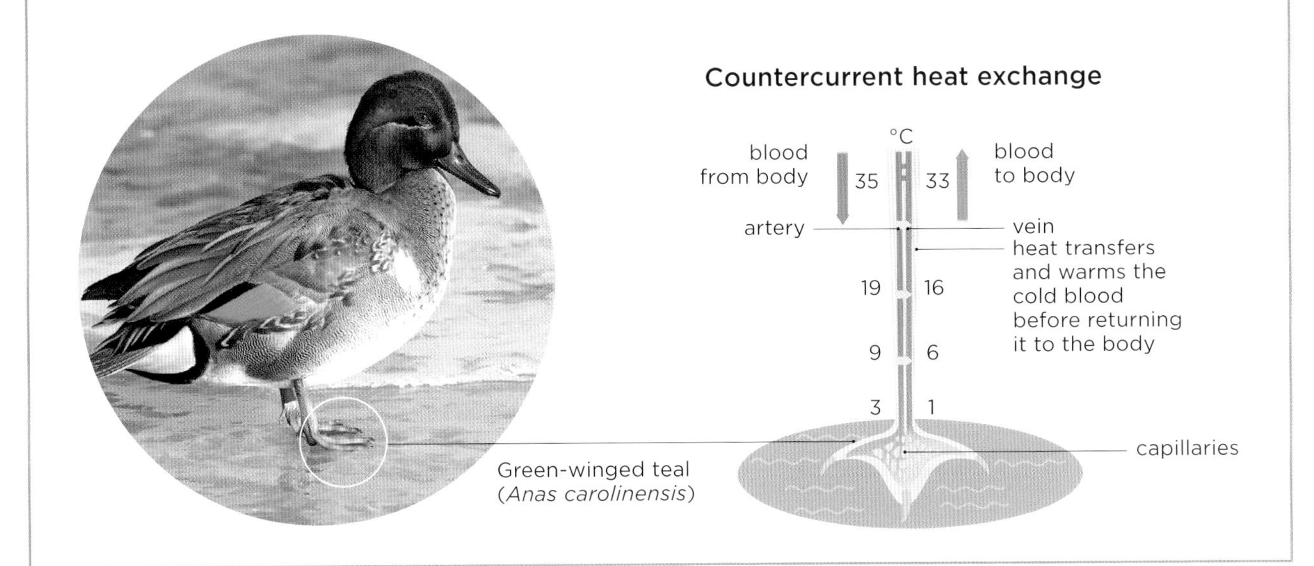

Green-winged teal (*Anas carolinensis*)

Immune system

Birds need to defend themselves against infection and disease caused by parasites and pathogens. The immune system is a complex, interacting network of cells and other factors that identify and destroy pathogens that invade the body, and in birds, as in other vertebrates, it has several branches. Innate immunity refers to systems that are general and provide immediate, but non-specific, defences. Acquired, or adaptive, immunity provides defence against specific pathogens that have been previously experienced. Unlike mammals, avian acquired immune responses include both cell-mediated and humoral (antibody-dependent) components.

Immune system cells

A range of different blood cells contribute to avian immune defences, including various white blood cells: heterophils, lymphocytes, basophils, eosinophils and monocytes. These cells interact with non-cellular, antibody-dependent defences (humoral components of the immune system). Heterophils are the most common white blood cells and play an important role in innate immunity, both through ingesting pathogens such as bacteria (phagocytosis) and contributing to defensive inflammation. Lymphocytes exist in two forms, B and T lymphocytes, which contribute to antibody-driven humoral immunity and cell-mediated immunity, respectively. The functional roles of basophils, eosinophils and monocytes are less understood, and although these are thought to play a role in immunity, mysteries remain as to how they contribute to innate and acquired responses.

Immune cells are primarily produced by the thymus gland, located in the neck, and the bursa, located near the cloaca. Other glands such as the spleen, the Harderian gland (behind the eye), tonsil-like glands in the intestine, and other lymphoid tissue are also involved in the immune system.

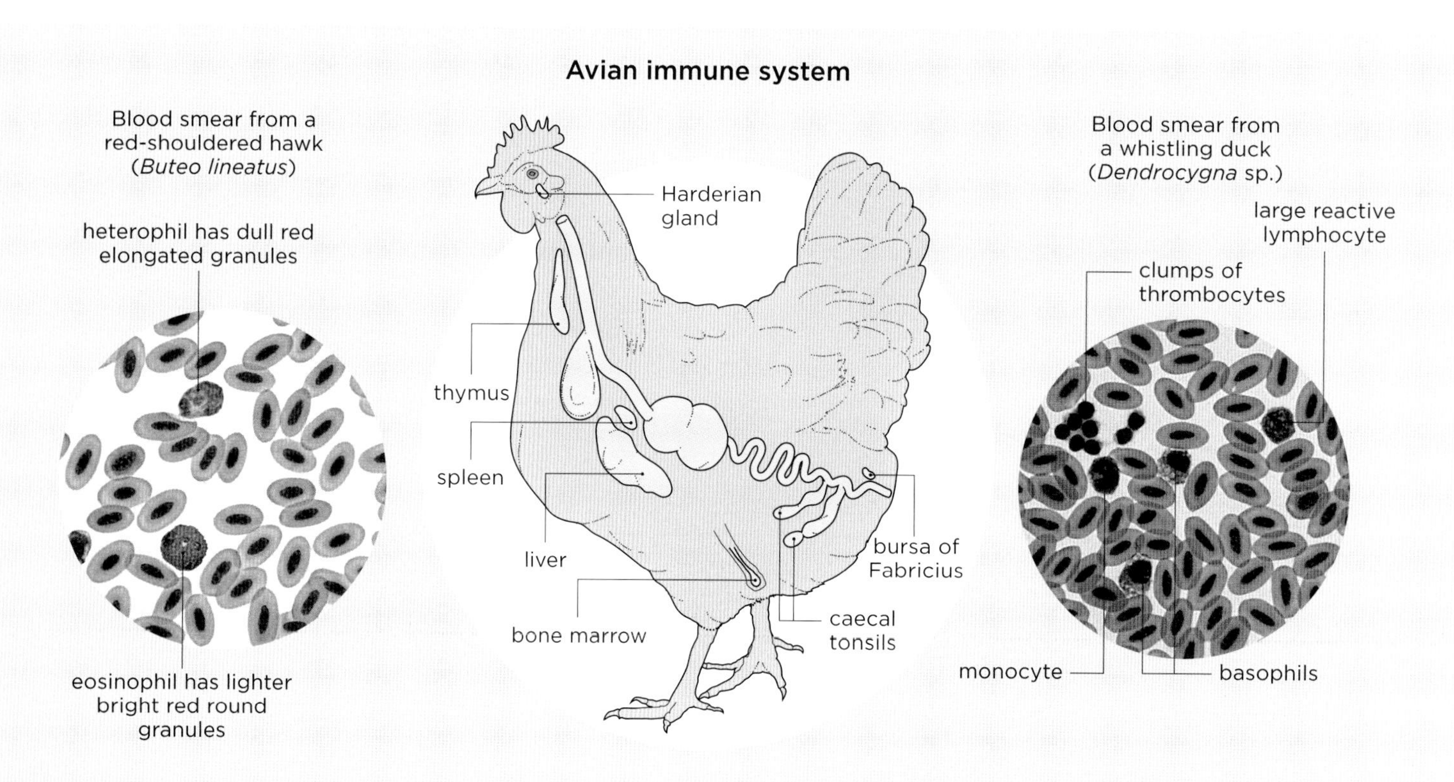

MAKING HISTORY

The two major arms of the acquired immune system are T-dependent (cell-mediated) and B-dependent (humoral-mediated) immunity. These two systems were identified in birds prior to their discovery in any other vertebrate animal. The bursa of Fabricius, a small pouch on the lower gut in birds, was first described in the seventeenth century by the Italian anatomist Hieronymus Fabricius. In the 1950s, following a series of experiments on birds in which the bursa was removed, scientists confirmed that cell- and humoral-mediated immune responses are carried out by distinct cell populations. These studies demonstrated that cells from the bursa, subsequently named B-cells, control antibody production. Others showed that cell-mediated immunity depended on the thymus, and these cells became known as T-cells. Later work demonstrated that mammalian B-cells are produced in bone marrow, but the B-cell designation, acknowledging their discovery in the bursa, has remained.

Hieronymus Fabricius

◄ The immune system comprises several organs throughout the body, which produce multiple kinds of immune cells as well as non-cellular immune factors such as antibodies. Various types of white blood cells can be distinguished visually with a microscope.

Ecoimmunology

The growing field of ecoimmunology takes an evolutionary approach to understanding how various immune responses interact with other traits to improve survival and reproductive success. Studies of birds have been at the forefront of this field and have focused on some long-standing questions. For example, do birds use attractive signals like plumage or courtship to advertise their ability to cope with pathogens? Do they select mates based on genetic compatibility for genes that encode antibodies? How do migratory birds cope with exposure to a greater variety of pathogens when they travel? Do the energetic costs of migratory flight put them at a greater risk of infection? The evidence to answer these questions is mixed, in part due to the fact that the immune system itself is so complex and multifaceted.

▼ Some ornamental features, such as the fleshy red wattles and comb of the red junglefowl (*Gallus gallus*) rooster, are thought to indicate how strong a male's immune system is and whether he has had a recent infection.

Thermoregulation and energy costs

The daily energy requirements of a bird and the energy value of its food largely determine how much it must eat each day. A bird's metabolism continues even when it is at rest and not digesting, and ambient temperatures are relatively benign. Energy use during such periods of inactivity is known as the basal metabolic rate, and for birds (and all other animals) this increases with body mass in a very predictable way. The total energy requirements of a bird include the additional energy it needs for production (e.g. growth, reproduction), thermoregulation and activity.

Pros and cons of endothermy

Birds and mammals are the only groups of animals that are endothermic, producing heat to keep their internal environment at a high and fairly constant temperature. Bird body temperatures (39–43°C) tend to be slightly higher than those of mammals, and also quite high relative to average air temperatures over most of the globe. But why did birds evolve a strategy of spending such a large amount of energy to maintain high body temperatures? The reason seems to be that by having a constant high body temperature birds become relatively independent of ambient temperatures, and hence could fill new niches and habitats. Cold reptiles (ectotherms), for example, are usually quite inactive, whereas birds living in the same chilly ambient temperatures can continue to forage, fly and defend territories. The trade-off is that birds expend orders of magnitude more energy maintaining their high body temperature than reptiles. Also, regulating a body temperature that is on the high side means that only occasionally will air temperatures be higher than body temperature. This is important, because when birds must cool down in the face of hot ambient temperatures, the only way they can do so is through evaporative cooling, which expends precious body water.

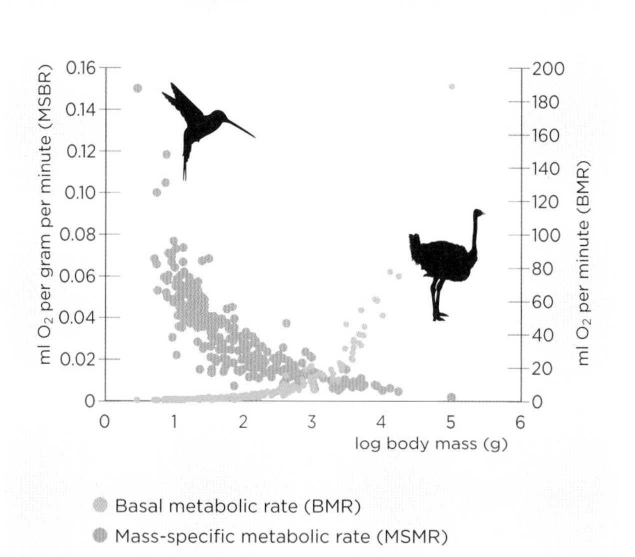

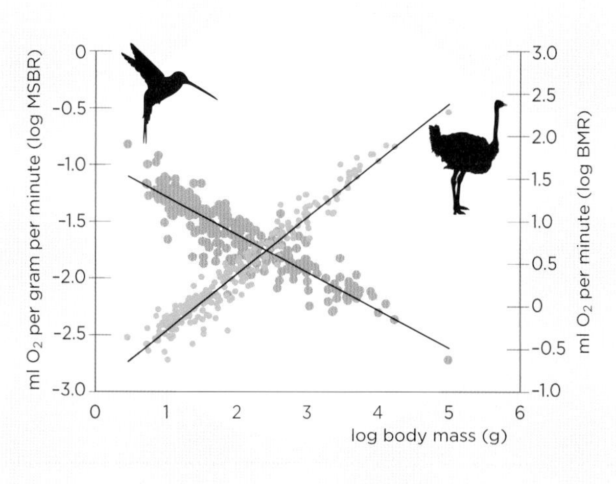

▶ The cost of living in birds is closely related to their body mass. Thus, ostriches expend more energy per unit time (basal metabolic rate, or BMR) than hummingbirds. However, hummers expend much more energy per gram of body mass, which helps to explain why they are such busybodies. Log-transformation (bottom figure) makes both relationships linear.

Heat gains and losses between a bird and its environment

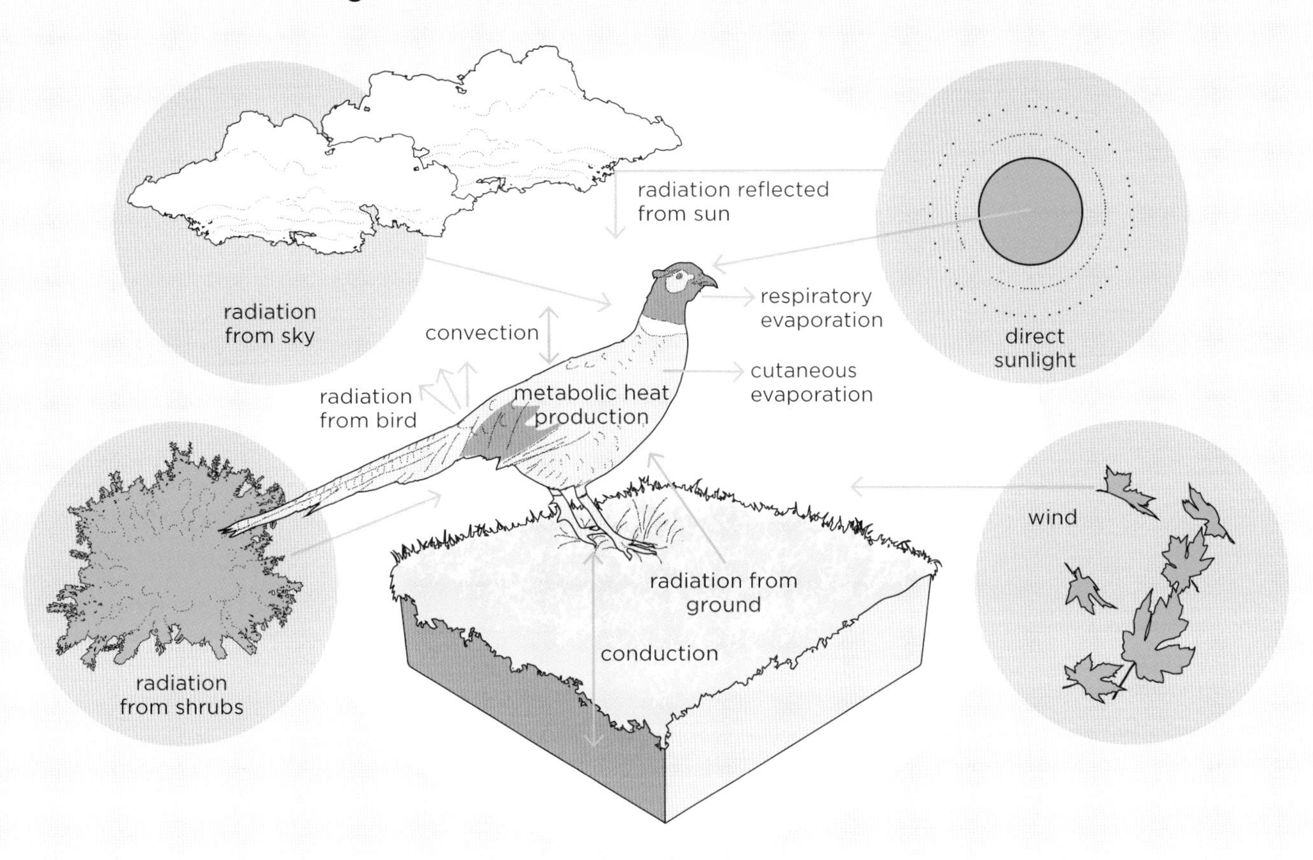

▲ Birds gain heat through direct and reflected radiation from the sun, and lose heat through radiation and breathing. The wind and ground can provide or dissipate heat depending on their temperature.

▼ Hummingbirds are energy powerhouses even when at rest. They expend more energy per gram of body mass than the majority of other (larger) birds and live a very fast lifestyle.

Staying warm in cold places

For a bird's body temperature to remain constant, heat loss must equal heat gain. Birds that live in cold places do not undergo major seasonal adjustments in insulation that are common in mammals, which often add both hair and fat as winter approaches. And nor do birds truly hibernate like mammals, which involves significant lowering of the body temperature to very low levels close to freezing point. Birds do add some fat insulation during winter, they puff up their feathers to stay warm and they try to find warmer places out of the wind, and some more social birds such as penguins huddle together to retain heat. However, only a few types of birds (e.g. hummingbirds, some poorwills and nightjars, mousebirds and some swifts) can lower their body temperature for extended periods of time (more than 24 hours) to conserve energy, a physiological state known as torpor. The primary way birds stay warm in cold places is by increasing their metabolism to increase heat production. However, they then have to recoup the substantial energy costs of this strategy by eating lots.

Staying cool in hot places

How do birds live, breed and stay relatively cool in the world's hottest, driest, low-productivity deserts, such as those found in the Arabian Peninsula? Not only are birds present in these extremely challenging environments, but they are often active at the hottest times of the day – unlike most desert rodents, which retire to cooler underground borrows. It was previously thought that birds were pre-adapted to desert life because of their high body temperatures. However, such temperatures are actually dangerously close to the thresholds at which molecules like proteins start to change their structure and lose their function, leading to death. Thus, staying cool in such hot places is essential.

The only physiological way that heat can be lost when ambient temperatures are greater than body temperature is through evaporative cooling, from water loss across the skin or respiratory surfaces. Birds can increase the efficiency of respiratory evaporative water loss by panting or through gular flutter (vibrating the throat with the mouth open). Even though birds do not have sweat glands, they still lose significant water across their skin, which can contribute to evaporative cooling. Cutaneous evaporation often accounts for half or more of total evaporative heat loss when birds are not active and surrounded by warm, dry air. Birds that live in hot places have evolved adaptations to reduce the cost of thermoregulation, including decreasing their metabolic rate to limit heat production, having efficient kidneys or salt glands to limit unnecessary water loss (see page 129), and decreasing water loss across the skin and respiratory surfaces whenever possible.

Given that birds lose heat by increasing evaporative cooling across their skin and lungs, staying cool in hot places increases their water requirements. Andrew McKechnie and Blair Wolf have predicted that the expected increases in global temperatures associated with climate change will substantially increase the water requirements of birds, to the point where their survival rates will significantly decline by 2080 and the frequency of catastrophic mortality events will increase.

THE THERMONEUTRAL ZONE

The range of ambient temperatures over which a bird's rate of metabolic heat production does not change is known as the thermoneutral zone. Metabolism remains constant because the bird relies on behavioural, insulatory or circulatory mechanisms to minimise heat loss at essentially no energy cost. On the cold side of the thermoneutral zone, a bird can face into the sun to warm up, tuck its beak under a wing or fluff up its feathers. Conversely, on the warmer side of the thermoneutral zone it can increase heat loss by increasing the blood supply to its skin.

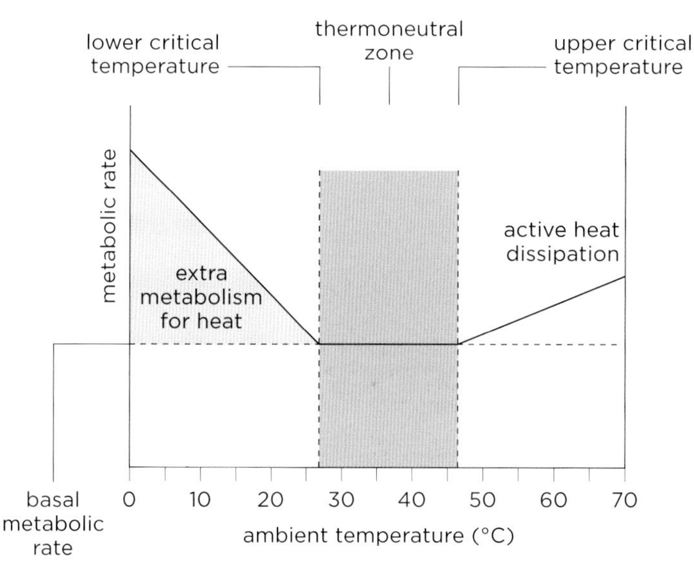

▲ Birds use no additional energy for thermoregulation and are in the 'thermoneutral zone' when outside temperatures are benign. Otherwise, birds must use more and more energy to stay warm when cold, or cool when hot.

▲ As in all birds, great blue herons (*Ardea herodias*) open their bills and flutter their throat (called 'gular fluttering') to increase heat loss on very hot days.

◀ Birds like this cormorant stay cool on hot days by posturing, panting, increasing the surface area of their body exposed to the wind and standing on cooler surfaces.

▼▼ When temperatures drop, birds can huddle together to stay warm, as seen in these European bee-eaters (*Merops apiaster*).

Digestive and excretory system

The digestive system of birds is relatively simple and similar to that of mammals. Food passes from the mouth to the oesophagus (where some birds have a crop for storing food) and the stomach, and then through the small intestine and large intestine before exiting via the cloaca. A remarkable feature of this system, especially in birds that migrate long distances, is the ability to rapidly and reversibly change the size of key organs – including the flight muscle, liver, intestine and heart – in response to changes in demand.

Chemical and mechanical digestion

The breakdown of proteins, fats and carbohydrates in food takes place primarily in the stomach and small intestine using both mechanical and chemical processes. In most birds the stomach is divided into a proventriculus, where chemical digestion is initiated, and a gizzard, where ingested food is mechanically mixed and ground (this therefore replaces the function of teeth). Many birds (e.g. geese, songbirds, penguins, pigeons) regularly swallow small, hard objects such as grit or pebbles to aid in the grinding of food in the gizzard.

Digestive enzymes are proteins produced by cells of the stomach, intestine and pancreas to regulate the rate of chemical breakdown of food compounds. Like all enzymes, digestive enzymes are, to some degree, specific to a certain substrate or reactant molecule, and their activity is sensitive to temperature and pH in particular.

The absorption of the end products of digestion in the foregut primarily occurs in the bird's small intestine. Digestive enzymes secreted from the liver and pancreas are delivered into the first portion of the small intestine, and in the latter portion digestion is completed and absorption takes precedence. The products of digestion (amino acids, fatty acids, glucose and fructose) are absorbed across the intestinal wall and then move via blood circulation to the bird's tissues and cells.

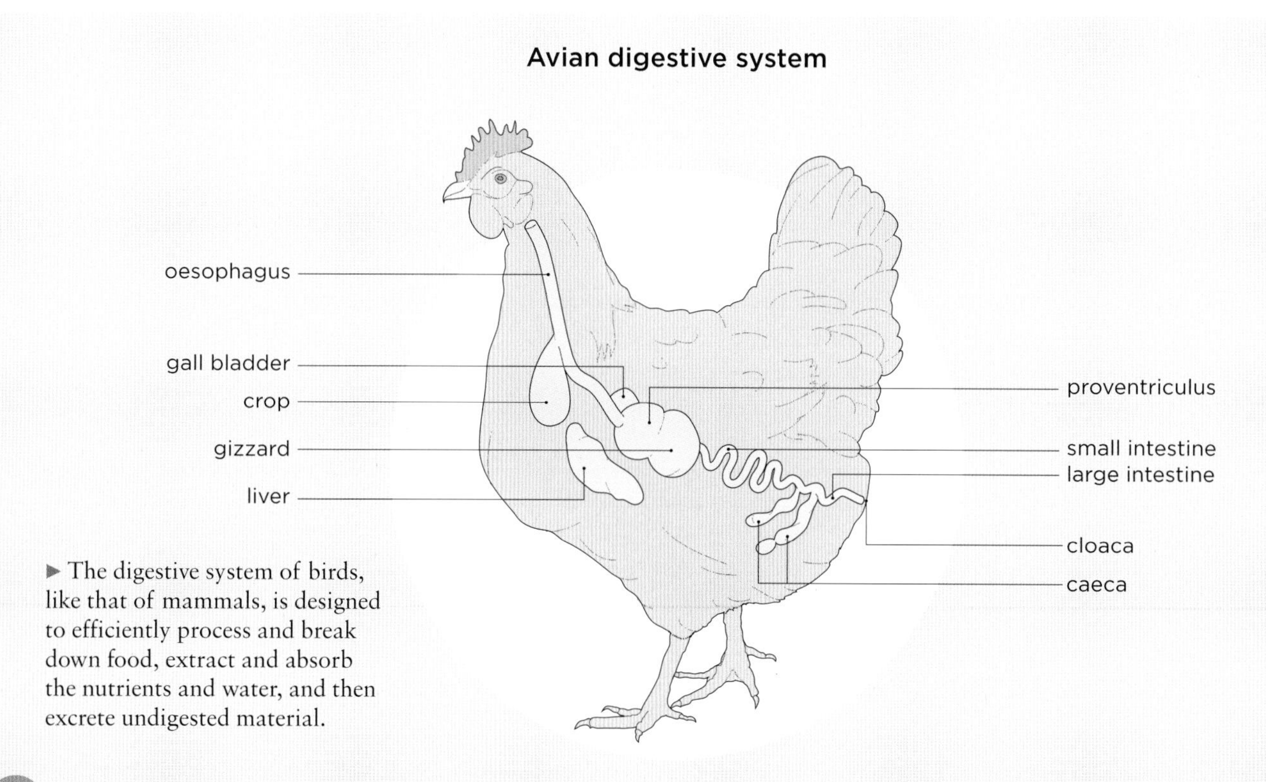

▶ The digestive system of birds, like that of mammals, is designed to efficiently process and break down food, extract and absorb the nutrients and water, and then excrete undigested material.

Avian herbivores

Cellulose is the primary structural material in plants and one of the most abundant biological compounds on Earth. But birds and other vertebrates do not produce cellulose-digesting enzymes and so cannot break cellulose down unaided. The evolution of enlarged portions of the foregut and/or hindgut to house symbiotic microbes (primarily bacteria and some protozoa) was crucial in the radiation of herbivorous mammals, because it provided the appropriate conditions for the microbial fermentation of cellulose. However, less than 3 per cent of living bird species have evolved a similar feature, perhaps because birds are relatively small and have high metabolic rates, and fermentation is a relatively slow process.

With the exception of the Amazonian hoatzin (*Opisthocomus hoazin*), the only living example of a foregut-fermenting bird, all avian herbivores harbour microbes in the hindgut (intestines, caeca or colon), where plant material is digested albeit much less effectively and efficiently than in mammalian herbivores. In general, the rarity of avian herbivores suggests that the nutritional challenges associated with eating leaves and other high-cellulose material are formidable and not often compatible with a bird's high-energy lifestyle involving a lot of flying.

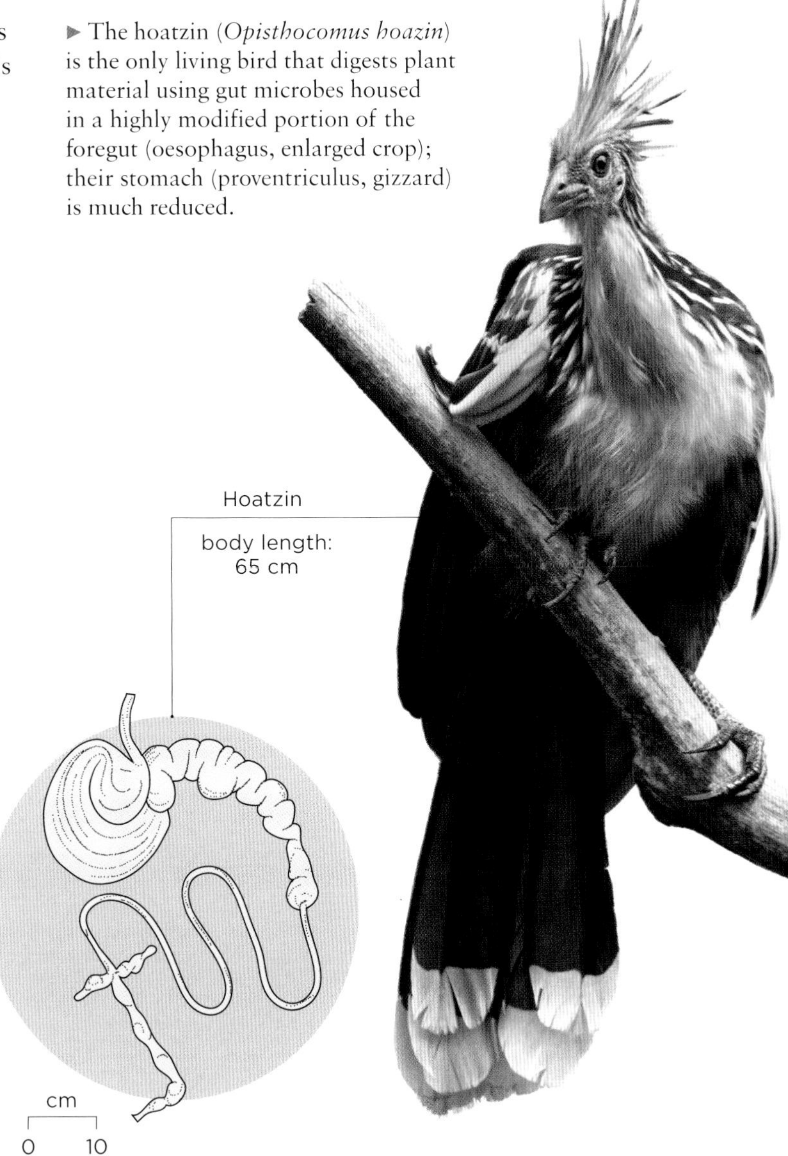

► The hoatzin (*Opisthocomus hoazin*) is the only living bird that digests plant material using gut microbes housed in a highly modified portion of the foregut (oesophagus, enlarged crop); their stomach (proventriculus, gizzard) is much reduced.

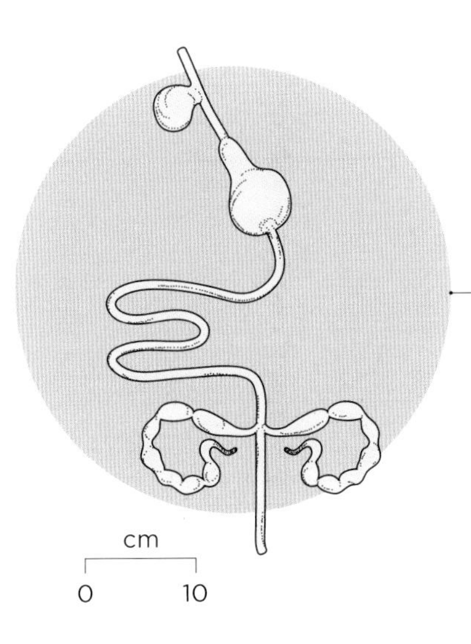

▲ It has been estimated that more than 25 per cent of the energy needs of ruffed grouse are provided by microbial fermentation in the hindgut.

▼ The ruffed grouse (*Bonasa umbellus*) is an avian herbivore that uses gut microbes housed in its very enlarged paired caeca to digest plant material.

The endocrine system

The endocrine system comprises a network of specialised endocrine glands (e.g. the pituitary and adrenal glands) and diffuse cells in the gut and fat that release potent organic chemical signals called hormones. Hormones are transported in the circulatory system to distant target cells in the animal's body, changing their function. While the nervous system mainly regulates short-term processes such as movement, the endocrine system regulates longer-term processes such as growth and development, energy balance and reproduction.

Endocrine networks

The brain and central nervous system are constantly receiving information from the body's sensory systems, including environmental information (e.g. light, temperature, social signals) and internal information from the physiological system (e.g. body temperature, blood nutrient levels). All of this information is integrated by the animal's neural circuitry and converges on the hypothalamus at the base of the brain. The hypothalamus transduces or changes neural signals into chemical, hormonal signals via neurosecretory cells. These cells receive neural input but synthesise and secrete 'releasing hormones' or 'release-inhibiting hormones' that target the pituitary gland – a small, lobed endocrine organ hanging just below the hypothalamus. This stimulates the pituitary

Hypothalamic–pituitary–gonadal–liver axis regulating reproduction in a female bird

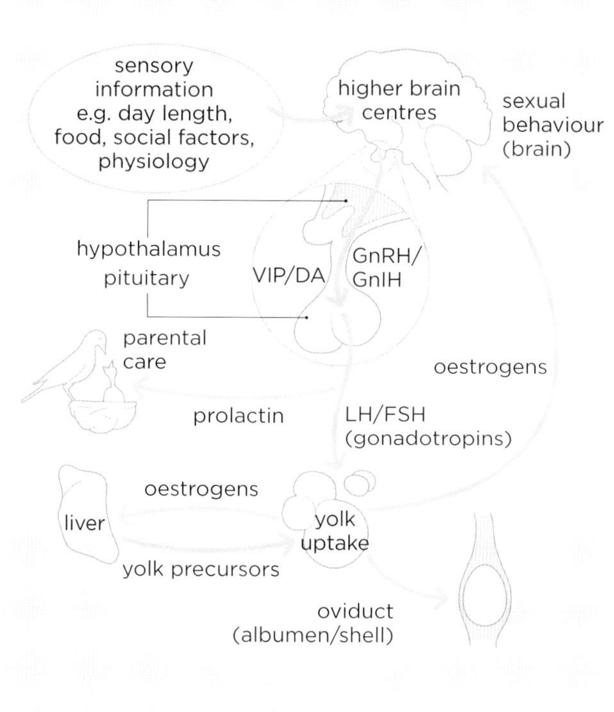

VIP	vasoactive intestinal peptide
DA	dopamine
GnRH	gonadotropin-releasing hormone
GnIH	gonadotropin-inhibiting hormone
LH	luteinizing hormone
FSH	follicle-stimulating hormone

◄▲ Growth and early development from an embryo to a hatched chick are regulated by the hypothalamic–pituitary–thyroid axis, and by growth hormones and many other hormones.

Deposition of fat under hormonal control

▲ Hormones regulate seasonal variation in body mass, as seen here in gentoo penguins (*Pygoscelis papua*), with a thin bird (left) that lacks fat deposits, and two fat birds (above) that have large subcutaneous fat deposits.

to produce a wide range of specific hormones that are released into the circulatory system. Neurosecretory cells in the hypothalamus therefore sit at the top of a cascade, or neuroendocrine axis, that regulates the function of peripheral cells, tissues and organs via the pituitary. Peripheral endocrine glands (e.g. the thyroid, gonads) synthesise and secrete their own hormones, which can feed back to the hypothalamus and pituitary, or change the function of other non-endocrine target cells – for example, promoting muscle growth, altering metabolic rate, or initiating gonad development.

The three main neuroendocrine networks, or axes, in birds are each defined by a suite of key organ(s) that communicate with specific regions of the brain, and that regulate different aspects of physiology. The hypothalamic–pituitary–thyroid axis regulates growth and metabolism, the hypothalamic–pituitary–adrenal axis regulates metabolism and stress, and the hypothalamic–pituitary–gonadal–liver axis regulates reproduction. The digestive system is also a major endocrine organ, releasing 'brain–gut' hormones that regulate the digestive process locally in the gut, but that also integrate with the hypothalamus to regulate feeding behaviour, food intake and energy expenditure.

Stress and growth hormones

Corticosterone is the main glucocorticoid hormone in birds, and is released from the adrenal gland. It is often referred to as a stress hormone, because plasma levels of corticosterone increase rapidly within three to five minutes of a stress-related event, such as a predator attack or a researcher catching the bird (although many other hormones also change in response to stress). Increases in plasma corticosterone concentrations redirect behaviour and physiology towards immediate survival – for example, increasing locomotor activity, inhibiting reproduction and enhancing immune function. At baseline levels corticosterone plays a fundamental role as a metabolic regulator of energy homeostasis during 'predictable' life-history events such as growth, migration and reproduction.

Growth hormones, thyroid hormones and growth factors regulate growth and development in birds, although these pathways are much less well understood than in mammals. Similarly, many of the 'brain–gut' hormones (e.g. ghrelin, leptin, insulin) known to regulate hunger, rate of food intake and meal size in mammals have been identified in birds and likely play similar roles. However, little work has been conducted on these potential regulatory mechanisms in birds beyond work in poultry.

The reproductive system

Like most vertebrates, birds do not maintain their reproductive system and gonads (the egg-producing ovary and sperm-producing testes) in a fully developed state year-round. Instead, the gonads are fully regressed during much of the year (the non-breeding phase), and grow and become functionally mature during the breeding phase. In birds breeding at mid- to high latitudes, this is triggered by environmental cues such as increasing day length and temperature. At the end of the breeding season, the gonads regress through a process called photorefractoriness. In males, the hypothalamic-pituitary-gonadal axis regulates testis function, and in females the hypothalamic-pituitary-gonadal-liver axis regulates ovary and oviduct function.

Seasonal reproductive cycles

Most birds have seasonal breeding cycles, with courtship, mating, egg-laying and rearing of young restricted to a particular time of year (often spring and early summer) when environmental conditions (e.g. food, temperature) are optimal for successful reproduction. These breeding cycles reflect underlying cycles of reproductive physiology, with the gonads of both sexes maturing and regressing annually. However, the timing of gonadal development – and the environmental cues critical for seasonal gonadal development – differ between the sexes.

Male reproductive system

Male birds have internal paired testes suspended from the upper wall of the body cavity just above the kidneys. Testis size varies seasonally, representing as little as 0.005 per cent of body mass in non-breeding males, but increasing 500- to 1,000-fold during breeding. In male European common starlings (*Sturnus vulgaris*), testis volume increases from less than 10 mm^3 in midwinter to 500 mm^3 during breeding in early April.

Each mature testis has thousands of coiled seminiferous tubules, containing germ cells (responsible for sperm production, or

Seasonal variation in testes size in male common murre or guillemot

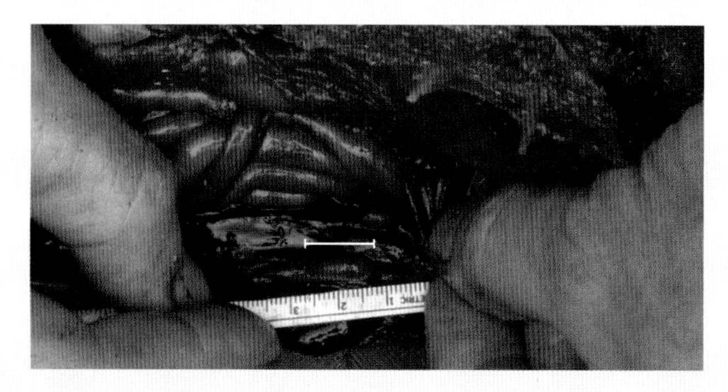

▲► A male non-breeding common murre (*Uria aalge*) with fully regressed testes (left), and during the breeding season with fully developed testes (right). (Testis shown by white bar.)

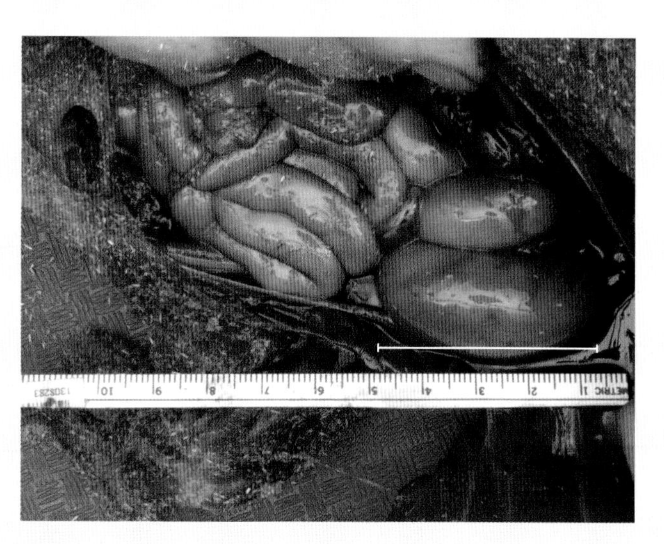

▶ A male green-winged teal (*Anas carolinensis*) with his large corkscrew-shaped penis extended.

spermatogenesis) and Sertoli cells (which assist in sperm maturation). Sperm are generally short and relatively simple in non-passerines and longer and spiral-shaped in passerines, but our understanding of the adaptive significance of this diversity remains limited. Interstitial or Leydig cells lying between the tubules secrete steroid hormones (e.g. testosterone), which control sexual behaviour (e.g. song, copulation) and the development of some, but not all, secondary sexual characteristics (e.g. plumage, bill colour). The seminiferous tubules connect to form a long epididymis and deferent duct, which carries mature sperm in seminal fluid to the cloaca, the common external opening for sperm transfer and excretion of urine and faeces.

Male birds lack a true intromittent organ, or penis. During courtship the male's cloaca swells, forming a cloacal protuberance, and in most species copulation and sperm transfer involves simple cloacal contact, with the male everting his cloaca into the opening of the female's oviduct. Some ratites (e.g. ostriches), ducks and geese have a more developed and erectile phallus, or pseudopenis, which is used in mating, but this lacks the internal urethra for passage of urine that occurs in mammals.

Male reproductive system

equidistant point grid overlaying testes cross section showing seminiferous tubules

IT

ST

interstitial tissue

seminiferous tubules

right adrenal gland

left testis

caudal vena cava

right kidney

right ureter

right ductus deferens

rectum

right seminal glomus

vent

▶ The male reproductive and urogenital system, with enlarged testes during breeding, and (inset) the ultrastructure of the testes, showing the seminiferous tubules (ST), where sperm formation occurs, and steroid-hormone-producing interstitial tissue (IT) comprising Leydig cells.

Variation in sperm morphology

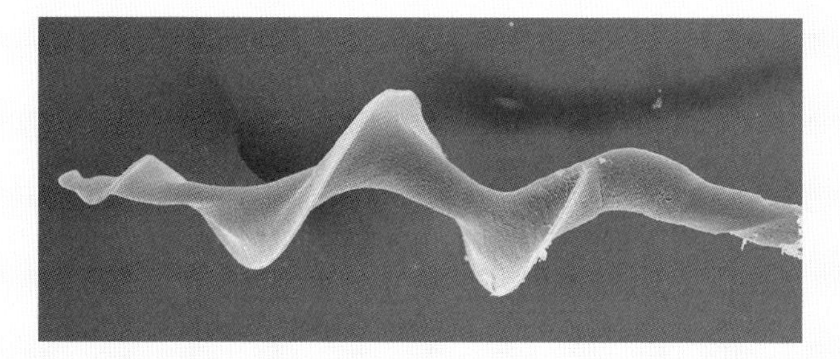

▲ Helical sperm of the bluethroat (*Luscinia svecica*).

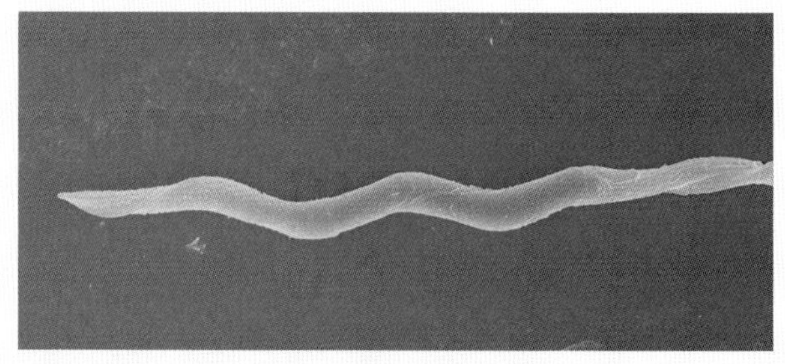

▲ Straighter sperm of the goldcrest (*Regulus regulus*).

Physiological control of testis function

In male birds, light levels appear to be a sufficient environmental cue for full gonadal maturation – long artificial photoperiods alone stimulate testis maturation and spermatogenesis in males of most wild birds when they are held in captivity and isolated from natural cues. So, testis maturation occurs gradually over several weeks or months, from midwinter to just before breeding, in response to increasing day length – in other words, it is mainly photoperiod-dependent. In birds, detection of light controlling seasonality involves extra-retinal photoreceptors in the brain not the eyes (unlike mammals)

As the days get longer following the winter solstice (21 December in the northern hemisphere), neurosecretory cells in the male's hypothalamus produce gonadotropin-releasing hormone (GnRH), with a concomitant decrease in gonadotropin-inhibiting hormone (GnIH). This stimulates endocrine cells in the pituitary gland to release two gonadotropin hormones, luteinising hormone (LH) and follicle-stimulating hormone (FSH), into the circulation. In turn, LH and FSH regulate steroid hormone production and spermatogenesis by the testes.

Female reproductive system

In many ways, the female reproductive system in birds is much more complex than that of males. Not only do the ovary and oviduct have to function together to form a complete egg, with yolk, albumen and shell, but the female's liver synthesises yolk lipids and proteins (hence, females have a true hypothalamic–pituitary–gonad–liver axis).

Avian embryos have two bilateral ovaries attached to the internal body wall at the top of the kidneys. Most adult female birds retain only a single functional ovary, with the loss of the second one having occurred far back in avian evolution – the egg-laying dinosaurs (maniraptorans), which lived more than 125 million years ago, had only one ovary. For unknown reasons, small raptors (Falconiformes) and the kiwi retain two functional ovaries. An ovary consists of many oocytes, cells that will give rise to ova or eggs, embedded in a matrix of connective tissue. Production of oocytes (oogenesis)

◄ Zebra finches are a valuable model species since they breed readily in captivity; females of many wild species do not become reproductively mature, or lay eggs, when held in standard, small cages typically used by researchers.

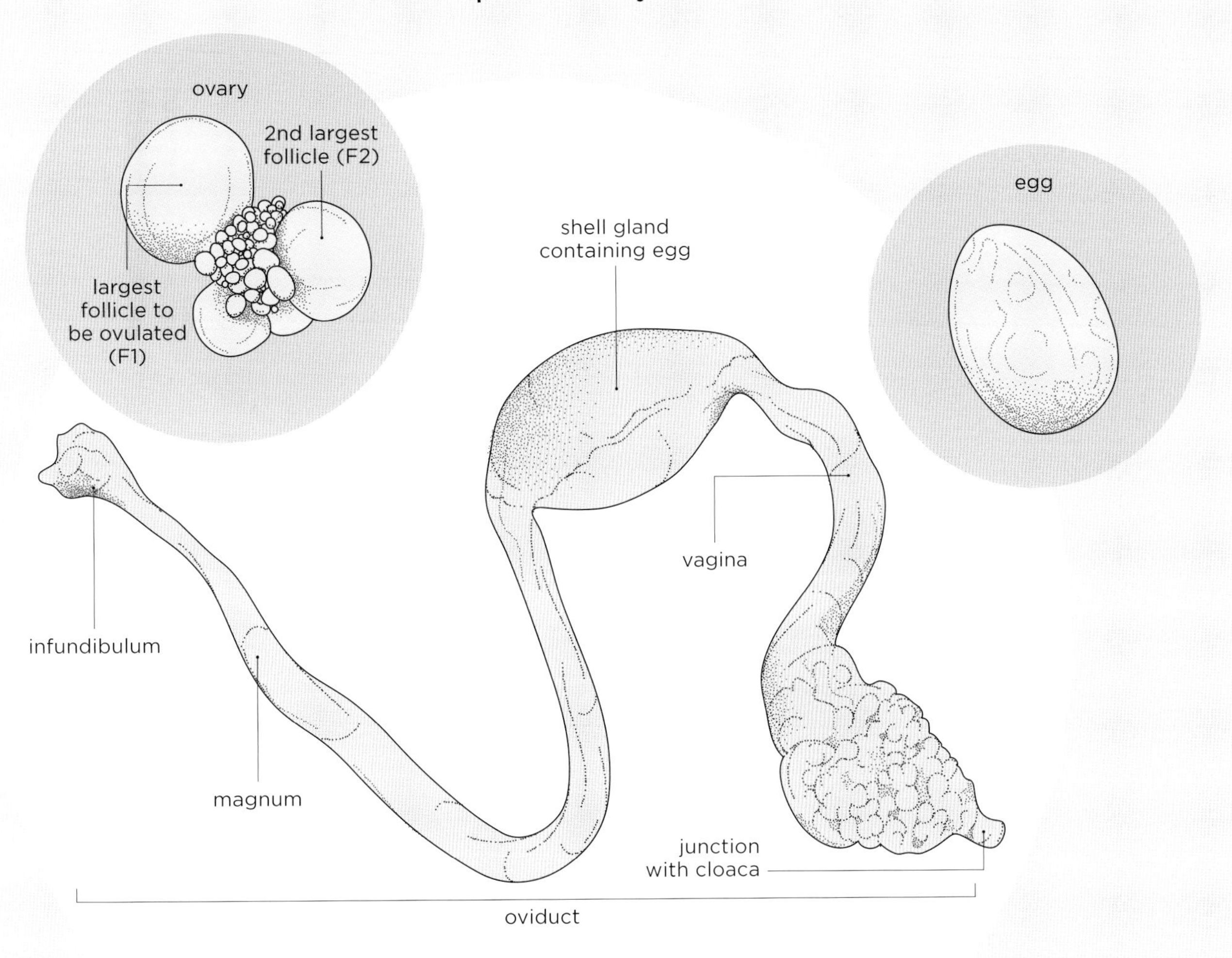

▲ The female reproductive system during breeding, showing the ovary, with a hierarchy of developing yolky follicles, and the oviduct, containing an ovulated follicle undergoing albumen and shell formation in the uterus.

is completed at hatching, with the ovary containing around 450,000 primary oocytes, although in most species as few as one to 20 of these will mature and be laid as eggs.

At the start of each breeding attempt, a small number of follicles are 'recruited', destined to become yolks. These start to develop into mature ova and contain the germinal disc region of cells that ultimately forms the embryo. Follicle development involves a variable phase of rapid yolk development, lasting 3–25 days depending on the species, in which lipid- and protein-rich yolk precursors are taken up from the circulatory system to form the yolk. Specialised cells (called granulosa and thecal cells) proliferate and surround the follicle during oocyte maturation, synthesising and secreting hormones (e.g. progesterone, oestrogens, growth factors) required for yolk development, oviduct function and control of female sexual behaviour. Yolk precursors are produced by the liver in response to stimulation from ovarian oestrogens; this represents a major shift in the lipid metabolism of egg-laying females.

Mature follicles are released into the oviduct (equivalent to the mammalian fallopian tube) typically every 24 hours in smaller birds, with the number of ovulations matching the bird's clutch size. Oviduct size varies seasonally and through the laying cycle. For example, in the Australian zebra finch (*Taeniopygia guttata*) the mass of the oviduct increases more than tenfold over five days at the onset of egg production. At ovulation, the yolky follicle is captured by the infundibulum, the first part of the oviduct, and fertilisation occurs here within 30 minutes of ovulation, before albumen deposition prevents sperm reaching the egg. Females can store sperm – often from multiple males – in special tubules in the oviduct. Sperm can remain viable for days to weeks and are released at the time of ovulation for fertilisation, so females do not need to mate at a specific time around ovulation. Females might even be able to control the sperm they use, effectively controlling paternity of their offspring!

▲ Females laying multi-egg clutches have a hierarchy of developing yolky ovarian follicles, these ovulating at 24-hour intervals, starting with F1, then F2, etc.

► Albatrosses lay a single-egg clutch and develop a single large yolky follicle in the ovary.

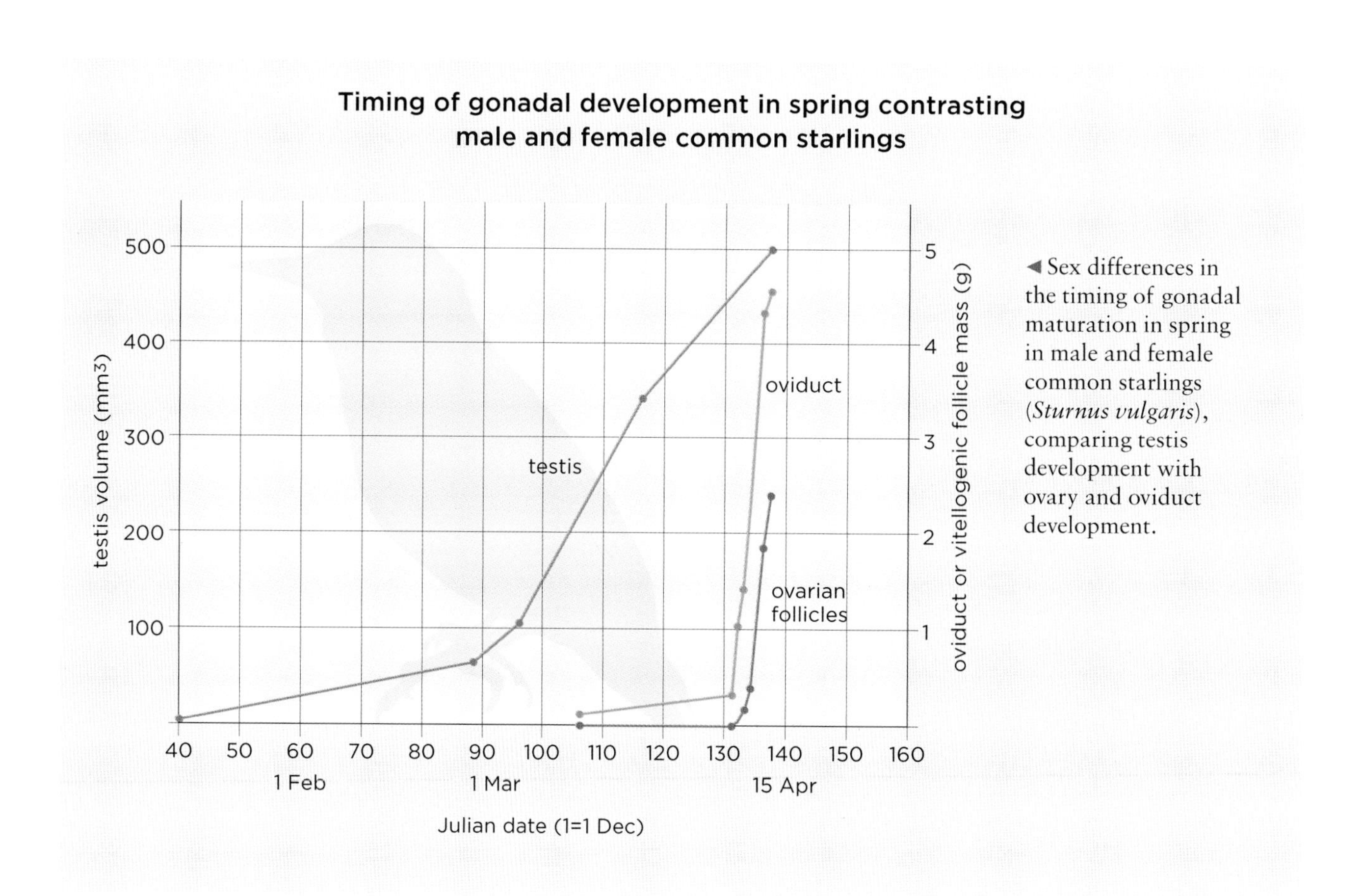

◄ Sex differences in the timing of gonadal maturation in spring in male and female common starlings (*Sturnus vulgaris*), comparing testis development with ovary and oviduct development.

▲ Shorebirds, like this red-wattled lapwing (*Vanellus indicus*), develop a limited number of yolky follicles and lay a fixed clutch size of three or four eggs.

Each yolky follicle then takes about 24 hours to move down the oviduct, where albumen, shell membranes, the shell and a cuticle are added, completing egg formation. Birds' eggs can be highly and variably coloured by red/brown (porphyrins) and blue/green (biliverdin) pigments, but the sometimes intricate mechanisms that produce these patterns are still not well known. Oviposition, or egg-laying, occurs most commonly once a day (although birds can skip days), and at a similar time of day, with the fully formed egg being expelled by muscle contraction via the female's cloaca.

Physiological control of ovary function

It has been known for more than 60 years that, in contrast to males, females of most free-living avian species held in captivity will not undergo complete ovarian development: yolk formation and egg production is very unusual in captivity (except when females are kept in large aviaries at low density). This suggests that the environmental cues regulating female reproduction may differ from those influencing males, and recent studies are confirming this. Increasing day length in spring clearly stimulates the female's hypothalamus, leading to GnRH release, and this leads to LH and FSH release from the female's pituitary. However, the ovary does not respond immediately to higher circulating levels of LH (unlike the testis), and there appears to be some level of 'peripheral' control in females at the level of the ovary and/or liver. Consequently, ovarian development and yolk formation do not occur gradually over weeks or months (as with testes development), but over just five to 20 days immediately before the first egg is laid. Final maturation of the female reproductive system before egg-laying is thought to be influenced by additional environmental cues (called supplementary factors), including temperature, food availability, and social information such as the presence of a male partner and a nest.

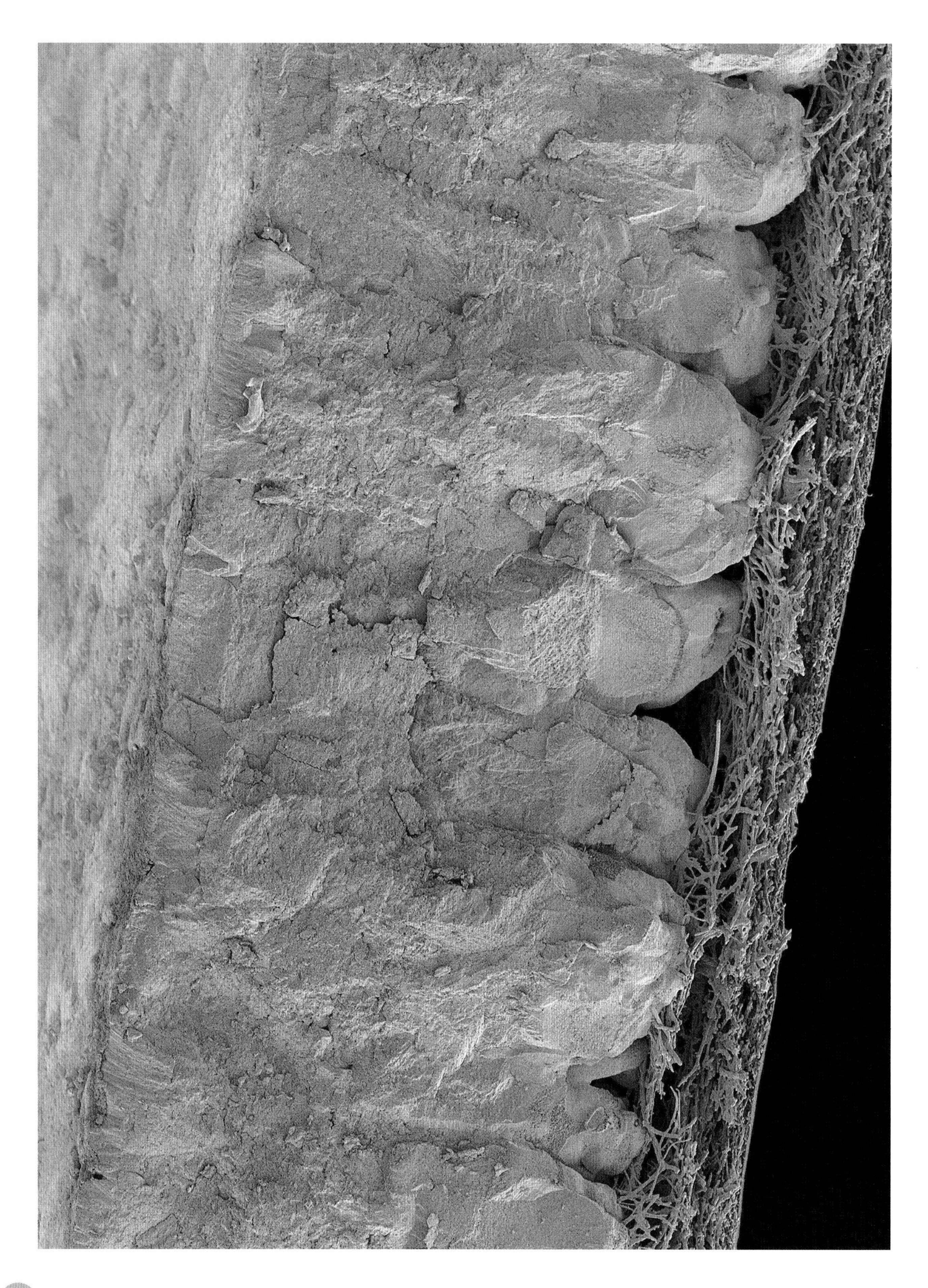

THE STRUCTURE OF A BIRD'S EGG

Eggs are sealed capsules that contain all the nutrients and energy used during embryonic development. The yolk supplies protein and energy-rich lipids for growth, as well as antioxidants for protection against oxidative stress, and the albumen provides additional protein and water, as well as having antibacterial properties. The shell provides physical protection, and is important in regulating water and gas exchange via large numbers of intricate channels or pores that connect the embryo with the external environment. Eggshell is an amazing substance: strong enough to avoid cracking during incubation and protecting the developing embryo, but weak enough to allow the chick to break out of the egg at hatching.

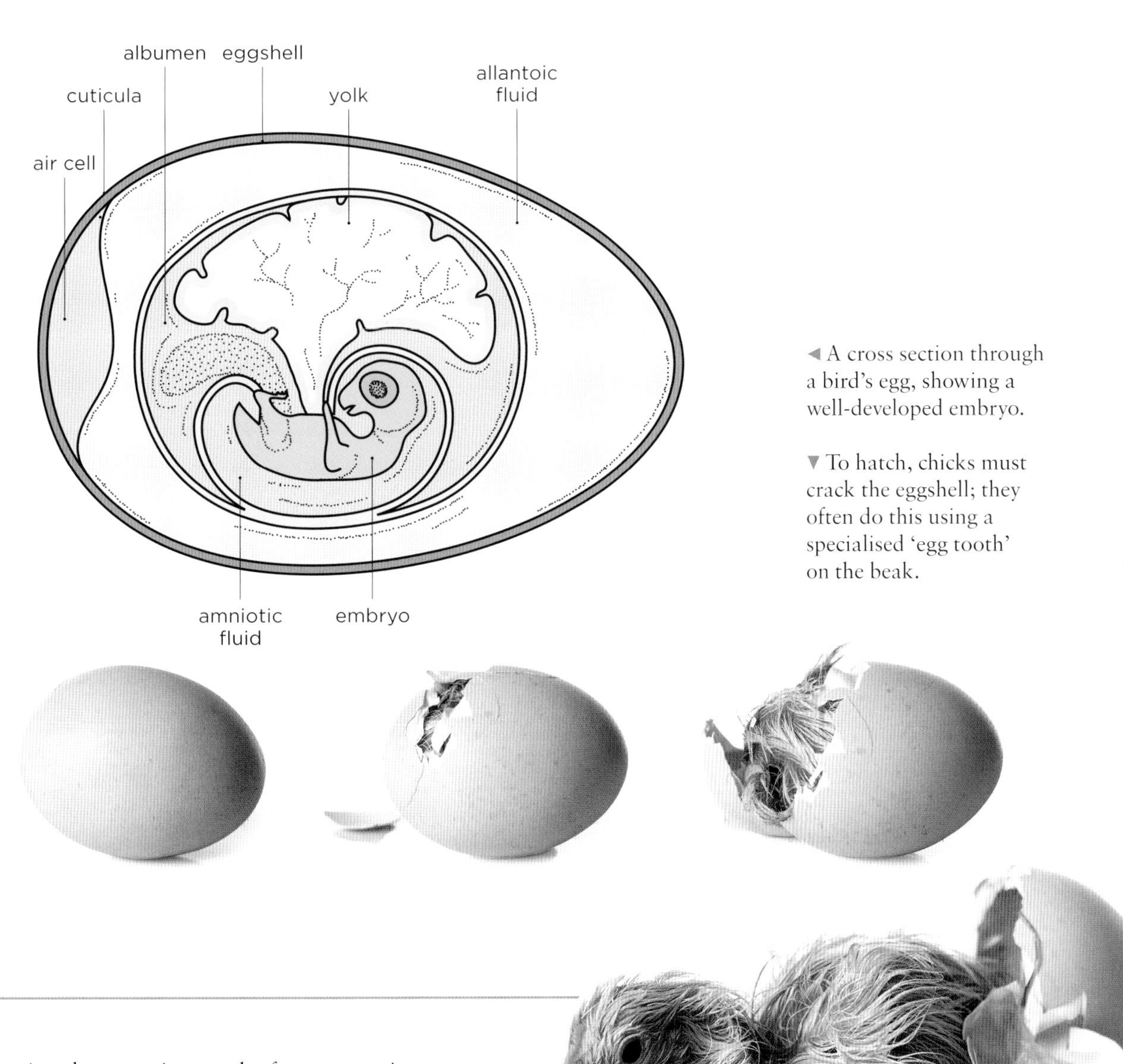

◄ A cross section through a bird's egg, showing a well-developed embryo.

▼ To hatch, chicks must crack the eggshell; they often do this using a specialised 'egg tooth' on the beak.

◄ Scanning electron micrograph of a cross section of the eggshell of a chicken (*Gallus gallus domesticus*), showing the hard, calcified layer (left) and an inner membrane of protein fibres.

CHAPTER 5

Getting Around

Although many animal species engage in gliding or other aerial forms of locomotion, birds, bats and insects are the only living animals capable of powered flight. This is the most energetically demanding form of animal locomotion, and meeting these demands has shaped the evolution of many aspects of bird anatomy, physiology and behaviour.

Powered flight has several necessary components – wings, an engine, a flight stroke, control systems, small body size, lightweight design and an energy source (fuel) – but different species of birds meet these requirements in different ways using different flight styles. Birds do, however, share one common specialisation for powered flight: their forelimbs alone are used for flight, while their hindlimbs are used for walking, running and jumping forms of locomotion. These dual locomotory capacities contributed to the ability of birds to exploit a diverse range of environments and to inhabit every continent on Earth.

◀ White-vented plumeleteer (*Chalybura buffonii*).

Aerodynamics

Powered flight requires the generation of forces to overcome gravity and drag. Overcoming the gravity is an acute problem for flying animals. Solid ground or the buoyancy of water counteract the force of gravity for walking and swimming animals, but the air provides little resistance to gravity for those that fly. Wings must provide enough lift to prevent a bird from falling to the ground, and they must also provide enough thrust to overcome drag forces in order to move the bird forward.

Overcoming drag

Drag is the force that acts in opposition to the forward movement of an object through air or water (see diagram opposite). There are three main types of drag: pressure drag, friction drag and induced drag. Flying birds have adapted to overcome these drag forces in various ways. Pressure drag is generated by low pressure forming behind an object as it moves through air, and can be reduced by having a streamlined fusiform shape. Friction drag is created by shear forces between air molecules as they flow over a surface, and increases with velocity. Friction drag can be reduced by body and wing shape, and by having smooth feather surfaces. Induced drag is created during flapping flight. On the downstroke of a wing, air is pushed downward to create lift, but some of this lift is in a backwards direction, creating drag. Induced drag can be reduced by having long, narrow wings (typical of petrels and albatrosses), and having open slots between primary flight feathers (as in some eagles and vultures; see page 161).

Producing lift

Gravity is the force pulling objects towards the centre of the Earth (see diagram opposite). Birds overcome both gravity and drag primarily through the use of their wings. During forward motion the body and tail can also provide lift, but most of the aerodynamic force to power flight is generated by the wings. Lift can be intuitively understood by thinking of the forces acting on your outstretched hand on a windy day, or if you carefully hold your arm out of the window of a moving vehicle. The wings of a gliding bird, or of a fixed-wing aircraft, generate lift – as does any solid object moving through a fluid medium. Lift is generated perpendicular to the wing due to lower air pressure forming above it, and wings are effective at generating this force thanks to their aerofoil shape. This causes differential airflow across the upper and lower surfaces of the wing, thereby increasing the pressure differential across the wing.

Aerodynamic lift

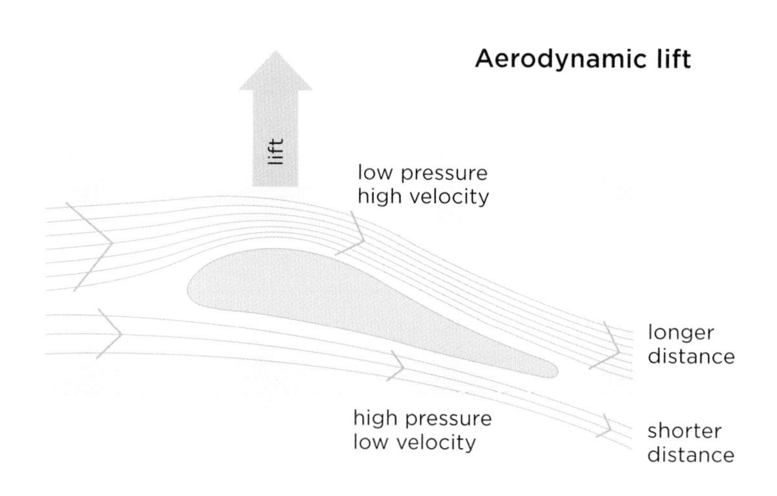

◄ Lift can be produced by wing shape. As a cambered wing passes through air, the air passing over the upper surface must travel at a higher velocity, creating lower pressure. The difference in air pressure between the upper and lower surfaces of the wing produce lift.

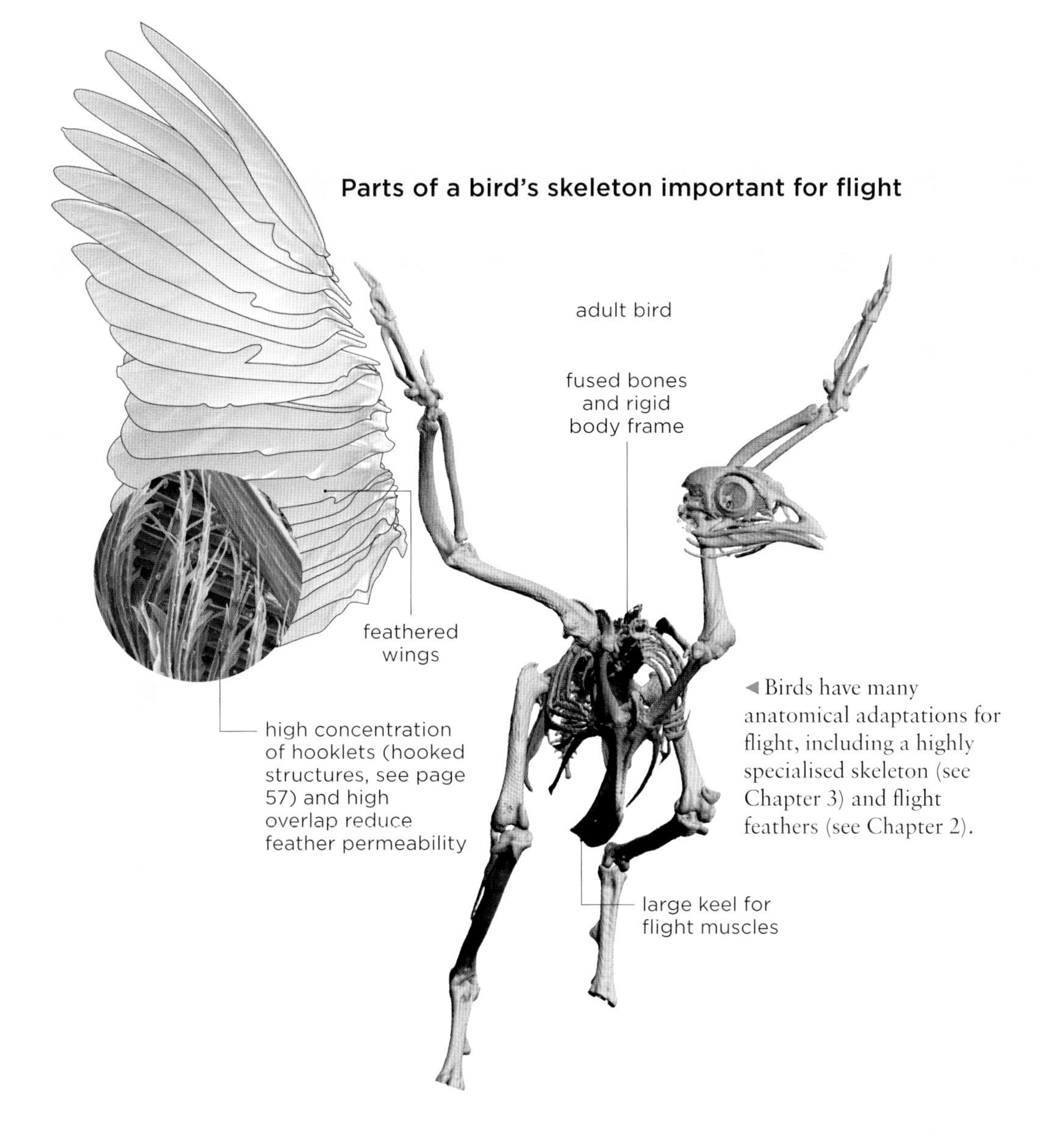

◄ Birds have many anatomical adaptations for flight, including a highly specialised skeleton (see Chapter 3) and flight feathers (see Chapter 2).

The amount of lift generated by a wing depends on several factors, including wing shape (aspect ratio), wing curvature (camber), feather structure, and orientation of the wing relative to the direction of motion (angle of attack). The wings of a flapping bird also generate lift, but the airflow due to forward motion of the body (as in gliding) and downstroke of the wing combine to create lift that acts both upward and forward, counteracting both gravity and drag. Lift and drag production vary across species depending on wing and feather shapes, and also from moment to moment within an individual bird as it flaps its wings. Birds rely on complex neuromuscular control to continuously change the shape of their wings during flapping in order to modify the size and direction of these forces, thereby staying aloft.

◄ The four forces experienced by any flying animal or aircraft.

Wings and tails

Different birds have different flight speeds and styles depending on their habitat, lifestyle and behaviour. Some need high manoeuvrability to navigate their way through a cluttered environment, while others are adapted to soar over the open ocean. And while some must generate massive force for quick acceleration and take-off, others need to keep their energy expenditure during flight to a minimum. These and other factors have led to the evolution of diverse wing and tail shapes.

Wing shape

The aerodynamic forces of lift and drag vary depending on wing size and shape. Aspect ratio is the measure of wing shape, and is calculated as the wing length (wingspan) divided by the wing width (wing chord). The long, pointed wings of swifts and albatrosses have high aspect ratios, providing greater efficiency (i.e. more lift relative to drag), and thus the birds require less energy for flight. In contrast, the shorter, wider wings of sparrows and titmice have low aspect ratios, which facilitate rapid take-offs and manoeuvring through a cluttered environment (see also pages 32–3).

In some birds such as hawks and eagles, the outermost feathers (the primaries) on their wings have wide spaces or slots between them. These open

Two extremes of wing aspect ratio

▲ The long, tapered wings of this northern royal albatross (*Diomedea sanfordi*), soaring near New Zealand, have a high aspect ratio, making them more efficient.

▲ Most sparrows, such as this Eurasian tree sparrow (*Passer montanus*), have relatively broad, short wings with a low aspect ratio, allowing manoeuvrability.

slots change wing shape to reduce the formation of vortexes at the wing tips and thus reduce the induced drag and increase lift.

The aspect ratio of the wing is usually considered in tandem with the wing loading, or the ratio of bird weight to wing area. This is because heavy birds with the same aspect ratio as lighter birds will have relatively reduced flight performance.

▲ Large raptors and other soaring birds can modify their wing shape by adjusting the gaps between the outermost primary flight feathers, as seen in this brahminy kite (*Haliastur indus*).

Forked tails improve lift

▲ Forked tails, such as that of tree swallows (*Tachycineta bicolor*), can be used as visual displays, but they also provide more lift compared to a wedge-shaped tail of equivalent length.

Tail shape

Besides the wings, the tail (and the body) can be important in generating more lift than drag – in fact, they provide up to 15 per cent of the lift required to overcome gravity. On first glance, bird tails would seem to provide drag and be costly to carry through the air. However, they actually provide substantial lift in many species. Aerodynamic modelling suggests that slightly forked tails that become triangular when fanned, as in American tree swallows (*Tachycineta bicolor*) and common terns (*Sterna hirundo*), provide high lift relative to their drag. In addition, tail feathers can effectively extend the length of the body during flight, contributing to a long fusiform shape that reduces pressure drag and the width of the wake behind a flying bird. Increasing the length of an object relative to its width creates a more aerodynamic shape by reducing pressure drag, and accounts for the shape not only of birds but of rowing shells, canoes and missiles.

Wing strokes

Aeroplanes have fixed wings that provide lift, and engines or propellers that provide thrust. In contrast, birds must provide both lift and thrust with their wings, and do so by stroking the wing up and down, essentially rowing through the air. As birds flap their wings, they continuously control their shape, camber and attack angle to modify aerodynamic forces.

Downstrokes and upstrokes

The power stroke is the downstroke of the wing that provides lift and propulsion, and is generated primarily by the large pectoralis flight muscles in the chest. During this stroke the wing can be swept down and forward, or down and backward, and is also rotated at the shoulder and other joints to adjust the angle of attack. These aspects of the downstroke vary during take-off, forward motion and hovering. The downstroke is followed by an upstroke, which serves as a recovery phase. Upward motion of the wing is powered by contraction of the supracoracoideus flight muscle (see pages 94–5) and adjusted by using other muscles controlling the upper limb. Compared to the downstroke, the upstroke is more variable across species and during different types of flight.

◄ This multiple-exposure composite image reveals the undulating upstrokes and downstrokes of the wings of a Mediterranean Audouin's gull (*Ichthyaetus audouinii*), creating lift and propulsion.

▼ Birds can change the shape of their wings as they manoeuvre, as in this western barn owl (*Tyto alba*) preparing to land.

▲ American kestrels (*Falco sparverius*) can use different wing gaits depending on their flight speed, from 60 km/h during rapid flight to almost stationary as they scan for prey.

Flight gaits

Variation in the upstroke creates different flight gaits, including vortex-ring gait and continuous-vortex gait. In the vortex-ring gait, lift is produced only during the downstroke of the wing, and drag is minimised during the upstroke by either rotating the wing tips backward or opening the primary flight feathers like a venetian blind. Continuous-vortex gait wing strokes create lift during both the downstroke and upstroke by keeping the wings partially or fully extended. Although lift is provided on both strokes, most is produced during the downstroke.

Some birds can use both types of gait, depending on their wing shape. At slower speeds all birds generally use vortex-ring gait during flight. At higher speeds, birds whose wings have a low aspect ratio (e.g. sparrows or pigeons) continue using vortex-ring gait, but those whose wings have a high aspect ratio (e.g. falcons) switch to a continuous-vortex gait. For example, American kestrels (*Falco sparverius*) use continuous-vortex gait at high speeds, and vortex-ring gait at intermediate speeds (by opening the primary feathers) and slow speeds (via wing-tip rotation).

Power for flight: physics meets biology

Newton's third law of motion states that for every action there is an equal and opposite reaction. For a flying bird, the downward force of gravity is counteracted by an opposing upward force of lift, and the forward force of thrust is counteracted by a backward force of drag. The good news for birds is the energy required to move a given distance during flight is quite low compared to swimming or running the same distance. The bad news, however, is that gravity causes them to use much more energy per unit of time.

Generating power

When hawks soar, they usually combine the lift generated by the aerodynamic shape of their wings with that from rising warm air currents. Flapping flight is more complicated than fixed-wing flight (e.g. soaring or gliding like an aeroplane), in that a bird's flexible wings essentially paddle through the wind to get aloft – a very energy-expensive manoeuvre. In this, lift generated by the aerodynamic shape of the wings is combined with the forward thrust and additional lift gained by flapping to sustain forward flight over longer distances.

The power requirements for walking and running increase with speed, whereas those of birds in flight are more complicated owing to the additional difficulty of staying aloft. Specifically, at relatively slow speeds birds must work harder to stay aloft than when flying at moderate speeds. The resulting graph that plots power requirements against flying speeds, which forms a shallow U-shaped curve (see below left), indicates that there is a flight speed at which birds can optimise their power requirements.

▼ Whooper swans (*Cygnus cygnus*) are among the heaviest living birds (15–23 kg) that can still fly. The extra power required for take-off for these very large, heavy birds must be augmented by running.

Optimising power use when flying

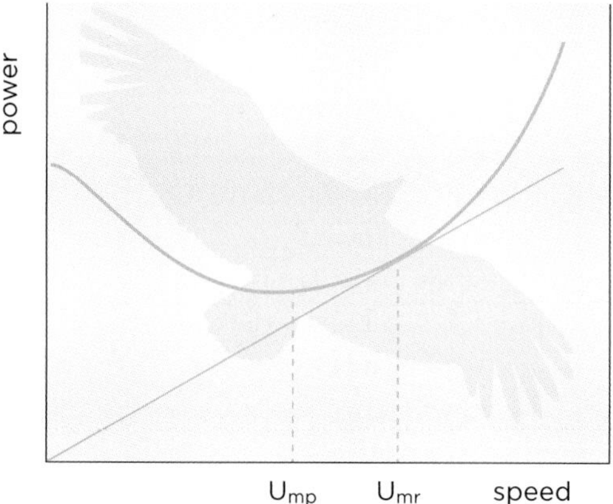

▼ Power required at take-off is high. As the bird picks up speed, the power required decreases to a minimum (U_{mp}), then increases again. Best 'fuel economy' is achieved at U_{mr} (maximum range).

▲ *Argentavis magnificens* is an extinct species from the Upper Miocene (*c.* 6 million years ago) that is believed to be the largest flying bird (with a wingspan of 7 m and weighing 70 kg) that ever lived. Given the laws of physics, it was likely a soarer incapable of powered, flapping flight.

Heavy birds

An interesting corollary of the laws of physics applied to bird flight is that, as the weight of a bird increases, there comes a point at which the downward force of gravity cannot be overcome and the bird cannot lift off the ground. Among the heaviest living birds that can fly under their own power for a sustained period are the New World trumpeter swan (*Cygnus buccinator*; 10–14 kg), and Old World mute swan (*C. olor*; 15–23 kg) and great bustard (*Otis tarda*; 15–20 kg). However, most similarly heavy birds, including the largest of the vultures (11–15 kg), fly mainly by gliding and soaring with the assistance of thermals, and use flapping flight only for initial take-off. The heaviest-known ancient bird that could fly was *Argentavis magnificens* (*c.* 70 kg), from the Upper Miocene (around 6 million years ago) of Argentina, although it was probably too large to be capable of continuous flapping flight and instead presumably flew by gliding and soaring like modern-day vultures. The heaviest extant birds, including the ostriches (100–150 kg), are relegated to a running form of locomotion.

Power for flight: the necessary metabolism

Given that flying is a relatively expensive form of locomotion, birds have evolved ways to lighten their load in order to reduce the effective force of gravity, and evolution has shaped their wings and feathers to maximise lift. In addition, they have specialised their fuel use, using fat as their primary source of energy for powering sustained flight.

Fuel for flight

Roughly 90 per cent of the energy birds use during long-distance flight comes from fat, with the remainder from protein and some carbohydrate stores. Some proteins and carbohydrates are also metabolised during such flights to provide necessary nutrients and water, rather than to meet the bird's primary energy needs directly. In contrast, a fast-running mammal (including a human) uses primarily carbohydrates (mostly glycogen stored within muscles) as fuel. Fat is the best fuel for weight-economising birds, however, because when stored it contains eight to ten times more energy per unit of wet mass (37 kJ/g) than alternative fuels such as proteins and carbohydrates (4–5 kJ/g).

However, the use of fats as fuel during intense exercise such as flying is difficult because fat metabolism requires more oxygen, and fats must be chaperoned by fatty acid transporters from where they are stored in the body to the site of oxidation in the muscle cells. Birds increase the number of these fatty acid transporters (as well as oxidative enzymes) during migration, and it appears that this response is essential in enabling birds to rely on fats as their primary fuel while flying long distances. As noted in Chapter 4, the increased oxygen needed for fat metabolism is provided by a flow-through respiratory system that is well coordinated with the circulatory system.

A robust antioxidant system

One of the other consequences of oxidising fats as fuel during flight is the greater potential for damage from the resulting reactive by-products. At the most basic chemical level, metabolising fats to produce energy involves the conversion of oxygen into water. During this process, some reactive intermediates (reactive oxygen species or free radicals) escape and can then react with, and damage, other useful

Model for the antioxidant capacity of a bird

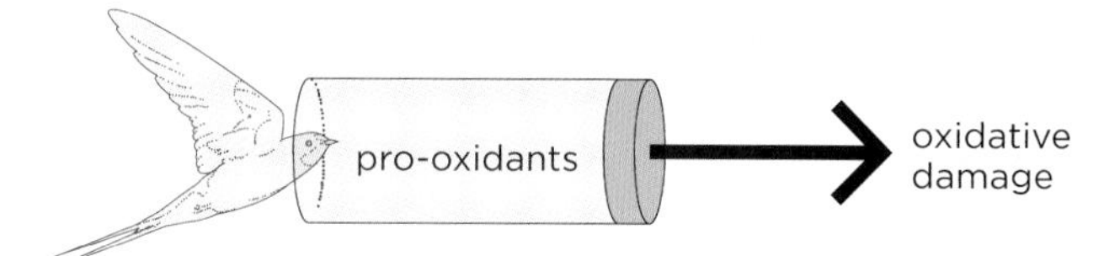

▲ Antioxidant capacity (visualised as a filter) intercepts pro-oxidants such as free radicals produced during exercise to reduce oxidative damage.

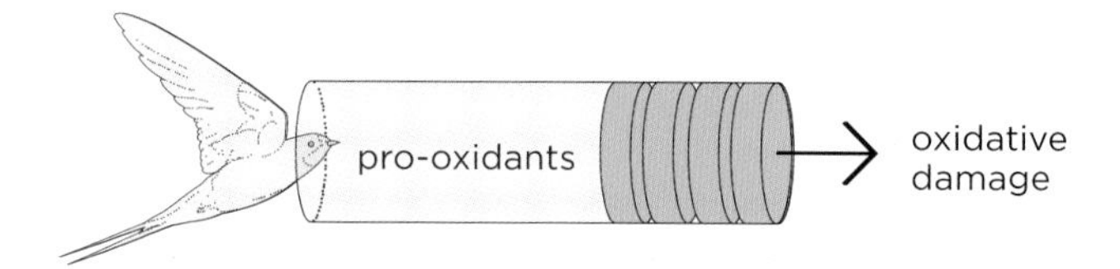

▲ A bird with less antioxidant capacity may suffer more oxidative damage (top), compared with one with a greater antioxidant capacity (bottom).

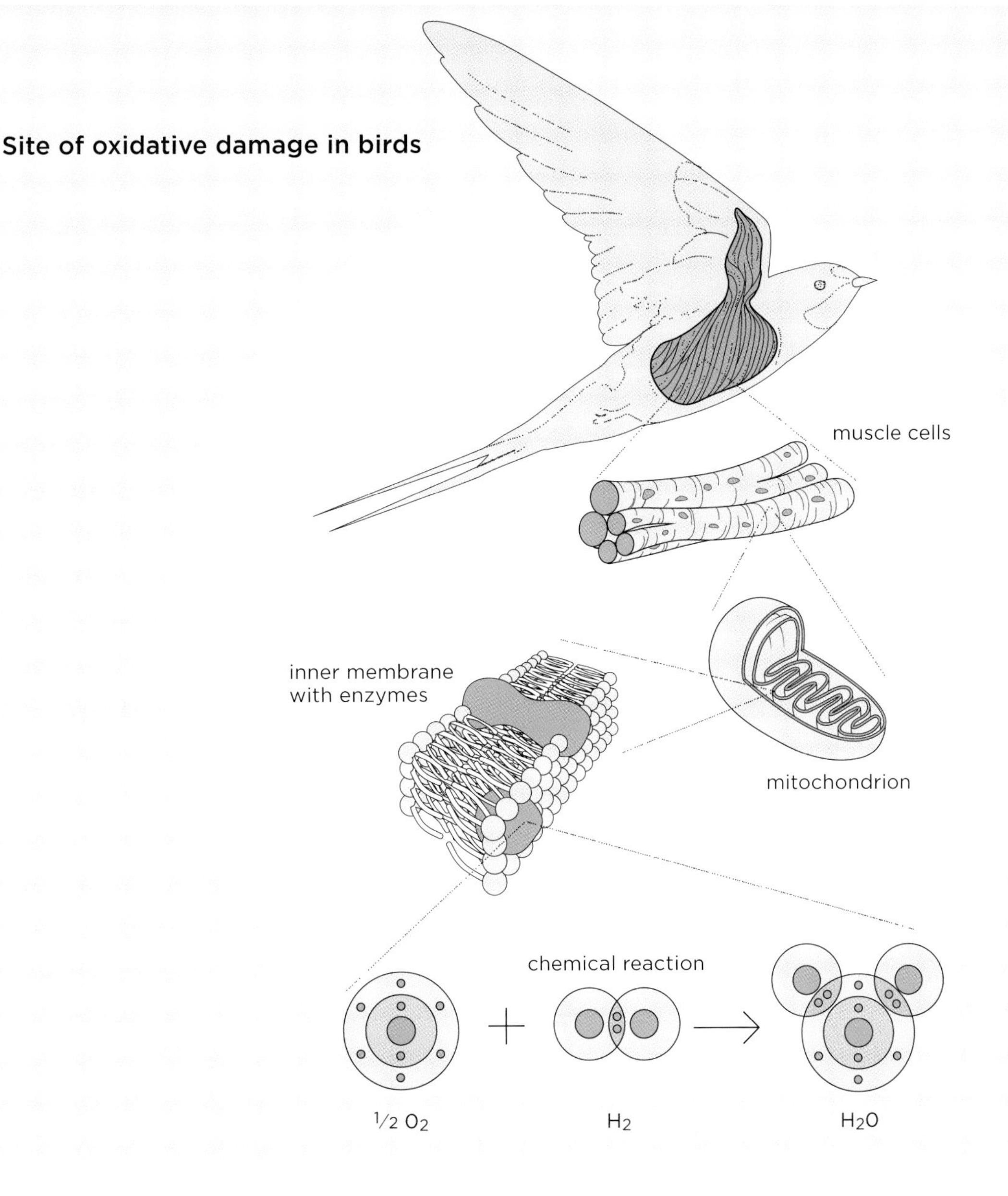

▲► Chemical reactions taking place in the mitochondria generate energy by metabolising food to convert oxygen into water. However, some oxygen escapes this process and these 'radicals' can chemically react with lipids to produce other types of damaging radicals.

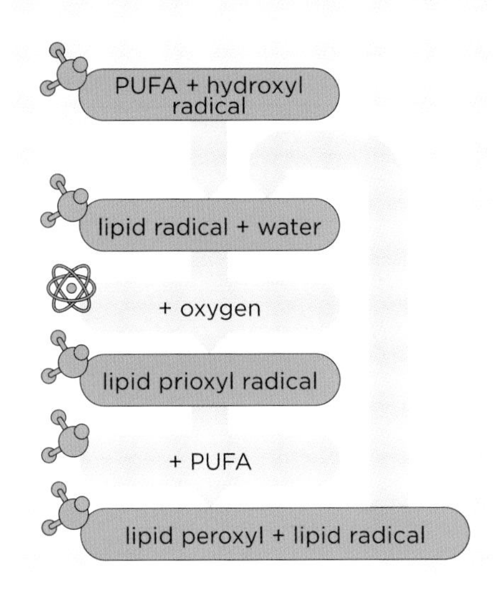

molecules (e.g. proteins, lipids, DNA). However, the antioxidant system of birds can convert or quench these free radicals and thereby avoid oxidative damage to these useful molecules. Thus, flying birds must have a robust antioxidant system – including antioxidant enzymes such as superoxide dismutase, and metabolites such as uric acid – as well as eating dietary antioxidants (e.g. carotenoids).

Flying at altitude

Birds that fly at altitude must overcome challenges not faced by terrestrial animals. Air pressure is lower at high altitudes, which reduces drag, but temperature and the partial pressure of oxygen are also lower, creating metabolic and thermal issues for high-flying birds. Although most birds stay within a couple of kilometres of the Earth, some species are known to fly at extreme heights and have physiological adaptations allowing them to do so.

High-flying migrants

Species that fly at extremely high altitudes include Eurasian whooper swans (*Cygnus cygnus*), which have been observed by an airline pilot at 8,230 m over the Atlantic Ocean, and Rüppell's griffon vulture (*Gyps rueppelli*), an individual of which was sucked into a jet engine at 11,552 m above the Ivory Coast in Africa. However, bar-headed geese (*Anser indicus*) can most regularly claim to be the highest-altitude migrants on the planet – during their migration through the Himalayas en route from their nesting grounds in Tibet to their wintering grounds in India, some individuals fly at altitudes close to the height of Mt Everest (8,848 m). Although most migratory birds fly at much lower altitudes, understanding how bar-headed geese fly at such high altitudes (where atmospheric oxygen levels are very low) provides insights into how the respiratory and circulatory systems of birds is pre-adapted for highly efficient oxygen uptake.

▲ High-flying whooper swans (*Cygnus cygnus*) above the clouds – the species has been recorded at an altitude of 8,230 m.

► Rüppell's griffon vulture (*Gyps rueppelli*) can soar at very high altitudes and is considered to be the highest-flying bird.

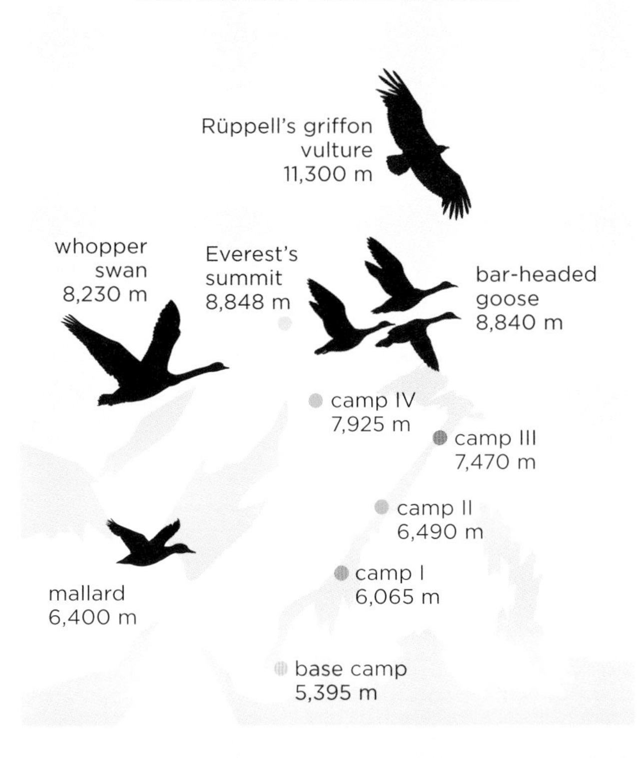

▲ Impressive records of high-flying birds compared to the altitude of Mt Everest and its base camps. The highest flyers have special adaptations, including increased lung capacity and a unique type of haemoglobin.

Adaptations for high-altitude flying

Like other birds, bar-headed geese produce more haemoglobin, more red blood cells and more myoglobin (the haemoglobin equivalent in muscle cells) as they prepare for migration. These changes enhance their blood's capacity to bind and transport oxygen. However, the geese differ from other species in several exceptional ways that allow them to perform unusually well at high altitudes. For a start, they have relatively larger lungs and a special type of haemoglobin that allows them to extract more oxygen from each breath. This enables these geese to increase the oxygen-carrying capacity of their blood without unduly increasing its viscosity and potentially reducing circulatory efficiency. Bar-headed geese also have more blood capillaries in their heart and flight muscles than do other geese. In addition, the capillaries penetrate deep within the muscle, allowing them to deliver oxygen throughout key tissues under low-oxygen conditions. The birds also have a special form of the enzyme cyclooxygenase, which is involved in energy production within their muscle cells, increasing the efficiency of their oxygen use.

▼ Bar-headed geese (*Anser indicus*) must cross the Himalayas, the highest mountain range in the world, to reach their breeding grounds in Mongolia.

Flight styles: flapping, gliding and soaring

Flapping of wings is usually necessary to generate enough lift for take-off, but once they are airborne birds vary in how much they flap and the nature of that flapping. Some birds soar for hours with virtually no wing flapping, while others flap continuously. And some birds have a bounding flight, while others flap and glide intermittently.

Continuous flapping

Continuous flapping flight is most common in species with a high wing loading (a high body mass relative to wing surface area), such as ducks and gamebirds. In some cases (ducks and divers) continuous flapping results in high-speed flight with low manoeuvrability, which requires long take-offs and landings. Some species can sustain continuous flapping for long periods. For example, the bar-tailed godwit (*Limosa lapponica*) can flap continuously for more than 11,000 km during its migration to and from Alaska and New Zealand. In other cases (pheasants, turkeys and tinamous) continuous flapping can be sustained for short bursts only and is mainly used to escape predators or move from the ground to a perch.

Flapping and gliding

Many species do not flap continuously, but instead intersperse periods of flapping with periods of gliding on extended wings (called flap-gliding) or periods of bounding with wings closed (called flap-bounding). These non-flapping periods likely save the birds energy through brief rests and/or allow them to adjust their speed by briefly pausing their forward thrust. Flap-bounding is common among many passerines and woodpeckers, and likely saves energy when the birds are travelling near their maximum speed. Flap-gliding is common in small to medium-sized birds such as common swifts (*Apus apus*) and common starlings (*Sturnus vulgaris*), and likely also saves energy when the birds are travelling at slow or moderate speeds.

Flap-bounding flight

▶ Finches exhibit a steep, roller-coaster flight, whereas woodpeckers generally fly in a pattern of moderate rises and falls.

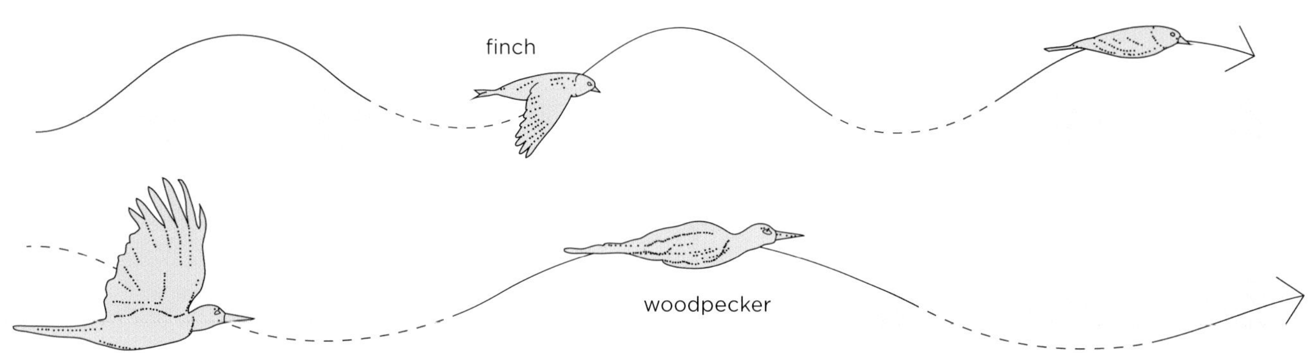

▲ Great white pelicans (*Pelecanus onocrotalus*) are expert at ridge soaring along shorelines with minimal flapping using rising coastal air currents.

HOVERING

Though hovering is most often associated with hummingbirds, many birds can hover, at least transiently. Hovering is a special form of flapping flight in which thrust is directed downward to keep a bird aloft without moving forward. Many birds hover briefly during landing, and kingfishers and ospreys may hover over water before diving to catch fish. Wind-hovering species such as Old World common kestrels (*Falco tinnunculus*) and American white-tailed kites (*Elanus leucurus*) hover while hunting by facing into the wind and flapping their wings, allowing them to survey the ground below for prey. However, hummingbirds are the most specialised hovering birds. They rely on a high attack angle and keep their wings extended during the upstroke, with the whole wing rotated in a similar way to the wing-tip rotation seen in other birds. As a result, the downstroke and upstroke combine with a figure-of-eight pattern made by the wings.

Gliding

Gliding is non-flapping passive flight with wings outstretched. The wings and body provide aerodynamic lift, and the rate at which the bird sinks towards the ground depends on the lift-to-drag ratio. Expert gliders can produce very low glide angles and hence stay aloft for extended periods, whereas other species may glide only briefly during landing. Non-avian gliders such as flying squirrels, flying snakes, flying fish and flying frogs can glide moderate distances. However, some birds intersperse gliding with periods of flapping to regain altitude, allowing them to remain aloft almost indefinitely. For example, frigatebirds can remain on the wing for weeks, flapping to gain altitude when necessary and then gliding. If air conditions are favourable, they can glide for extended periods without flapping, a flight mode known as soaring.

Soaring

Soaring is a form of gliding in which birds make use of rising air currents to counteract gravity and provide sufficient lift to stay aloft for long durations without flapping. There are three types of soaring: thermal soaring, ridge soaring and dynamic soaring. In thermal soaring, birds spiral upward within rising columns of warm air to gain altitude, then glide towards their goal. Hawks and vultures may glide from one rising thermal to another, spiralling upwards within each before gliding to the next. During migration, large numbers of birds may aggregate in warm air columns, forming so-called 'kettles' of spinning groups.

In ridge soaring, birds use rising air along ocean waves or ridges of land to stay aloft, such as when North American brown pelicans (*Pelecanus occidentalis*) soar along waves or coastlines. In dynamic soaring, birds alternately tack into prevailing winds to gain altitude and glide downwind to gain speed. Seabirds such as albatrosses and petrels use this form of soaring, taking advantage of the vertical gradient of wind speeds over the open ocean.

Flight displays

In addition to being a form of locomotion, flight may be used in visual and acoustic communication. Many species of birds use flight as part of complex visual displays, combined with vocalisations or non-vocal sounds produced by feathers themselves. Flight may also be used to accentuate ornamental feathers, or to produce sounds that are carried great distances.

Manakin dances

Flight is incorporated into the complex visual displays, or dances, of the Central and South American manakins, which hop, hover and fly during courtship. Manakin males court females at specific sites, and use a combination of complicated movements and sounds to attract them. These display flights may even be coordinated between multiple individuals, as seen in the dance of lance-tailed manakins (*Chiroxiphia lanceolata*), in which the males alternately hop and hover over one another. Golden-collared manakins (*Manacus vitellinus*) court females by leaping from stem to stem in an arena, and loudly snapping their wings together over their back while in mid-leap.

Diving display flights

In other flight displays, birds may dive from high altitude and then sweep upward, showing off plumage features or making sounds as they do so. One magnificent example is the American Anna's hummingbird (*Calypte anna*), the males of which dive towards perched females. As they pass over the female, they orient their iridescent throat feathers to create a flashing pattern, and then swoop upward while spreading their tail feathers, which vibrate to create a loud sound.

Vocalising in flight

Some species sing or call during flight displays. The Old World rollers are so named for their rolling flight displays. For example, the African purple roller (*Coracias naevius*) rolls back and forth around its flight axis during long diving display flights that are combined with mating calls. Eurasian skylarks (*Alauda arvensis*) perform extended song flights, which may also last up to a half an hour. The males sing throughout the climb, level flight and descent of this display, and adjust their flight speed to maximise their endurance.

▼ These blue manakins (*Chiroxiphia caudata*) collaborate during their complex jump-hover flight displays, with an alpha male alternating with subordinates to perform in front of a female.

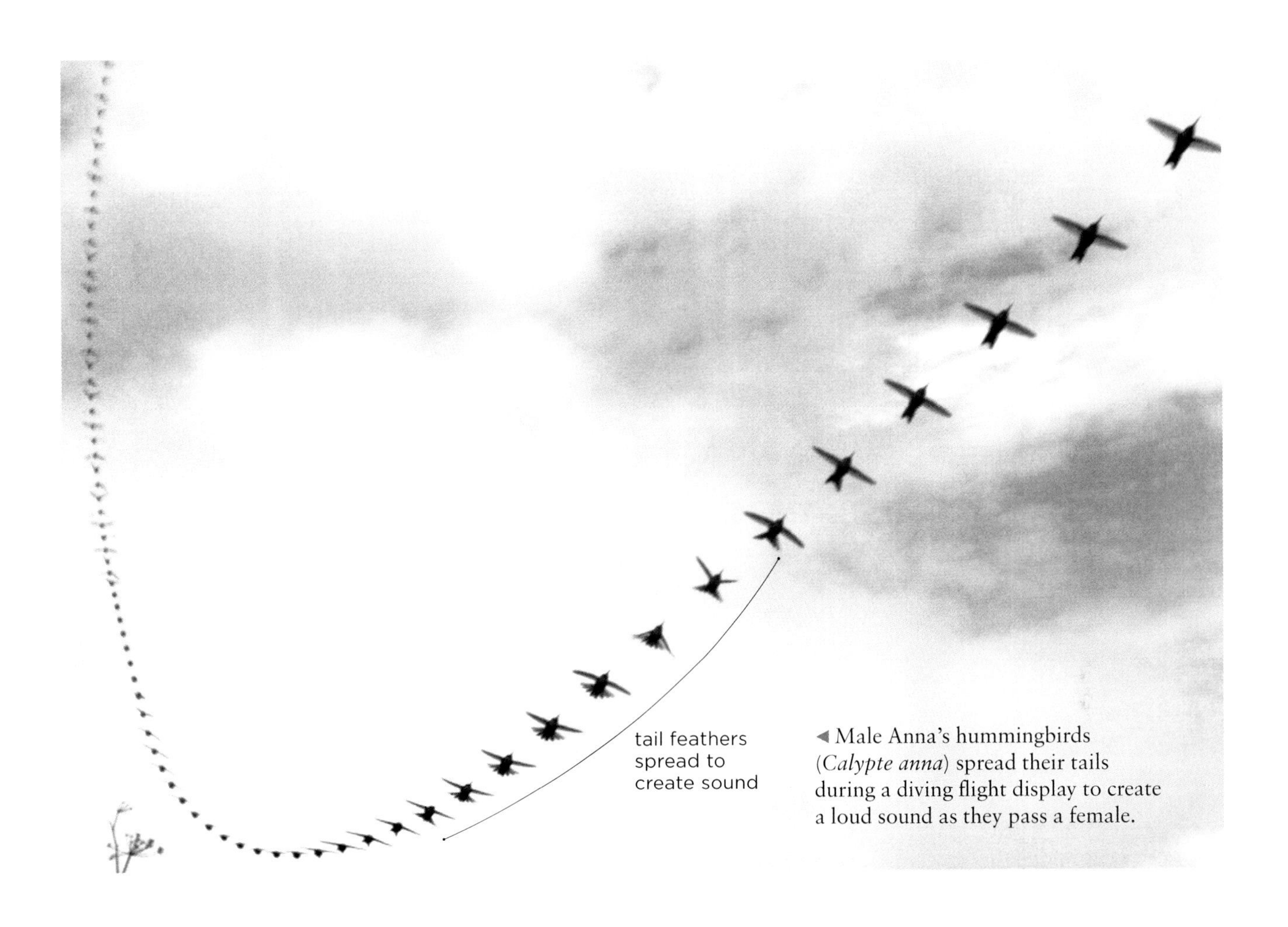

◄ Male Anna's hummingbirds (*Calypte anna*) spread their tails during a diving flight display to create a loud sound as they pass a female.

SOUNDING OFF

Like Anna's hummingbird (see main text), some birds make sounds in ways other than vocalisations during their flight displays. For example, Latham's snipe (*Gallinago hardwickii*) breeding in Japan have elaborate displays that incorporate sounds produced by their feathers. American common nighthawks (*Chordeiles minor*) also perform a flight display that incorporates feather sounds. Around dusk, they fly to a height, then rapidly descend and swoop within a metre of the ground. At the bottom of this swoop they adjust their wings to produce a loud, low-pitched booming or whooshing sound.

► Male Latham's snipe (*Gallinago hardwickii*) perform elaborate flight displays to females. This sound spectrogram illustrates the short, loud calls (short vertical markings) and longer intermittent buzzes produced by their feathers.

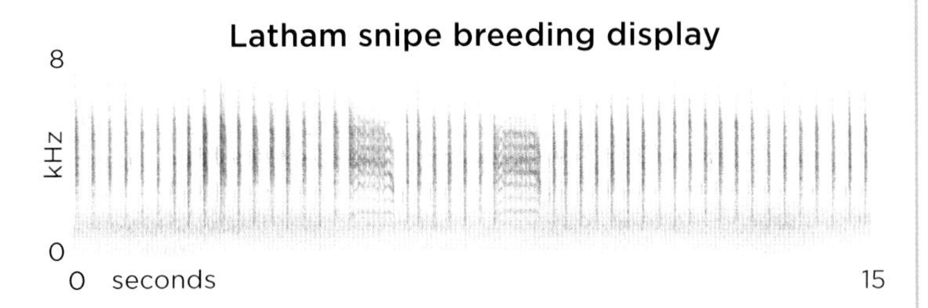

Sleeping on the wing

The majority of birds are diurnal, being active during the day and sleeping at night. However, when these birds migrate, they depart soon after sunset and often fly throughout the night. This switch to being nocturnally active is thought to have several potential benefits, including making it easier to avoid predators and providing more favourable flying conditions (the atmosphere is cooler and less stratified at night). However, flying throughout the night and then needing to feed and refuel during the day makes it very difficult to find time to sleep!

Avian sleep versus mammalian sleep

Birds, like many other animals, sleep in the sense that their nervous system is relatively inactive, their muscles are relaxed, their eyes are usually closed and they themselves are largely unresponsive to the outside world. In both mammals and birds, sleep is categorised by certain patterns of brain activity: rapid eye movement (REM) sleep, with recognisable low-amplitude, fast-changing brainwaves; and non-REM sleep, with high-amplitude, slow-changing brainwaves. Active sleep periods are short in birds, unlike in most mammals, and in some avian species sleep is interrupted frequently so that the birds can check for potential predators.

Birds can also sleep with only one-half of their brain at a time, a phenomenon known as unihemispheric sleep. The visual systems of birds are crossed in relation to the brain halves; that is, optic nerves from each eye go to the opposite sides of the brain. By alternating the sleeping half of the brain throughout the night, birds can watch out for predators with one eye open while resting the

▶ Migratory warblers, including these garden warblers (*Sylvia borin*), adjust their sleeping behaviour (with the head tucked or untucked) depending on their condition and thereby trade off energy costs and predation risk.

other half of their brain. The advent of miniaturised wireless technology for recording the brain activity of flying birds has led to the remarkable discovery that birds such as the Old World alpine swifts (*Tachymarptis melba*) can fly for days at a time and sleep aloft using this unihemispheric sleep system. Great frigatebirds (*Fregata minor*) also use unihemispheric sleep in flight, but in addition they sometimes sleep with both brain hemispheres – so, it appears that the birds do not need to be awake to maintain the aerodynamic control of flight. That said, researchers have found that frigatebirds sleep for only 0.7 hours per day during flights that last up to ten days, whereas on land they sleep for 12.8 hours per day.

◀ Alpine swifts (*Tachymarptis melba*) can sleep while flying, resting one half of their brain while the other remains active.

The costs of sleeping safely

Garden warblers (*Sylvia borin*) migrate in spring from Africa to Europe, and they usually rest and refuel along the way. Researchers studied these long-distance migratory birds during a stopover on an island in the Mediterranean after they had been flying more than ten hours the previous night. Following the long nocturnal flight, the warblers must recover sleep and also refuel while avoiding being eaten themselves by predators. In a series of cleverly designed field experiments, the researchers discovered that the birds modify the depth of their sleep and resting posture so as to remain alert to predators, while at the same time conserving their energy. Specifically, leaner warblers sleep more deeply, saving more energy, but are more vulnerable to predators compared to fatter warblers, which sleep less deeply yet are better able to escape predators. In short, these migrating songbirds modify their sleep behaviour in such a way as to strategically trade off sleep-mediated energy conservation with antipredatory vigilance.

Flock formations

Birds may fly individually, in large groups or in organised formations. Flocks of birds may consist of family groups, groups of unrelated birds of the same species or mixed-species flocks. Birds may fly together in a flock for self-defence, as many eyes have a greater chance of spotting predators, and to reduce the costs of flight.

Flocking

Most birds that migrate at night fly solo, even though hundreds of thousands of individuals may be aloft on any given night during a species' migration. Some birds, however, are more social and fly in groups. In the Americas, Harris's hawks (*Parabuteo unicinctus*) hunt in organised groups to capture prey, while other species fly in groups to avoid becoming prey. Mixed-species flocks of birds occur in a variety of ecosystems but are common in Neotropical forests. Individuals may benefit from being in a flock through increased foraging efficiency by increased detection of predators.

▶ Harris's hawks (*Parabuteo unicinctus*) of the Sonoran Desert are unusually social for a hawk species, hunting in coordinated groups.

▼ Massive flocks of common starlings (*Sturnus vulgaris*) gather at dusk prior to roosting, forming visually stunning murmurations.

Types of flock formation

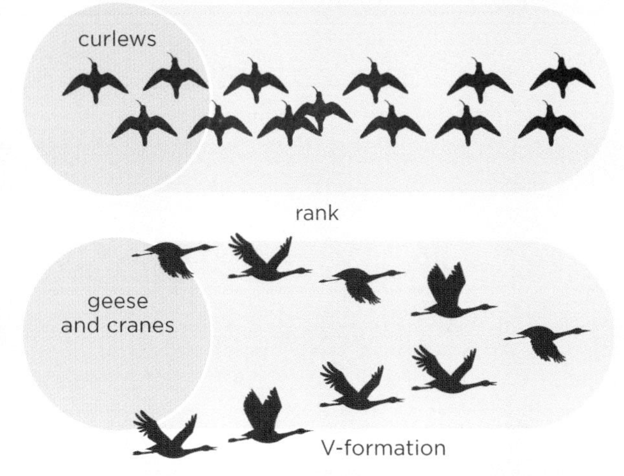

▲ Species vary in whether their flocks are loosely organised or more coordinated. The V-formation provides aerodynamic benefits to birds within the flock.

Some birds flock seasonally, keeping to themselves in the breeding season but then joining groups in the non-breeding season. Flocking can also vary over a daily cycle, with massive gatherings of birds coming together at dusk, just before they descend to communal roosts. Such dusk flocking occurs in red-lored Amazon parrots (*Amazona autumnalis*) in Panama and in many corvid species. However, it is perhaps most famous in common starlings, whose dusk flockings (called murmurations) can include thousands, or even hundreds of thousands, of individuals.

Flying in formation

When birds fly together in a flock, they may be able to take advantage of the air movement created by those around them to gain some aerodynamic benefit. Consider the classic V-shaped groups of flying geese, pelicans and cormorants. In a study of endangered northern bald ibises (*Geronticus eremita*) raised at Salzburg Zoo in Austria, birds were fitted with transmitters that could record the precise movements of their wings while flying freely in V-formation behind an ultralight aircraft. This demonstrated that each bird positions itself with respect to adjacent individuals, and the birds time the flapping of their wings to maximise the capture of 'upwash' and minimise their exposure to 'downwash' (the helpful and unhelpful air from their neighbours, respectively). Much like the cyclists in a peloton slipstream behind one another, so birds in organised flock formations may reduce flight costs by flying close to one another.

▼▼ Australian budgerigars (*Melopsittacus undulatus*) can form massive flocks of tens of thousands of birds when in search of water.

Getting around on land

Present-day birds evolved from a walking (bipedal), featherless lizard-like animal(s) into an ancestral feathered, flying bird. While most contemporary birds fly, some have secondarily lost this ability, including the familiar ratites (ostriches, cassowaries, kiwi, rheas and emu, *Dromaius novaehollandiae*) and penguins, as well as more than 60 other species from a wide diversity of families. These birds rely on walking or running on land, and swimming at sea, to get around.

Wings of flightless birds

All flightless birds except one group (the very large, herbivorous New Zealand moas, which were hunted to extinction around 600–700 years ago) retain the forelimb wing structure of their flying ancestors, albeit in a reduced form. These birds use their wings for non-flying purposes, including while running (e.g. ostriches) or for swimming (e.g. penguins), or quite extensively for courtship and displays. In the extreme, wings may be reduced to an almost vestigial state, as in the kiwi.

NO ESCAPE

Kiwi, South Island takahē (*Porphyrio hochstetteri*) and kākāpō (*Strigops habroptila*) are New Zealand birds that evolved flightlessness in the absence of predators. Unfortunately, however, their populations are now very low owing to predation by rats, weasels and other non-native species that have been introduced.

◄ The North Island brown kiwi (*Apteryx mantelli*) of New Zealand, like all kiwi, has highly reduced wings (inset) and is a walker, not a flyer. This flightlessness evolved in the absence of predators.

▲ The emu (*Dromaius novaehollandiae*) is the second-largest living bird (35–40 kg) and is flightless. Its relatively small wings are used for balance during running and for displays.

Too large to fly

The energy expenditure (kilojoules per gram of body mass) required for flight is very high compared to that for running, and these energy costs increase with body weight – this means that at some point birds are too heavy to fly. Accordingly, the heaviest flying birds today are relatively light compared to flightless running birds. In general, the evolution of flightlessness in birds is associated with larger body size, as well as island environments with few or no predators and reduced competition (see box).

Ostriches are the largest living birds and among the fastest bird runners. Interestingly, if one considers the energy required for flying versus running per unit distance travelled – essentially like considering petrol mileage in vehicles – then running is more expensive than flying. That is, a running bird will expend more energy than a similar-sized flying bird to move a given distance. This advantage in fuel mileage for flying birds enables their long-distance migrations, but flight still requires many physiological and anatomical adaptations, as has been discussed earlier in the chapter.

Power and energy requirements of locomotion of different groups and species

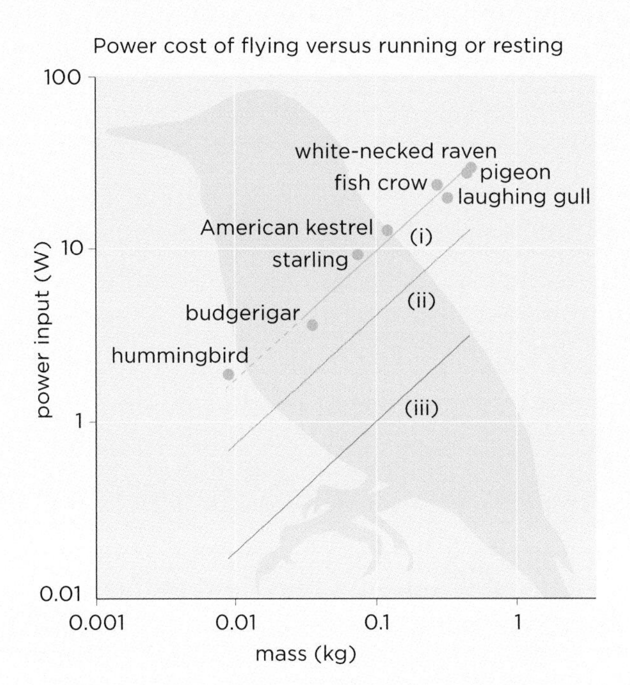

▲ For a given body mass, birds in flight (i) require more power (energy per unit time) than a running mammal (ii) and much less when simply at rest (iii).

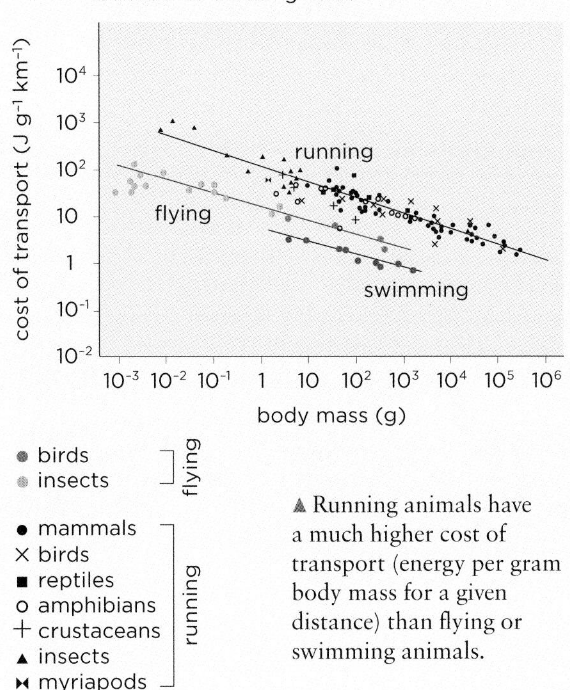

▲ Running animals have a much higher cost of transport (energy per gram body mass for a given distance) than flying or swimming animals.

Swimming and diving

In addition to flying, running and perching, many birds use their wings and legs for aquatic locomotion. Swimming is an important or primary form of locomotion for many birds such as waterfowl, loons (divers), cormorants, pelicans and penguins. Convergent evolution has resulted in several common features among such birds, including webbed feet with narrow foot bones, posteriorly oriented hips and legs, smaller and sturdier wings (for species that flap underwater), wide bodies to increase flotation, and oily feathers.

Swimming

Some species are so specialised for swimming and diving that they are very awkward on land (e.g. loons or divers) or have lost the ability to fly (e.g. penguins). Other species, including Atlantic puffins (*Fratercula arctica*), are adept at both swimming and flying.

Swimming on the water's surface involves overcoming drag, and although swimming speeds are slower than typical flying speeds, water is more viscous than air. Swimming at the surface creates a wave in front of the animal, limiting the speed at which it can travel. At slow speeds birds engage in paddling, such that propulsion is provided by pushing their webbed feet through the water. To achieve higher speeds, birds hydroplane to reduce the drag induced by water. Smaller (lighter) birds generate enough forward propulsion by paddling to hydroplane and lift their bodies above the surface of the water. Larger birds such as South American steamer ducks cannot generate sufficient speed to hydroplane by paddling alone, and instead use their wings as oars to steam over the water.

At higher speeds, birds can flap their wings with enough power to lift their body out of the water while they run across its surface, a behaviour known as taxiing, skittering or paddle-assisted flying. Taxiing differs from steaming in that the wings do not enter the water, and the recovery stroke of the feet is above the water. A few landbirds, including North American bald eagles (*Haliaeetus leucocephalus*) and western ospreys (*Pandion haliaetus*), occasionally swim slowly using their wings as oars, such as when they catch a fish with their talons that is too large to lift out of the water by flight.

◄ Atlantic puffins (*Fratercula arctica*) are adept at flying, swimming on the surface of the ocean and extended diving.

▼ Flightless Fuegian steamer ducks (*Tachyeres pteneres*) 'steam' over the water, using their wings as oars.

► Atlantic puffin (*Fratercula arctica*) diving.

▲ American dippers (*Cinclus mexicanus*) are an aquatic songbird, catching their food on the bottom of fast flowing streams by diving or running along it.

Diving

Birds sometimes dive below the surface of the water to avoid predators or to forage, a behaviour that has been a way of life for some species since the Cretaceous period. Fossils of Hesperornithes, a group of theropod dinosaurs that colonised the oceans in the Mesozoic era (66–22 million years ago), indicate that they were likely poor at flying, if they flew at all, but they appear to have been adapted to foot-propelled swimming and diving. Diving is present in many different taxa of modern birds. Dippers are passerines that can run along the bottom of streams or rivers and occasionally also swim underwater using their wings. They forage like this even in turbulent streams during winter, gathering invertebrate prey from the streambed. Some species (e.g. kingfishers and terns) dive into the water from a height to capture fish and return to the surface quickly. These birds may hover over the water until they spot prey, and then plunge-dive from a great height. Other species (e.g. loons or divers, and penguins) may dive from the surface and then spend longer periods of time underwater on foraging trips.

Underwater propulsion

Swimming underwater has an advantage over flying in that birds have natural buoyancy and so do not need to generate lift. However, water is many times more viscous than air, so much greater hydrodynamic force is required for propulsion, or thrust. Propulsion for diving is generated by either the wings, in which the birds essentially fly underwater (e.g. penguins), or the feet (e.g. grebes, loons or divers, and cormorants). Foot propulsion can involve oscillating the feet horizontally or vertically, depending on the species. Ducks and cormorants kick their feet up and down like a paddle wheel, but loons (divers) and grebes kick their feet from side to side like a kayak paddle.

Puffins and penguins flap their wings underwater for propulsion. While they are underwater they need to generate downward force with their wings to prevent themselves from floating upward, as well as propulsion to move forward. Flying birds that use wing propulsion underwater commonly keep their wings partly folded while diving, as much less wing area is required to produce the same force in water than in air. In a few cases, such as the scoters, both the wings and the feet are used for propulsion while the bird swims underwater.

▼ Propulsion while swimming underwater can be provided by the wings or feet. African penguins (*Spheniscus demersus*, main image) and other penguin species 'fly' through the water by flapping their wings. In contrast, common loons or great northern divers (*Gavia immer*, inset) propel themselves by kicking their feet from side to side.

CHAPTER 6

Migration and Navigation

Once birds evolved the ability to fly, migration and globe-trotting became possible. The movement of birds across the Earth is most pronounced in autumn and spring, when many species that breed in temperate and higher latitudes move toward (in autumn) or away from (in spring) lower latitudes, where they spend the winter. Such globe-trotting has required the evolution of sophisticated navigation and timekeeping skills, to allow birds to find their way and set off at the right time.

Some birds travel thousands of kilometres to and from their regular breeding and wintering areas on a fixed schedule, whereas others travel only when environmental conditions deteriorate and they are forced to find greener pastures. Variation in the timing, duration and direction of these usually seasonal movements for any given species occurs because the migrations are controlled by a combination of internal (genetic, physiological) and external (environmental) factors.

◄ Lesser snow geese (*Anser caerulescens caerulescens*).

Migration patterns

With few exceptions, almost every inch of our planet is overflown by a bird at some point in the year. During migration, most birds move along recognised flyways that have become defined over time by ecological barriers such as tall mountains, large deserts and oceans, and by prolonged glacial periods. Most major flyways are oriented north to south and are predominantly used in autumn and spring, when many species migrate between breeding and wintering grounds. In some areas, however, birds migrate in a wider variety of directions to benefit from wet or dry seasons, or for those species that breed on mountains, they simply move to lower altitudes during winter.

Obligate versus facultative migrants

The timing, duration and routes used by birds during their migrations all vary. However, our ability to track the movements of the full diversity of birds over the globe has become possible relatively recently with the development of very small tracking devices, many of which communicate via orbiting satellites. Some migratory birds traverse the full length of a given flyway (e.g. bar-tailed godwit, *Limosa lapponica*, seasonally move between western Alaska and New Zealand), whereas most move much shorter distances. Likewise, some migratory birds initiate and end migration on the same day each year; that is, timing of migration is principally controlled endogenously, or internally, and is quite independent of environmental conditions. Other migratory birds require certain changes in the environment (e.g. cold temperatures) to prompt their departure during a given season and hence the timing of their movements varies from year to year.

Like human snowbirds or grey nomads, who spend the winter in warmer places, most birds

Principal migration flyways of the world

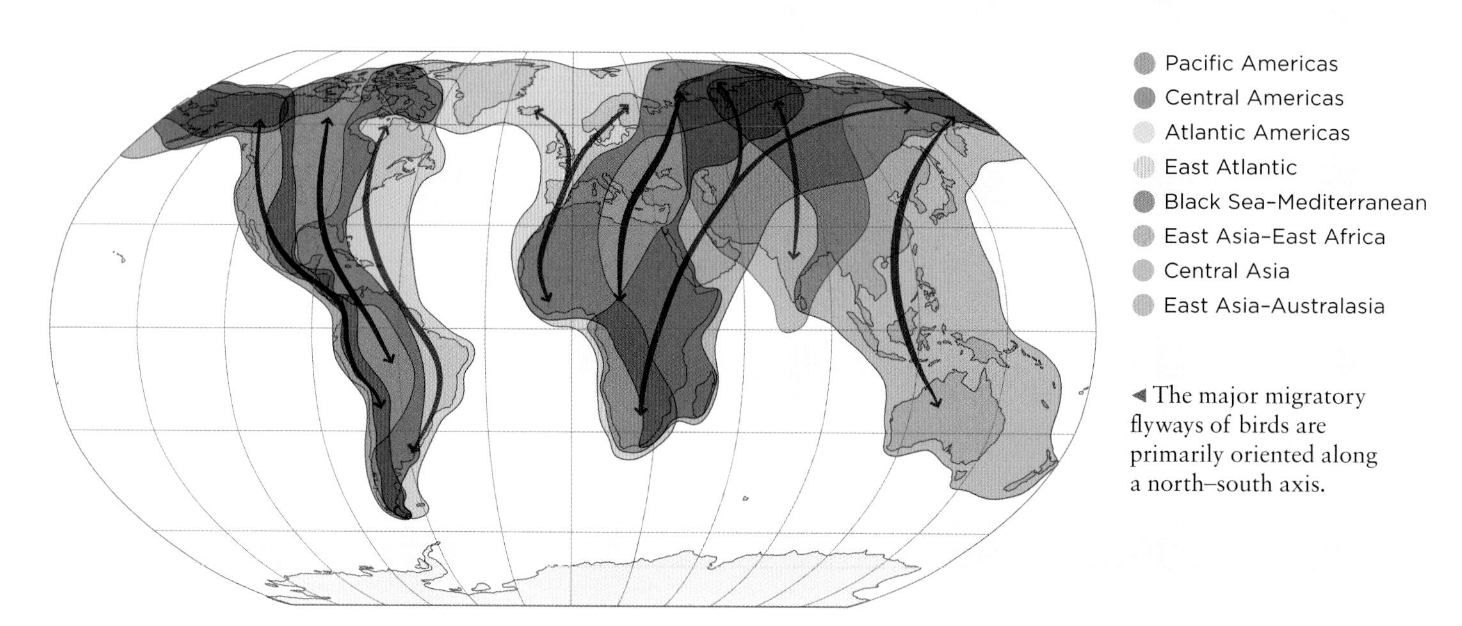

◀ The major migratory flyways of birds are primarily oriented along a north–south axis.

Eurasian distribution of the common blackbird

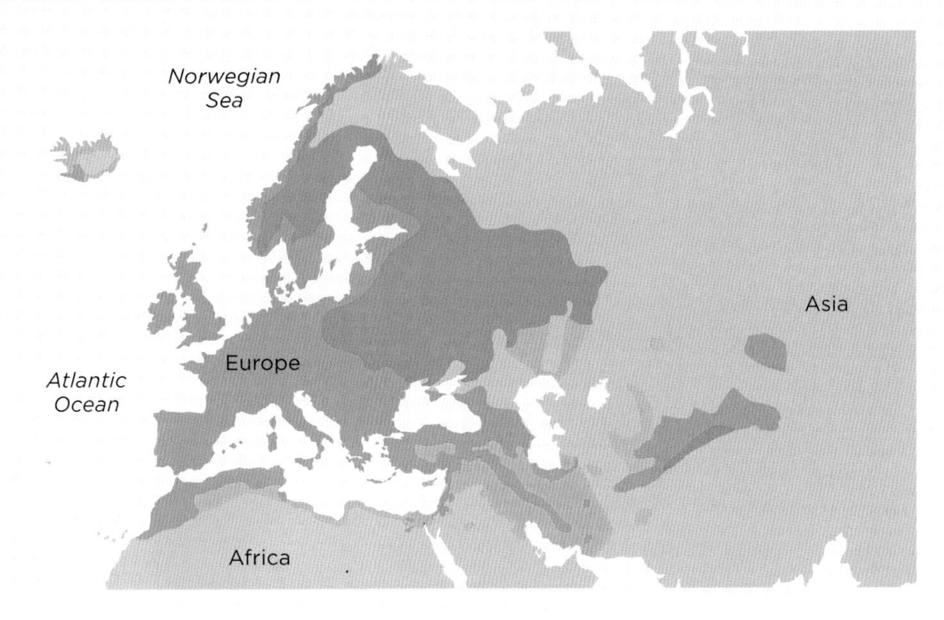

▶ The common blackbird (*Turdus merula*) is known as a partial migrant, in that some populations migrate between summer and wintering areas, whereas others do not migrate and instead remain resident year-round (purple shading).

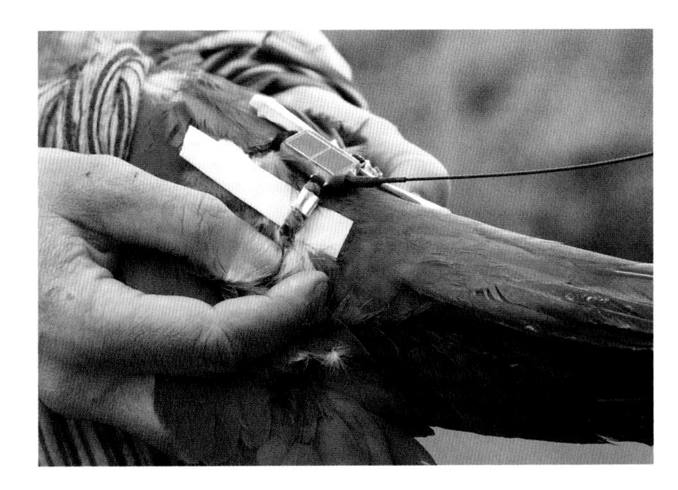

◀ Innovations in transmitter technology allow the tracking of individual birds as they traverse the globe. Shown here is a transmitter with a small solar panel that charges a battery inside.

seasonally migrate to more benign areas on their own, even though many individuals may be moving from the same place at the same time. The general route taken by most migrating birds from breeding areas to/from wintering areas is inherited, although local environmental conditions (e.g. prevailing winds) and barriers (e.g. oceans for landbirds) influence the exact path an individual bird takes. Young birds migrating for the first time (usually well before they are a year old) can be found in unusual places, in part because the entire migration is across unfamiliar landscapes, and also because the inherited avian navigation system is not as precise as our Global Positioning System (GPS).

Birds that rely more on internal (inherited, physiological) regulation of migration are known as obligate migrants and tend to be much more predictable in the timing, duration and direction of their movements. Those that rely more on environmental factors are known as facultative migrants and are usually much less predictable in their patterns of movement. This mix of migration types can even be observed within a single species, in which case it is known as a partial migrant. For example, some common blackbirds (*Turdus merula*) are year-round residents in western Europe, but others are facultative or obligate migrants that breed further north and then move to more southerly non-breeding areas.

Common blackbird (*Turdus merula*)

Experimental determination of migration direction

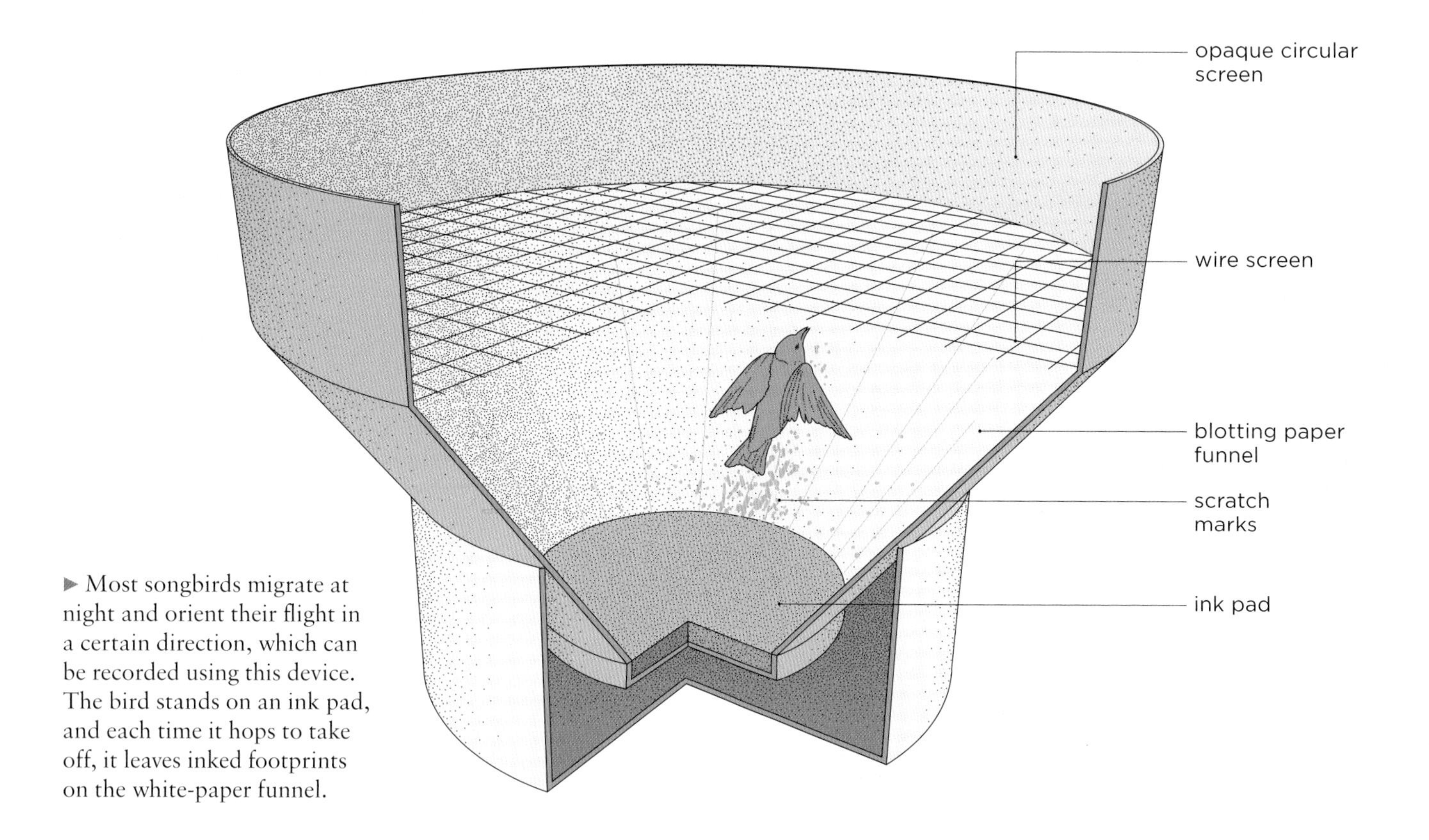

► Most songbirds migrate at night and orient their flight in a certain direction, which can be recorded using this device. The bird stands on an ink pad, and each time it hops to take off, it leaves inked footprints on the white-paper funnel.

Controls of bird migration

How can we determine the extent to which the timing, duration and direction of migratory movements is controlled by internal, inherited factors versus external environmental factors? Documenting where birds go throughout the year has been a long-standing activity for bird researchers, but only recently have we learned about the physiological mechanisms that control bird migration. A particularly revealing set of experiments was conducted by the German ornithologist Peter Berthold in the 1970s and 1980s on the Eurasian blackcap (*Sylvia atricapilla*). These common warblers breed throughout Europe and include populations that are long- and short-distance migrants.

The researchers captured birds from a variety of different populations and kept them under normal daylight conditions in cages. Conveniently, when blackcaps and many other small songbirds are brought into captivity, they respond to natural changes in photoperiod just like free-living birds. For example, during shorter days in autumn they became active at night (a behaviour known by the German term *Zugunruhe*, meaning 'migratory restlessness'), and the extent of nocturnal activity corresponds to the duration of migration for that population. In addition, small songbirds like blackcaps preferentially orient their night-time activity in the direction in which they travel, and by placing the captive birds in large funnels (called Emlen funnels, after their designers) at night and recording the way they jumped and fluttered, the researchers could determine their preferred direction of travel.

The experiments revealed that blackcaps from populations that migrate in a southwesterly direction to cross the Strait of Gibraltar oriented their movements to the southwest, whereas birds from populations that migrate in a southeasterly direction to the Middle East oriented their movements in that direction. Remarkably, when blackcaps from these two populations were cross-bred, the hybrids migrated on average in an intermediate direction (directly south), which suggests that there is an important innate, inherited genetic component to their migration strategy. This has been confirmed in dozens of species and seems an essential guide for birds during their annual migrations.

Influence of environmental cues

Besides the innate component to migration determining the direction of travel, the other important finding from the blackcap experiments and many other similar studies is that photoperiod (the duration of daylight, from sunrise to sunset) provides a proximate, predictable environmental cue, instigating key hormonal changes that control the timing and preparation for migration (in addition to reproduction; see pages 148–53). Other cues supplement the information from day length, including temperature, food supply and social cues (e.g. the activity of conspecifics). In summary, the timing of seasonal movements such as the migration of birds, and much of the variation between species in the timing of migration, is the product of an interplay between photoperiodically induced physiological changes, coincident changes in certain other environmental cues (e.g. food, temperature), and an evolved endogenous annual rhythm. In addition, the direction and duration of migration is also inherited to various degrees, depending on the species and population.

Eurasian blackcap
(*Sylvia atricapilla*)

Migration pathways of Eurasian blackcaps in Europe

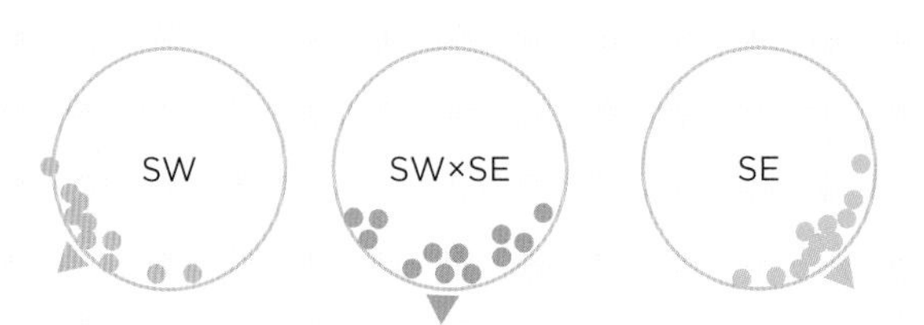

▲ Eurasian blackcaps (*Sylvia atricapilla*) inherit a tendency to migrate in different directions depending on which population they are from. The coloured dots show preferred direction of travel of certain individuals, while the triangle indicates the average direction of travel of each group – in this case, either southwest (SW) or southeast (SE), or an intermediate direction (middle circle) when individuals from the two populations are cross-bred.

Migration connectivity and conservation

The fact that a species is migratory influences its populations and conservation. Most birds return annually to the same breeding areas, especially when they have bred successfully; faithfulness to wintering areas is less common. If individuals of a given species or population consistently use the same breeding and wintering areas, then this coordinated movement results in strong migratory connectivity. In contrast, weak migratory connectivity occurs when individuals from one breeding area have different wintering grounds, resulting in substantial mixing of birds and less population integrity.

Delineating migratory populations

The pattern of site faithfulness and population connectivity of migratory species has consequences for how that species is managed. A classic example of the importance of accurately delineating populations for conservation purposes was the delayed recognition of 'migratory' and 'resident' Canada geese (*Branta canadensis*) along the Atlantic Flyway. Genetically distinct populations of these birds cannot be differentiated by appearance alone; rather, individuals from different breeding areas must be tracked to assess the extent of their movements. Prior to the 1990s, Canada geese in the eastern USA were managed as one large population, even though biologists recognised that there were breeding Canada geese throughout eastern Canada and all the way down the eastern coast of the US. Unfortunately, differential survival, breeding success and hunting pressures led to steep declines in the 'migratory' Canada geese that mostly bred in eastern Canada and wintered along the east coast of the US, while sedentary 'resident' populations that bred along the east coast of the US exploded to nuisance levels. Recognising this complex population structure, wildlife biologists changed the timing of the hunting season (an early September start to the season targeted the 'residents' and occurred before the arrival of the 'migrants'), which led to reductions in the harvest of 'migratory' Canada geese and the recovery of their populations.

Strong to weak migratory connectivity

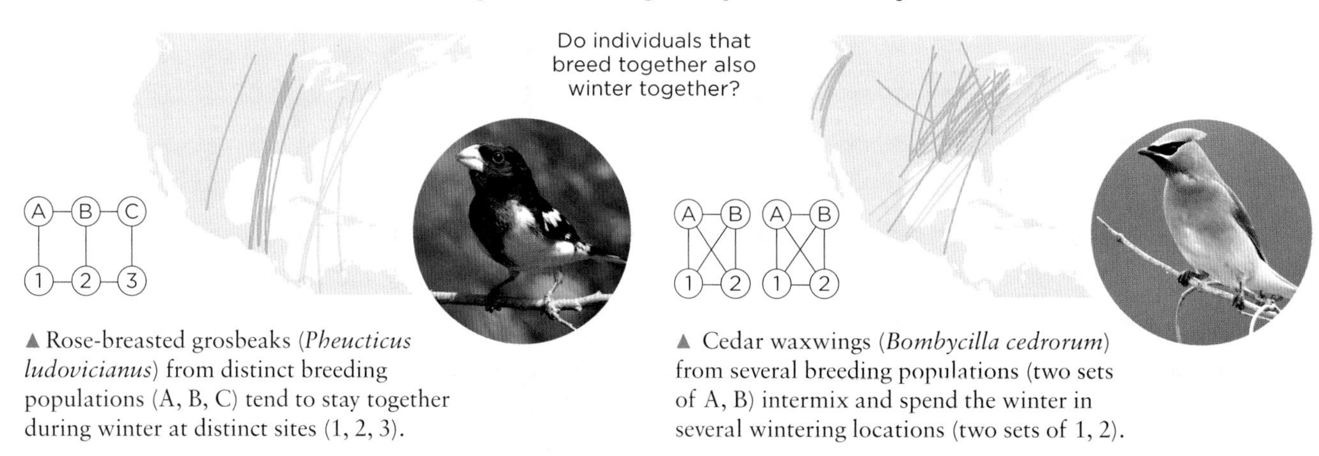

▲ Rose-breasted grosbeaks (*Pheucticus ludovicianus*) from distinct breeding populations (A, B, C) tend to stay together during winter at distinct sites (1, 2, 3).

▲ Cedar waxwings (*Bombycilla cedrorum*) from several breeding populations (two sets of A, B) intermix and spend the winter in several wintering locations (two sets of 1, 2).

Managing migratory populations

Determining the key breeding and non-breeding areas of migratory birds has also led to the recovery of several threatened species. For example, the Kirtland's warbler (*Setophaga kirtlandii*) is a small songbird that was included on the very first US list of endangered species in 1967 under the Endangered Species Preservation Act of 1966. At that time there were fewer than a couple of hundred breeding pairs of Kirtland's warblers in the jack pine (*Pinus banksiana*) forests of upper Michigan and Wisconsin, and little was known about the reasons for their endangered status and where they went during the non-breeding season.

Initial efforts to help the Kirtland's warbler focused on enhancing breeding success, primarily by burning the jack pine forests to improve the extent and quality of the breeding habitat, and removing brown-headed cowbirds (*Molothrus ater*), a notorious parasitic species. Female cowbirds lay their eggs in nests of other species, including those of Kirtland's warblers, often at the expense of the host's own young. Although these management efforts enhanced the breeding success of the Kirtland's warbler, the population did not increase, suggesting that it was regulated by events during the non-breeding season. Not until the species' wintering areas in the Bahamas were discovered and also successfully managed did the population recover. Today, there are an estimated 2,000 breeding pairs, although the species depends on continuing management to establish young forest habitat and to reduce cowbird populations.

▲ Kirtland's warbler (*Setophaga kirtlandii*) is an endangered migrant with very limited breeding and wintering areas, and a strong migratory connectivity. Conservation relies on management in both areas.

▲ Green-winged teals (*Anas carolinensis*) from many breeding populations intermix substantially during winter.

Managing and protecting migration stopovers is also important for conservation, especially at key sites just before or after geographic barriers such as oceans or deserts, where birds are highly concentrated – for example, at Tarifa in Spain on the Strait of Gibraltar, at Falsterbo in southwestern Sweden and at Eilat in southern Israel. Illegal killing, habitat degradation, wind turbines and other threats can have disproportionately large impacts on migratory birds when they are concentrated in such sites.

In sum, the conservation of migratory birds requires coordinating efforts across their breeding, migration and wintering areas – no small task given that they move across almost every inch of our planet.

Finding home: navigation

Bird movements range from short-distance foraging trips to globe-spanning migrations that cross geographic barriers such as deserts, oceans and mountain ranges. Despite the massive scale of some bird movements, they are not random. After migrating hundreds or even thousands of kilometres, individual birds often return to breed within metres of their prior nesting location.

Philopatry

A high degree of site fidelity is known as philopatry, and most birds exhibit breeding-site philopatry. For example, males of numerous northern temperate-zone songbird species return to breed within metres of the same location each year, purple martins (*Progne subis*) winter in the Amazon basin and return to the same North American nesting colonies each spring, and in migratory Australian golden whistlers (*Pachycephala pectoralis*) 66 per cent of males return to the same nest site and 100 per cent pair with the same female. Most species exhibit very high philopatry to their first breeding site, but migratory common linnets (*Linaria cannabina*) breeding on the remote island of Heligoland in the North Sea return with almost perfect fidelity.

Although wintering philopatry is less studied, some species such as the Asian brown shrike (*Lanius cristatus*) are known to return to the same overwintering sites each year. Achieving such feats requires an accurate navigation system.

▼ Australian golden whistlers (*Pachycephala pectoralis*) exhibit remarkable philopatry. Males return to the same breeding territories and mate with the same female.

▲ Ultralight aircraft such as this have been used to teach migration routes to captive-reared cranes and geese.

First-time flyers

Birds have an amazing ability to navigate, finding their way to a known location from a variety of other locations. Especially revealing is the autumn migration of birds born in the summer, which must travel hundreds or even thousands of kilometres to a location they have never visited before. In some species that migrate in flocks during the day, such as geese and cranes, younger birds may learn migratory routes by following more experienced birds. Conservationists have tried to exploit this in order to teach captive-reared birds to migrate. For example, captive-reared whooping cranes (*Grus americana*) that imprinted on the humans who raised them were trained to follow ultralight aircraft to their wintering grounds in Florida. However, although these cranes did appear to learn their migratory route, their rearing in captivity and imprinting on humans may have affected other behaviours and the practice has been discontinued.

Many other species migrate individually and often at night, conditions that make it unlikely that young birds learn migration routes from older, experienced birds. As described above, studies have found that Eurasian blackcaps, independent of experience, become motivated to migrate in their first autumn for a set period of time, and the direction of their nocturnal migratory restlessness in captivity corresponds to that of their population of origin (see page 191). Another remarkable example of innate control of migration is provided by Eurasian common cuckoos (*Cuculus canorus*), which lay their eggs in the nests of other species, leaving the foster parents to raise their young. Despite being raised by parents of a different species, young cuckoos migrate to wintering areas in central Africa, irrespective of where their foster parents may travel.

These studies demonstrate that many birds have innate behavioural mechanisms that motivate them to fly in particular directions for particular durations on their first autumn migration. In subsequent years, birds often then use their learned experiences to navigate to and from their breeding grounds.

Navigation

Many birds exhibit true navigation: the ability to move to a known location from a variety of starting positions. In one study, white-crowned sparrows (*Zonotrichia leucophrys*) that were relocated thousands of kilometres from their wintering grounds in California one year returned to it the following winter. On a shorter scale, birds that are displaced tens or hundreds of kilometres from their nesting or breeding sites will readily find their way back.

This navigation ability exists naturally in all species studied to date and has been exploited in homing pigeons (*Columba livia domestica*), bred specially for racing and to carry messages. Navigation requires a sense of direction (compass) and a sense of location (map). Much of our knowledge of how birds navigate results from experimental studies of homing pigeons, wherein researchers manipulate the sensory input to the pigeons and determine how this affects their ability to return home. The main conclusion is that birds have multiple ways of sensing direction, and that these overlapping systems provide them with robust mechanisms for orientation and navigation. Virtually every sensory modality has been shown to play a role in navigation in birds, from olfaction to vision, hearing and magnetoreception.

Maps and compasses

Cognitive maps are learned mental representations of the spatial layout of the environment. At least some birds are able to learn the arrangement of landmarks and features in their world and use a form of cognitive map. Studies of food-storing birds such as Clark's nutcrackers (*Nucifraga columbiana*) in the mountains of North America and willow tits (*Poecile montanus*) in Scandinavia demonstrate that birds use the spatial arrangement of visual cues to find hidden food. Remarkably, homing pigeons use olfaction to form a map-like

◄ Clark's nutcrackers (*Nucifraga columbiana*) can store, and then later retrieve, up to 100,000 food items.

Triangulation by Clark's nutcracker

▲ Clark's nutcrackers (*Nucifraga columbiana*) appear to use landmarks to triangulate the location of food caches that may be buried in the snow.

representation of their environment. The spatial arrangement of volatile compounds and the direction of prevailing winds provide information to the pigeons about their location within their home range.

Birds exploit a variety of cues to orient themselves, including visual references such as coastlines, mountain ranges and even man-made features like roads. Beyond visual landmarks, olfactory cues may be used by petrels and other birds to orient themselves toward food sources in the open ocean or to nesting burrows. Infrasound (very low frequency sound waves) can travel great distances and may provide birds with information about distant shorelines, mountain ranges or storm systems. Birds can also use cues from the Earth's magnetic field to determine direction. This field varies with respect to polarity (north versus south), intensity and inclination (angle of the field relative to the Earth's surface). In some cases, variation in the strength of the magnetic field may be used for a map sense for homing, but magnetic inclination is the cue most likely used as a compass, through magnetoreception in the beak and retina (see pages 120–1). The magnetic compass sense of nocturnally migrating songbirds appears to require calibration each night, with birds observing polarised light at sunrise and sunset in order use magnetic cues to orient themselves effectively.

▼▼ White storks (*Ciconia ciconia*) migrate in flocks numbering in the hundreds or thousands.

Celestial cues for navigation

Throughout human history we have used the sky for navigation. The positions of the sun and the stars provide us with compass directions, and when these are combined with the time of day we can calculate latitude and longitude. Birds have been using these same celestial cues for navigation since before the evolution of humans, relying on the sun as a compass during the day and attending to the stars at night.

Sun compass

Birds use sky-based cues to determine direction. If the time of day is known, the position of the sun provides a solar compass. Experimental studies have used mirrors to 'move' the sun's direction, or have phase-shifted birds using artificial lights to make light phases (dawn and dusk) earlier or later relative to natural cycles. These studies have clearly demonstrated that homing pigeons and common starlings (*Sturnus vulgaris*) use the position of the sun for orientation and, because birds do not have watches, they rely on their internal circadian clock to correct the sun's position relative to the time of day. Remarkably, pigeons need to learn the path of the sun in order to use it as a compass. Scientists know this because birds reared indoors that had only ever seen the position of the sun in the morning were unable to use the afternoon sun to orient themselves.

Star compass

Nocturnally migrating birds use a star-based compass. However, rather than learning particular constellations, songbirds appear to pay attention to the general rotation of the night sky in order to determine which direction is north. Experiments on young American indigo buntings (*Passerina cyanea*), in which the birds were housed in a planetarium, demonstrated that they observe the slow rotation of the night sky and learn the location of Polaris, the North Star or Pole Star, relative to other constellations. During their autumn migration, after they have gained this knowledge, they orient themselves away from Polaris to move southward. When indigo buntings were experimentally raised in a planetarium where the night sky artificially rotated around a different star (Betelgeuse), they oriented themselves away from this star during their first migration. It is likely that many other species of night-migrating birds use star compasses in a similar fashion.

▼ Many birds, including these common starlings (*Sturnus vulgaris*), use the position of the sun as a compass, correcting for the time of day.

▶ American indigo buntings (*Passerina cyanea*) observe the night sky to learn the position of the North Star. When migrating south, they head away from this star.

Physiological challenges of migration

Even if the navigation system of birds keeps them from getting lost, a complete migration for most species involves periods of flying and fasting alternating with periods of resting and refuelling at stopover sites, repeated multiple times. Successful migration therefore requires overcoming a series of physiological challenges, including finding a way to store and carry enough energy, nutrients and water for the long-distance flight, avoiding fatigue and other damaging by-products of intense exercise, and then somehow rapidly restoring the necessary reserves at stopovers for the next stage of migration.

Preparing for the long journey

Birds store fat in preparation for migration by eating lots of food – as much as two to three times more than when they are not migrating – and by being picky eaters. A bird that is ready for migration may have doubled its body weight, of which 50–60 per cent will be fat. As explained in the previous chapter, fat is the best choice in the trade-off between energy and weight saving because stored fat contains much more energy per unit of weight in comparison to either protein or carbohydrates. For this reason, many birds preparing for migration choose foods

that facilitate fattening (e.g. high-fat fruits and seeds) and that contain a certain fat quality (i.e. fatty acid composition). Specifically, changes in the diet of birds primarily determines seasonal changes in the quality of their stored fat. During migration, birds have a higher proportion of monounsaturated fats (like those in olive oil) relative to polyunsaturated fats, especially linoleic acid (an omega-6 fatty acid) and linolenic acid (an omega-3 fatty acid), both of which are 'essential' fatty acids for birds, meaning they cannot produce them and so must acquire them from diet. During their autumn migration, birds obtain these important omega-3 and omega-6 fatty acids by eating certain fruits, which has led to the habitat conservation and management practices of providing plenty of fruiting shrubs at important stopover sites used during migration.

◀ Red-winged blackbirds (*Agelaius phoeniceus*) gather in large flocks as they prepare for migration.

▲ Many songbirds, including the fieldfare (*Turdus pilaris*), eat plenty of nutrient- and energy-rich fruit during their autumn migration.

The dramatic increase in feeding (called hyperphagia) and fattening in preparation for migration is associated with dramatic increases in the size and function of digestive organs such as the gizzard and small intestine. These concomitant increases in food intake and digestive organs allow birds to maintain the efficiency of digestion and thus maximise nutrient and energy gain, at the same time loading their bodies with fuel for migration.

Although stored fat provides much more energy when metabolised than either protein or carbohydrates, compared to these alternative fuels the metabolism of fat provides very little water, requires more oxygen and produces more oxidative by-products that can cause damage at the cellular level. Birds therefore have to metabolise some protein to satisfy their water needs while migrating, and as noted in Chapter 5, their respiratory and circulatory systems are designed to satisfy the elevated requirements for oxygen for metabolism at this time. Migratory birds contend with potentially damaging oxidative by-products generated during increased metabolism and flight by building up their antioxidant system (see pages 166–7).

During a migratory flight

One of the more remarkable aspects of migration is that the vast majority of birds do not eat or drink during a migratory flight because they are aloft and typically away from sources of food and water. A single migratory flight can last from many hours to several days for small songbirds such as New World blackpoll warblers (*Setophaga striata*) and Old World northern wheatears (*Oenanthe oenanthe*), and as long as a week or more for some larger shorebirds such as the bar-tailed godwit (*Limosa lapponica*). As discussed earlier, fat provides the vast majority of fuel for such migratory flights, but many other aspects of a bird's body composition are modulated during migration.

Digestive organs are among the most costly in the body (only the brain has a higher metabolic rate); thus, it might be expected that during a migratory flight, when most birds are not eating or drinking, these organs would decrease in size and function. Indeed, the digestive organs exhibit remarkable flexibility, increasing in size as the need arises and birds eat large quantities of food in preparation for migration, then decreasing in size when they are fasting during flight. However, it has been shown that such remarkable flexibility in the digestive and other organs of the body (e.g. heart, flight muscle) are more closely related to intrinsic differences in the turnover rates of these tissues rather than their function (e.g. loss of digestive organs while fasting). Furthermore, some of this 'lost' digestive tissue is used as an important source of protein and metabolic water during a migratory flight.

Refuelling at stopover sites

The refuelling process at a given stopover site along the migration route is somewhat similar to the preparations for the migration journey – birds carefully select what to eat and consume large quantities of food. However, birds at stopover sites must contend with a few unique physiological challenges. First, as they arrive after a migratory flight they cannot immediately eat lots because the reduced size of their digestive organs (see above) constrains food intake. Thus, the digestive organs must be rebuilt before birds can feed at maximum capacity, which can take a few days depending on the duration of the prior migratory flight.

Second, birds that have just completed a migratory flight must contend with any oxidative damage that has occurred, while also rebuilding their antioxidant capacity in preparation for the next flight. Birds can increase their antioxidant capacity by increasing production of endogenous antioxidant enzymes (e.g. superoxide dismutase) and by increasing their consumption of dietary

◄ During migration, all birds – including these evening grosbeaks (*Hesperiphona vespertina*) – must consume large quantities of food to fuel their journey.

▲ Large-bodied, long-lived species such as this common crane (*Grus grus*) usually travel with their juveniles (rusty-brown heads) during their autumn migration, literally showing them the route.

antioxidants. During autumn migration, birds acquire dietary antioxidants such as vitamin E and anthocyanins from fruits, which also contain the specific types of fatty acids they need for fuel. The relationship between plants and the birds that eat their fruits is usually mutualistic, meaning that fruit consumption benefits both the bird (which receives nutrients and energy) and the plant (through seed dispersal), and many fruits are both widely available and easily consumed. Unfortunately, we know little about the source of dietary antioxidants for birds during spring migration, when fruits are much less abundant and birds must also begin preparations for reproduction.

Other bird movements

Although seasonal migrations between wintering and breeding grounds are often the most impressive and conspicuous movements birds make, they exhibit a variety of other movement patterns that span a range of geographic scales. These include natal dispersal (movement of young birds away from the nest location where they were hatched) and breeding dispersal (movement of adults between breeding attempts), as well as nomadic movements and irruptive behaviour.

Dispersal

In almost all species, young birds move away from the location at which they hatched, a process called natal dispersal. For most temperate-zone birds that breed in spring, this dispersal occurs in late summer and early autumn. Young birds roam tens of kilometres or more away from their parents' breeding territory, possibly in search of breeding sites for the following spring. When these birds begin their first breeding attempt, it is typically at a site some distance from their hatch location. The range of natal dispersals typically has a skewed distribution, with most birds moving a short distance from their hatch site and relatively fewer dispersing greater distances. In addition, there is a sex difference in most species, with females dispersing further than males. For

▼ In Australia, male lovely fairywrens (*Malurus amabilis*) are homebodies, settling close to where they were born. Females disperse further, settling several kilometres from their natal home.

▲ Female Eurasian black kites (*Milvus migrans*) are more likely than males to move nesting locations between breeding seasons, thus showing higher breeding dispersal.

example, in lovely fairywrens (*Malurus amabilis*) of tropical Australia and New Guinea, males disperse only 0.1–0.4 km, whereas females disperse 2–5 km. This sex difference of greater natal dispersal by females has been reported for numerous species and is in direct contrast to the pattern typically seen in mammals, where males disperse further than females.

In addition to natal dispersal, birds may engage in breeding dispersal, or movement between breeding attempts, although most move only short distances. In territorial songbirds, dispersal between breeding seasons may result from competition for the highest-quality territories. As with natal dispersal, females often have greater breeding dispersal distances than males. For example, female black kites (*Milvus migrans*) are also more likely to move breeding territories than are males. Other factors leading to breeding dispersal are failure of a nest due to weather or predators, or the death of a mate.

Dispersal may result in the range expansion of a population or species. When rapid colonisation of a new area occurs, adults typically disperse into an unoccupied region, and further generations of offspring then continue to disperse into unoccupied breeding sites. North American western bluebirds (*Sialia mexicana*) nest in cavities and tend to breed in clusters of suitable habitat, often tree stands damaged by forest fires and then excavated by woodpeckers. Male western bluebirds that are more aggressive tend to disperse further and thereby colonise new areas of suitable habitat.

Prospecting

Prospecting involves exploratory movements by birds as they search for suitable habitats. Some birds prospect for suitable breeding sites well before they actually breed – it may even occur during the hatch year, with young birds roaming and exploring habitats before they migrate south. In other species such as Old World collared flycatchers (*Ficedula albicollis*), some individuals are non-breeding owing to a lack of available nest sites. The non-breeders pay careful attention to nearby breeding flycatchers and are more likely to breed in subsequent years at sites where others have successfully raised young. Similarly, black-legged kittiwakes (*Rissa tridactyla*) are more likely to return to a site to breed following their own nest failure if they observe that their neighbours' nests are successful. Thus, prospecting is not random and involves birds gathering information about the suitability of locations for future reproductive success.

breed at sites with sufficient cones for a period of time, and then move on to search for new sites with abundant cones. In Africa, red-billed quelea (*Quelea quelea*) similarly live in nomadic flocks searching for unpredictable grass-seed crops following periods of rainfall that vary from year to year. Nomadism is thus most common in those species that rely for breeding on food resources that are temporally or geographically unpredictable.

Nomadism

In contrast to the predictable seasonal to-and-fro migration of most migratory birds, nomadic species engage in large-scale movements that may be unpredictable in both time and space. Nomadic species often move to track fluctuating resources, a behaviour that is most common in birds that breed at high latitudes (tundra and boreal forest), in high steppes or in arid tropical regions. Snowy owls (*Bubo scandiacus*) on the Arctic tundra forage on lemmings that have unpredictable population cycles, these varying both from year to year and place to place. As a result, the owls may move thousands of kilometres between breeding attempts to track these resources.

In central Australia, various bird species – ranging from songbirds to ducks – move nomadically to regions that have recently had rainfall, to take advantage of a flush of plant growth and associated insects. For example, grey teal (*Anas gracilis*) may fly hundreds or thousands of kilometres to regions that have had recent rain or flooding. How they can detect the change in conditions from such great distances is unknown. In North America, Europe and Asia, red crossbills (*Loxia curvirostra*) use their specialised crossed bills to pry open pine cones. However, the cones are not abundant on a regular seasonal schedule, and bumper crops emerge unpredictably at different locations (called mast years). Thus, crossbill flocks will

▲ Snowy owls (*Bubo scandiacus*) are irruptive migrants, appearing unpredictably in large numbers in more temperate zones during the winter, possibly driven by population cycles and food availability. If food sources remain high, owls may remain on their breeding grounds year-round, but when food supplies become scarce, they will move vast distances to find a new resource and overwinter.

▲ Red-billed quelea (*Quelea quelea*) can live in vast flocks. In some parts of Africa these flocks move nomadically in response to rains and can become crop pests.

Irruptions

Irruptions are dramatic increases in the local abundance of a nomadic species. Nomads can be thought of as skimming the cream, moving from location to location and thus benefiting from abundant food resources. Successful reproduction in such an area can lead to rapid population growth, and thus available food supplies can be depleted, with relatively less food per individual. In these circumstances, nomads or facultative migrants dramatically shift their home range, leading to an irruption. For example, in some years large numbers of Arctic-nesting owls and finches move southward for the winter in periodic irruptive migrations, much to the delight of birdwatchers who live in these southerly latitudes. Bohemian waxwings (*Bombycilla garrulus*) that breed in Scandinavia and Russia can also irrupt into the UK and Europe. Sometimes, these irruptions follow multi-year cycles – common redpoll (*Acanthis flammea*) abundance in the northeastern USA fluctuates on an approximately two-year cycle. By tracking food resources and population numbers at higher latitudes, many birdwatchers attempt to predict irruptions.

Migration risks: light at night

Birds die more often during migration than at any other time of the year – even when they are literally 'sitting ducks' during summer nesting, or when they are trying to stay warm in winter. High mortality during migration indicates that travelling long distances, which for many species takes place at night under the cover of darkness, is a risky business.

The risk of predation

Migrating birds face many challenges, from poor weather carrying landbirds out to sea, to predation by raptors such as peregrine falcons (*Falco peregrinus*) and Merlins (*F. columbarius*), which often time their own migrations to coincide with those of their bird prey. Recent studies have suggested that many migratory shorebirds – including American western sandpipers (*Calidris mauri*) – have changed their behaviour and even the regulation of their body mass in response to the recovery of raptor numbers (following the ban on the use of DDT in the US in the early 1970s) on shared migratory flyways.

To make matters worse, humans have modified environments on grand scales (e.g. climate change), constructed barriers (e.g. tall glass buildings, cellphone towers with cable tethers) and lethal attractants (e.g. city lights), and introduced predators (e.g. cats), all of which further increase mortality rates for birds during migration. And even though we humans know of many ways to reduce these avian mortalities, unfortunately we rarely take these preventative measures.

City lights and nocturnal migration

Given that birds use photoperiod, or day length, to time key events in their lives, such as when to breed and when to migrate, it might be expected that extensive city lighting at night would interfere or influence this. Indeed, this does appear to be the case, but the effect of artificial light at night on breeding birds is different than on migrating birds. For example, common blackbirds living in cities in Germany breed and moult as much as one month earlier than those living in nearby naturally dark forests, and blackbirds brought into captivity and exposed to dim city lighting at night also breed earlier. During avian migrations, city lights attract birds as they fly at night and can cause them to become disoriented, such that they collide with buildings and/or waste energy, leaving them vulnerable to other urban threats.

▼ Humans light up the otherwise dark night, creating a variety of problems for the billions of night-migrating birds.

BIRD-FRIENDLY GLASS

Given that millions of birds perish each year when they collide with windows or glass buildings, many bird-advocacy groups have funded studies to determine how glass can be modified to make it more apparent to birds so that they can avoid these lethal obstacles. Taking inspiration from spider webs, manufacturers have added ultraviolet (UV) patterns within glass and/or film coatings to the outside. Birds can see this wavelength of light, whereas we cannot, so such an addition might be visible to them but the glass looks no different to us. Coatings or decals can also be added to glass to increase its visibility to birds – this is most effective if the coatings or markings are spaced 5 cm apart horizontally and 10 cm apart vertically across the glass surface.

Given that most migration occurs over just a few nights each autumn and spring, conservationists have worked with some of the large cities along key migration flyways to turn off or dim the lights in buildings for short stretches. Scientists are also studying whether certain colours of artificial lighting affect birds less during night-time migration.

feather friendly glass

▲ Carcasses of birds found killed by window collisions on a single university campus over an eight-month period.

Migration risks: storms

Migratory birds pay close attention to prevailing winds and rain, and time their movements to take advantage of favourable conditions. However, they cannot forecast the weather along their entire migratory route, and at times they must cope with storms and even hurricanes.

Coping with storms

Many birds that migrate over land simply stop (fall-out) if they encounter bad weather. When such fall-outs occur along coastlines, thousands of birds, of many different species, may congregate – a boon for local birdwatchers. Similarly, islands in large bodies of water, including offshore oil platforms and large ships, can often provide refuge for large numbers of birds during bad weather.

▼ In stormy weather birds can be blown off course and out over the ocean. For those that cannot swim, human-made structures such as boats and oil rigs can provide a critical refuge.

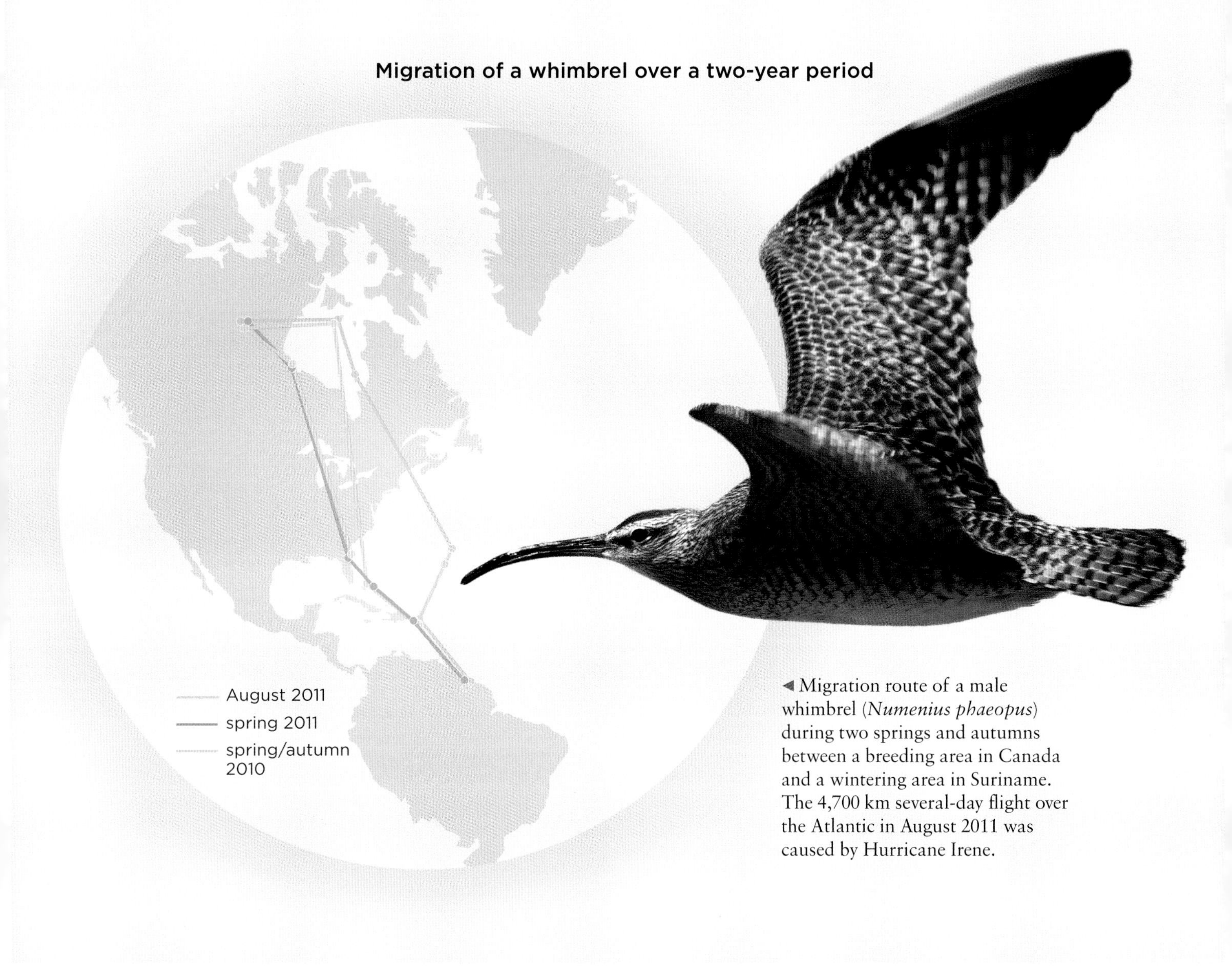

◄ Migration route of a male whimbrel (*Numenius phaeopus*) during two springs and autumns between a breeding area in Canada and a wintering area in Suriname. The 4,700 km several-day flight over the Atlantic in August 2011 was caused by Hurricane Irene.

Some birds try to remain aloft during storms and confront them directly. For example, a satellite telemetry study of whimbrels (*Numenius phaeopus*) breeding in Canada and wintering in South America documented how some of these strong-flying shorebirds deal with hurricanes in the Caribbean during their autumn migration. In 2011, one whimbrel named Chinquapin flew *into* Hurricane Irene's dangerous northeast quadrant and shot out the south side, while another bird named Goshen flew through the storm's outer edge. The previous year, Chinquapin successfully flew around the edge of Tropical Storm Colin, whereas a second tagged whimbrel perished in the same weather system. Smaller migrating birds can become trapped inside large storms. For example, a large flock of migrating chimney swifts (*Chaetura pelagica*) was swept up by Hurricane Wilma in 2005, the most intense hurricane ever recorded to date in the Atlantic Basin. The swifts rode the storm from the eastern US over to western Europe, much to the delight of birders there, who were very excited to see these New World vagrants.

An additional threat of climate change is that shifting weather patterns, caused by rising ocean temperatures and variation in the jet stream, result in more intense and more frequent storms. Thus, although birds' high mobility provide some capacity for them to cope with climate change, the migration patterns of some species will be affected and they may experience population declines as a result.

Climate change and migration

Climate change poses a threat to birds in several ways beyond increasing the frequency of major storm events such as hurricanes. As global temperatures increase due to human production of greenhouse gases, the leafing and flowering of plants, and the emergence of the insects that feed on them, are occurring earlier and earlier each year. Because many migrant birds rely on cues such as photoperiod to time their annual migrations, they may therefore arrive back at their breeding grounds too late to take full advantage of the peak abundance of food resources in spring.

Mismatched timing

The start of the breeding season in European pied flycatchers (*Ficedula hypoleuca*) in northern Europe has been advancing in recent years, but the emergence of the main insect species they depend on to feed their young has been advancing even more rapidly. The local insects are responding more quickly to the warmer local climate than the flycatchers because the birds cannot detect conditions on the breeding grounds from their wintering grounds in Africa. Large-scale studies of North American songbirds have demonstrated steady advances in spring migration timing, but whether these advances are sufficient to take advantage of advancing spring food resources is not yet known.

Range shifts

In addition to mismatches in timing of migration, climate change is driving range shifts in plant and insect species, which in turn is driving range shifts in breeding birds. Birds are highly mobile, and shifts in their range can be facilitated by natal and breeding dispersal. However, range shifts can also induce costs, such as when the suitable breeding and wintering habitats of a species move further apart. For example, it has been estimated that the migration distance between the wintering and breeding habitats of thrush nightingales (*Luscinia luscinia*) – in eastern Europe and western temperate Asia, and in sub-Saharan Africa, respectively – will increase by about 800 km, or five days' travel, in the forthcoming decades.

Thus, many species will need to increase the distance they migrate in addition to adjusting the timing of their spring migration if they are to cope with climate change. Another threat is that range expansions may become limited. For example, as cold-adapted plant and insect species move to higher altitudes in the mountains in response to warming temperatures, the high-altitude breeding birds that track these insect resources must also go higher; however, eventually even the mountain peak may become unsuitable and species will become locally extinct.

▲ Pied flycatchers (*Ficedula hypoleuca*) cannot detect warmer conditions on their breeding grounds prior to migration from their wintering areas, which creates a timing mismatch.

► Thrush nightingales (*Luscinia luscinia*) must increase the distance they migrate as their preferred deciduous forest breeding areas move north with climate change.

CHAPTER 7

Food and Foraging

Most birds are omnivores, eating a variety of animal and plant matter, although diets can vary seasonally. Many seed eaters switch to protein-rich insects during breeding to meet the needs of growing chicks, and migratory birds will switch to energy-rich fruits. These dietary generalists typically retain the basic avian body form, with a relatively non-specialised bill shape and foraging behaviour. However, many birds are specialised meat or fish eaters; others, more rarely, specialise on plant material; some scavenge for dead prey; and a few are extreme diet specialists, relying on a single prey type. There are even bone-crushing birds and vampire birds that feed on blood.

Dietary specialists have evolved a range of morphological, behavioural and sensory adaptations, including ultraviolet vision and echolocation, which allow them to target specific prey or forage in particular ways. Finding food can also be a social activity, with many birds feeding in flocks; a few birds use tools to obtain food; and social learning of foraging skills is likely widespread.

◀ Red-headed barbet (*Eubucco bourcierii*).

Who eats what, and when?

Birds must find and eat enough food each day, or over several days, to satisfy their nutritional requirements and maintain energy balance. However, food availability to meet these demands varies through the year, even in tropical regions, so diet often also varies seasonally. Reproduction imposes very different demands compared with the non-breeding period: parents must not only find more, and potentially different, food to meet their own elevated nutritional requirements for mating, incubation and parental care, but in many species they also have to obtain large amounts of additional food for their growing offspring.

Different diets

Most birds are omnivores, eating a variety of animal and plant matter, although seasonal variation in the relative amounts of these is common. Small mammals and birds form the main prey for many carnivorous birds of prey, and smaller raptors will take amphibians and reptiles, as well as insects. New and Old World vultures eat mainly carrion, as do a few other species such as Antarctic giant petrels. Fish form most of the diet in piscivorous species, in both marine (e.g. auks and penguins) and freshwater systems (e.g. diving ducks). And insectivorous species rely on insects or other invertebrates, either gleaned from the foliage or obtained by probing in soil or mud.

Plant-eating birds are common, relying on an abundance of fruits and seeds at certain times of year, while frugivorous (fruit-eating) species are most diverse in the tropics, where fruit can be available year-round. Nectar provides a rich source of energy for many small birds, especially hummingbirds, and species such as sapsuckers tap into the sugar-rich phloem of plants. A few bird species are extreme diet specialists – for example, American snail kites (*Rostrhamus sociabilis*) rely almost entirely on a single species of *Pomacea* snail, and lesser flamingos (*Phoeniconaias minor*) filter-feed on blue-green algae in alkaline lakes of East Africa.

◄ Snail kites (*Rostrhamus sociabilis*) feed almost exclusively on a single species of apple snail (*Pomacea* sp.).

The variety of avian diets

▲ Although most birds are omnivores, different species can feed on a wide variety of different foods, from fish, meat and insects, to plant matter.

Generalists and specialists

A long-standing dichotomy with bird diets has been between generalists, which eat a diverse array of foods, and specialists, which eat only a narrow range. However, this variation is now generally acknowledged to be a continuum, with the diets of individuals in different species or environments varying along a generalist–specialist spectrum. Individuals within the same species can often have different dietary preferences (i.e. they can show individual diet specialisation), which may arise due to sexually dimorphic feeding adaptations such as bill size, or because of variation in learned behaviours. Among shorebirds, ruddy turnstone (*Arenaria interpres*) individuals specialise in one of six alternative foraging methods, including probing, digging or overturning objects to find prey. Individual Eurasian oystercatchers (*Haematopus ostralegus*) also specialise in the prey they target, using prey-capture techniques such as probing in mud for worms or hammering or stabbing mussels.

Seasonal variation in diets

Bird diets commonly vary across the seasons, partly because of changes in the availability of foods through the year, and partly because of increased nutritional demands from specific activities – for example, higher protein requirements during chick rearing, or higher energy requirements during migration. In North American horned larks (*Eremophila alpestris*), seeds make up more than 60 per cent of the diet between July and December, but less than 20 per cent in April and May, when the birds are breeding and feeding chicks. Neotropical songbirds rely on insects while breeding in temperate regions, but eat more fruits on migration or while overwintering in the tropics.

Food scarcity during winter in extreme northern and southern latitudes causes birds to migrate, and many of these species time their migration to coincide with the abundance of specific foods. For example, the arrival of shorebirds at stopover sites often coincides with the mass spawning events of their prey. During spring migration, sandpipers stopping in the Pacific Northwest feed on the abundant herring roe available there at this time, and red knots (*Calidris canutus*) en route to the Arctic from South America refuel on the eggs of Atlantic horseshoe crabs (*Limulus polyphemus*) in Delaware Bay before continuing their journey.

▼ North American red knots (*Calidris canutus*) time their spring migration to coincide with the spawning of Atlantic horseshoe crabs (*Limulus polyphemus*) in Delaware Bay, where they feed on the abundant crab eggs.

Seasonal variation in diets

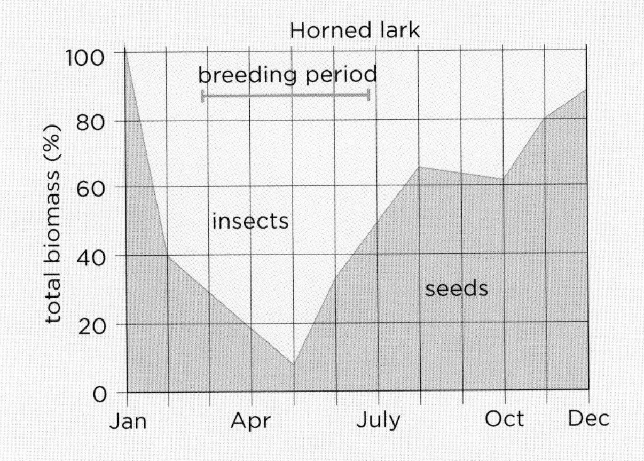

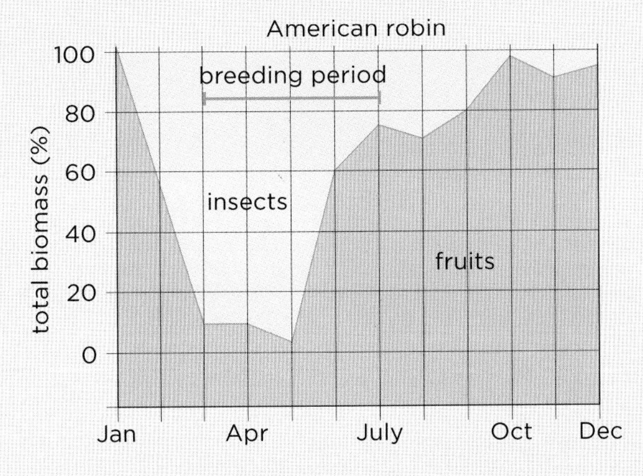

▲ Horned larks (*Eremophila alpestris*) switch from seeds to insects during breeding (top), while American robins (*Turdus migratorius*) eat insects during breeding but switch to fruits during migration and over winter (above).

ARE FEEDERS GOOD OR BAD?

In the UK, 40–50 per cent of householders feed birds in their gardens, dishing out an estimated 50,000–60,000 tonnes of bird food each year. Providing food in winter, when natural foods might be scarce, allows birds to survive food shortages, but it is not clear whether feeding year-round has a positive or negative effect on birds. In the US, species that use bird feeders most frequently tend to be doing just as well as, or better than, species that use them more sporadically. But some studies in the UK suggest that winter feeding can have a negative effect on subsequent breeding performance, advancing laying date but decreasing clutch size and breeding success. This may be because additional food provides 'incorrect' cues to birds for timing of breeding. Feeders also change the evolution of birds, selecting for longer bills that are better at reaching the food inside (e.g. as seen in great tits, *Parus major*, in the UK), and changing migratory behaviour (e.g. Eurasian blackcaps, *Sylvia atricapilla*, are now wintering in the UK).

Although adults birds can consume a variety of foods during their annual cycle, most growing chicks require a protein-rich animal diet. Few birds rely on frugivory for chick rearing, in part because few plants in temperate and higher latitudes fruit in spring and early summer, when most birds breed here. Even tropical frugivorous birds switch to insect diets to feed their chicks, due to the low nutritional quality of many fruits. However, some tropical species can select calcium-, iron- or protein-rich rich fruits to meet the specific demands of breeding and chick growth.

Only a few species do not show seasonal shifts in diets, one example being the North American house finch (*Haemorhous mexicanus*), which relies on seeds, buds, flowers, leaves and fruits for up to 97 per cent of its diet throughout the year. Similarly, while Australian zebra finches (*Taeniopygia guttata*) eat grass seeds year-round, during breeding birds eat sprouting seeds made available by the seasonal onset of rainfall, providing females with the specific building blocks (amino acids) required for egg production, and chicks with a more protein-rich diet.

Predators and scavengers

Raptors are consummate predators, armed with sharp beaks and talons for capturing and tearing up their prey. Large birds of prey such as eagles hunt during the day and are capable of taking quite large birds and mammals, including ducks, rabbits, monkeys and even antelope. Smaller birds of prey consume small mice and voles, small birds, amphibians, reptiles and insects. Owls catch prey at night, aided by acute visual and auditory sensory adaptations. Vultures, condors and kites, on the other hand, scavenge dead animals or carrion, which they can detect from very large distances.

◄ Peregrine falcons (*Falco peregrinus*) reach speeds of up to 320 km/h as they fall out of the sky to catch small birds in the air.

Carnivores: predators and scavengers

Birds of prey, or raptors, are apex predators, sitting at the top of many food chains. They feed mainly on other vertebrates, especially mammals and other birds. More than 500 species of raptors occur worldwide, including those that hunt primarily during the day (diurnal foragers, e.g. eagles, hawks, falcons) and those that are specialised for hunting at night (nocturnal foragers, e.g. owls). Birds of prey possess a range of morphological, behavioural and sensory adaptations for catching, handling and killing large prey. Hawks and owls snatch small mammals such as mice, voles and rabbits from the ground, while sea eagles and ospreys catch fish from the water's surface. Many birds of prey then kill their prey by crushing it with their powerful feet.

African crowned eagles (*Stephanoaetus coronatus*) will take bushbuck antelope (*Tragelaphus* spp.) up to six times their own mass and are consequently known as the 'leopards of the sky'. Harpy eagles (*Harpia harpyja*), with a wingspan of up to 2 m, can effortlessly catch sloths and monkeys larger than themselves among the tree tops in the rainforests of South America. Falcons have exceptional flight abilities, allowing them to take small birds directly from the air. Peregrine falcons (*Falco peregrinus*), for example, drop out of the sky at more than 320 km/h, instantly killing their prey on impact. Secretarybirds (*Sagittarius serpentarius*) of the African savannah are atypical raptors with long legs that hunt primarily on foot rather than from the air, specialising in snakes and other reptiles

Most raptors have sharp, curved talons or claws for grasping or killing prey, and a powerful curved beak for tearing flesh. Hawks and eagles have excellent eyesight that is two to eight times better than that of humans. Their large, forward-facing eyes are positioned on the sides of the head, providing them with powerful binocular vision and allowing them to detect medium-sized prey from at least 1.6 km away, or a 0.5 cm-long insect from the top of a 20 m-high tree or while hovering. Like humans, all birds – including raptors – have excellent colour vision, but most birds can also see into the ultraviolet (UV) range. In the case of Eurasian common kestrels (*Falco tinnunculus*), for example, it has been suggested that their UV vision allows them to detect trails of urine left by voles running through grass.

► Harpy eagles (*Harpia harpyja*) hunt sloths and monkeys in the treetops of the South American rainforests.

Nocturnal foraging

Owls have extremely acute vision, allowing them to hunt in low-light conditions at night – although many are crepuscular, foraging mainly at dawn and dusk. They can also fly almost silently, on broad wings with soft-edged feathers (see pages 66–7). The large eyes are framed by a feathered facial disc and provide excellent binocular vision, but because they face forward the birds can see only straight ahead. To compensate for this, owls have 14 additional bones in their neck, increasing articulation and allowing them to rotate their large, round head about 270° to scan for prey without having to twist their body. Owls also have an acute sense of hearing, and the feathered facial disc directs sound to the ears. The ears themselves are offset, one being slightly higher on the head than the other, helping to localise sound. In western barn owls (*Tyto alba*), one ear canal also points upwards and one down, so that each ear catches sound at a slightly different time. This provides the birds with an auditory map of their environment, allowing for pinpoint accuracy in locating prey (see also page 108).

Smaller owls eat insects or small mammals, but larger owls such as horned or eagle-owls will take large rabbits, hares and ducks. African fishing owls (*Scotopelia* spp.) and Asian fish owls (*Ketupa* spp.) specialise in catching fish, scanning slow-moving streams from overhanging trees and taking prey up to 2 kg in weight. In contrast to other owls, burrowing owls (*Athene* spp.) are often active during the day in the grasslands and deserts of North and South America. They eat mainly insects and small mammals, and are known to spread mammal dung by their burrow entrances to attract dung beetles, their preferred prey.

Scavengers

Rather than killing live prey, some raptors – including vultures, condors and kites – are scavengers, feeding on dead animals or carrion. These birds soar on large, broad wings high above the ground with the help of air thermals, and can cover up to 240 km a day in their search for carcasses. Old World vultures and condors mostly rely on acute vison to find food and will use social cues from other vultures and birds of prey to home in on a carcass. In contrast, New World vultures such as turkey vultures (*Cathartes aura*) have a strong sense of smell, and can detect carrion from 1.5 km away, even if the carcass is fully covered by leaves and vegetation. Many vultures have a bare, unfeathered head, perhaps so that they can avoid soiling their plumage when feeding deep inside a rotting carcass. Crows, some gulls and giant petrels also routinely obtain food by scavenging.

▶ The bare, unfeathered heads of some scavengers, such as this Rüppell's vulture (*Gyps rueppelli*), might help them avoid soiling their plumage when feeding on rotting carcasses.

RECONSTRUCTING DIETS USING OWL PELLETS

Many birds of prey, especially owls, regurgitate the indigestible, hard parts of their food (e.g. bones, hair, insect parts) as pellets. After feeding, it takes around six hours for an owl to produce a pellet and the birds will regurgitate one or two pellets after a night's hunting. Pellets can often be collected underneath trees or in barns where birds roost overnight – or even purchased online!

The pellets can be soaked in water for 24 hours and then gently teased apart, preferably with tweezers, to extract the remains of the owl's prey. Analysis of owl pellets from Homestead Cave, Utah, United States, has even been used to document human changes to the landscape over the last 100 years based on changes in the small mammal remains.

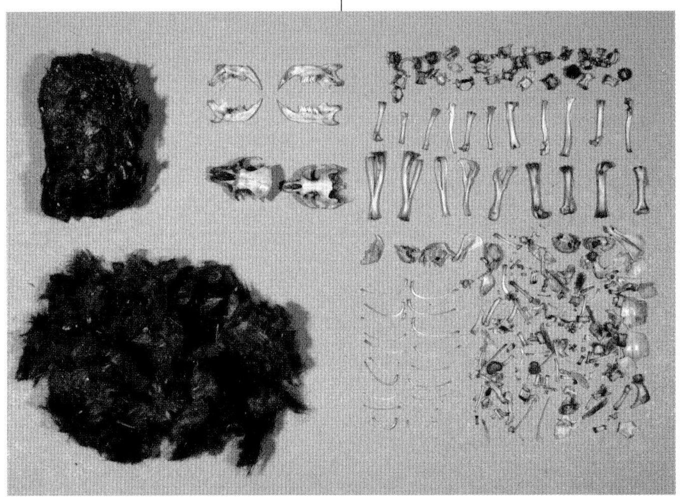

► Owls, such as this northern saw-whet owl (*Aegolius acadicus*), regurgitate pellets containing undigested parts of their prey; these can be collected from below roosts and dissected to reveal what the owl has been eating.

Insect hunters and fish eaters

In terrestrial habitats, many smaller birds forage for insect prey among the foliage of bushes and trees, sometimes sallying or hawking – launching themselves from a perch to take insects in flight. Others probe for worms and grubs in soil or mud, and aerial insectivores such as swallows take advantage of the abundance of flying insects. Aquatic birds catch fish and other marine prey in the underwater world, either by plunge-diving or 'swimming' underwater using flippers or foot propulsion.

Insectivores

Around 60 per cent of all birds are insectivorous, relying on insects and other small invertebrates for at least part of the year, and insects are essential in providing protein-rich food for growing chicks. It has been estimated that birds consume 400–500 million tonnes of insect prey annually, equivalent to the amount of energy required to run a large city the size of New York. Flight gives birds access to large numbers of small flying insects such as dipterans (true flies), and a slender bill allows them to catch and handle these insects easily. Aerial insectivores – including swifts, swallows, martins and nightjars – take insects on the wing in continuous feeding, spending more than 60 per cent of the day flying, compared to 10–20 per cent in more terrestrial insectivores. Other small birds catch insects in the air by hawking or sallying, in which they launch themselves from a perch, catch a hovering insect and then return to the same perch or a different one. This method of foraging is common among both Old World and New World flycatchers.

◄ Common swifts (*Apus apus*) and other aerial insectivores catch insects on the wing while in flight.

▼ Some insectivores, such as this great crested flycatcher (*Myiarchus crinitus*), feed by 'sallying' out from a branch, catching a hovering insect and then returning to the same branch.

Insectivorous birds living in forests and grasslands generally glean their prey from vegetation, picking individual insects from plants as they hop around. This foraging mode, in which the birds make short flights or jumps between branches or stems in their search for insects, and quick hovers to catch prey, is thought to be especially energy-expensive. Woodpeckers, nuthatches and creepers are agile climbers, moving up and down tree trunks as they glean insects from bark and branches. The birds have stiff tail feathers to provide them support on vertical trunks. Woodpeckers also hammer into trees and strip bark off with their beak to get at the insects below.

Many birds forage on the ground, picking through leaf litter or probing into the top turf layer of grass or pastures for soil-dwelling larvae. Several shorebird species that breed in terrestrial habitats or use agricultural areas, such as the European golden plover (*Pluvialis apricaria*) and Australian masked lapwing (*Vanellus miles*), also rely on insects obtained from the soil. However, during the non-breeding period most shorebirds occupy saltmarshes and mudflats, probing through mud or sand in search of invertebrate prey such as small snails and worms.

▼ Red-backed shrikes (*Lanius collurio*) store food, like this grasshopper, in a 'larder' by spearing the prey on plant thorns.

Barn swallow feeding insects to its young

▲ A barn swallow (*Hirundo rustica*) will both catch insects on the wing and then feed them to its newly hatched chick, all without landing.

Fish eaters: feeding underwater

Marine and freshwater birds catch aquatic prey, including fish, cephalopods and crustaceans, by swimming or diving underwater in seas, lakes or streams. Among seabirds, the wide-ranging pelagic gadfly petrels, storm petrels and snow petrel (*Pagodroma nivea*) pick prey from the water's surface while flying, appearing to dance on the water, while fulmars, gulls and shearwaters feed while swimming on the surface. American black skimmers (*Rynchops niger*) also pick prey from the water's surface, but with a unique fishing method: they fly close to the surface with the lower mandible in the water, and as soon as this comes into contact with a prey item the bill snaps shut.

Many other seabirds dive underwater, either from the surface or from the air (plunge-diving), allowing them to pursue prey deeper in the water column. While pursuing prey underwater, birds propel themselves using either their wings (e.g. penguins, auks, diving petrels) or their feet (e.g. cormorants, grebes, loons or divers, and some ducks). In general, wing-propelled birds dive deeper and for a longer period of time than foot-propelled divers. Diving birds have a streamlined body shape and excellent plumage waterproofing, and many have sharp, serrated (saw-like) bills to secure slippery prey.

With large wings designed for soaring over the ocean, albatrosses are the poorest divers among marine birds and feed mainly at the surface. The larger wandering albatross (*Diomedea exulans*) feeds within 1 m of the surface, but smaller albatrosses can reach down 5 m. Albatrosses cover thousands of square kilometres in their search for fish, squid and jellyfish, with foraging trips lasting four to five days and extending between subantarctic islands and the South American coast or the Antarctic Peninsula during breeding.

Gannets and boobies dive into the water from 30–40 m above the surface and at speeds of up to 85 km/h. Northern gannets (*Morus bassanus*) have sealed nostrils and air sacs under their skin (extensions of their respiratory system) to cushion their body, and they collapse their wings just before hitting the water to avoid injury. Gannets typically reach depths of around 10 m by plunge-diving, but they then propel themselves deeper using their wings and feet.

Penguins and auks dive underwater from the surface, but they also reach greater depths using their flippers or short, stiff wings for propulsion. Auks, including the circumpolar common murre or common guillemot (*Uria aalge*), can dive to 180 m, although most feeding dives reach 20–50 m and last 40–140 seconds. Many penguins routinely dive to depths in excess of 50 m. The smallest species, the little penguin (*Eudyptula minor*) of Australia and New Zealand, dives to 30 m in pursuit of fish, the medium-sized macaroni penguin (*Eudyptes chrysolophus*) dives to 50 m and takes mainly krill, and the largest species, the emperor penguin (*Aptenodytes forsteri*), routinely feeds at depths of 100–200 m on squid and fish (although the maximum dive depth recorded for this species is 565 m).

Penguins have many physiological adaptations to allow them to make such prolonged deep dives, including increased levels of myoglobin (the protein that stores oxygen in muscles). In addition, their heart rate is as low as three beats per minute during a dive, well below the resting rate of 60–70 bpm, and even their body temperature lowers, reducing energy costs and prolonging valuable oxygen stores.

▲ Black skimmers (*Rynchops niger*) fly close to the water's surface with the lower part of their bill in the water. As soon as this detects prey, the bill snaps shut.

▼ Snow petrels (*Pagodroma nivea*) dance among small ice floes, picking krill from just below the water's surface.

Freshwater habitats

A similar diversity of foraging modes is found in birds occupying freshwater lakes, rivers, wetlands and streams. Eurasian common kingfishers (*Alcedo atthis*) plunge from a perch above a stream to catch small fish, and in doing so inspired the design of Japan's high-speed bullet trains! The Afro-Asian pied kingfisher (*Ceryle rudis*) is one of the few species to hover above the water before diving. In the Amazon, streamlined anhinga (*Anhinga anhinga*) dive from the water's surface and pursue fish by foot propulsion, and Eurasian white-throated dippers (*Cinclus cinclus*) and South American torrent ducks (*Merganetta armata*) also dive underwater in fast-flowing mountain streams, turning over stones to obtain aquatic insects. Dippers have large feet to grip the bottom, essentially walking underwater, and they also have a translucent third eyelid, allowing them to see underwater even when this membrane is closed to protect the eye, and flaps that close the nostrils when they are submerged.

A wide variety of wading birds (e.g. herons, egrets) stalk prey through shallower water or stand motionless, employing a sit-and-wait strategy, before spearing their meal with their sharp, elongated bill. Aquatic prey often hide in relative safety in the shade of overhanging vegetation, and African black herons (*Egretta ardesiaca*) have taken advantage of this behaviour, fooling fish into coming within striking distance by 'mantling' or canopy feeding. In this, the birds spread their wings around their head like an umbrella, creating shade that attracts fish, which can then be easily snapped up. American green herons (*Butorides virescens*) fish using bait, holding a small insect in their bill tip just above the water or dropping a piece of food onto its surface to lure fish within range.

◄ Black herons (*Egretta ardesiaca*) use their wings to create a patch of shade that attracts fish, which they can then easily snap up.

STUDYING DIVING IN PENGUINS

Time-depth recorders (TDRs) are data loggers or mini computers that can be attached to the back of a penguin, and that record and store information on the bird's behaviour at sea. Modern TDRs can record when a bird dives and the maximum depth it reaches (based on pressure), as well as information on swim speed, light levels and sea temperature. The data can either be recovered by recapturing the bird, or via transmission to a satellite. Ingestible data loggers can be implanted in the bird's stomach. Because stomach temperature decreases when a penguin ingests cold seawater and prey, these stomach loggers record when the bird eats.

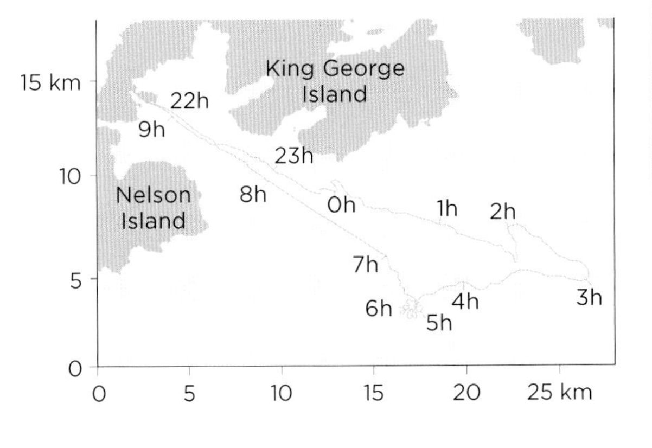

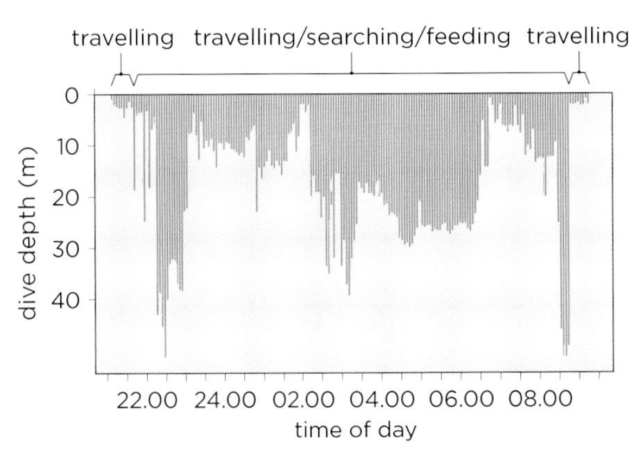

► A foraging trip by a chinstrap penguin (*Pygoscelis antarcticus*) at King George Island, showing movement and position of the bird over time (top) and changes in diving depth while foraging (bottom).

Plant eaters

Few birds rely exclusively on herbivory (a plant diet) or frugivory (a fruit diet), but many omnivorous birds include plant material in their diet, taking advantage of the abundance of seeds, nuts and fruits from late summer through to winter. Birds lack the enzyme required to break down cellulose, the main component of leaves, so feeding on plants requires special adaptations to their digestive system. Many species also take advantage of plant fluids, including sugar-rich sap and nectar. In return for their meal, birds provide plants with important services, including pollination and seed dispersal.

Seeds and nuts

Seeds are packed with nutrients intended for the developing seedling following germination, and as such they provide a rich food source for many birds. Granivorous birds such as sparrows and finches rely on seeds, grain and nuts, the most widely used plant foods. Some plants, such as oaks and beeches, produce very synchronous but highly variable numbers of seeds in different years, called masting, and this can have cascading effects on bird populations. For example, breeding densities of many woodland species in the UK are higher following good mast years.

To protect the plant embryo, seeds have a tough coat. Birds use their small, thin bill to dexterously remove this outer layer in order to access the nutritious seed inside. Scottish crossbills (*Loxia scotica*) and Clark's nutcrackers (*Nucifraga columbiana*) have specialised bills that allow them to pry seeds out of pine cones, and parrots and cockatoos that feed on very large seeds such as palm nuts use their large, powerful bills to crack the shells open to get to the seed inside. Once the bird has eaten the seed, it's tough, muscular gizzard crushes and breaks it, often using ingested grit and stones to aid the grinding process.

► Red-tailed black cockatoos (*Calyptorhynchus banksii*) have bills that are powerful enough to crack the tough shells of the nuts they eat.

▲ A common blackbird (*Turdus merula*) eating the red berries of a European yew (*Taxus baccata*), which are toxic to most mammals.

Fruits

Many plants enclose their seeds in protective tissue derived from the ovary, and these fruits or berries can be juicy, fleshy and highly nutritious, as well as being brightly coloured. Some plants use colours to indicate that their fruits are poisonous, but this toxicity may affect only mammals and many birds can consume 'toxic' berries – as in the case of thrushes, which eat red European yew (*Taxus baccata*) berries. Other plants have bright red or black fruits to attract avian frugivores.

Although fruits are widely available and easily consumed, they tend to be low in protein and, in addition, birds lack the enzyme sucrase, so cannot digest certain fruit sugars. Different bird species select different fruits – for example, thrushes tend to consume high-fat fruits, whereas waxwings prefer fruits with a high carbohydrate content. After removing a fruit's fleshy pulp, many birds regurgitate the seeds intact or these pass out undigested in their faeces – often a long way from the parent plant and thereby contributing to seed dispersal.

In temperate or higher latitudes few plants fruit in spring and early summer, when most birds breed. As a result, few species rely on fruits year-round, although the American cedar waxwing (*Bombycilla cedrorum*) is one exception. In the tropics, year-round availability of fruit is more common, and some tropical birds select specific fruits to meet the mineral and energetic demands of breeding. However, year-round reliance on frugivory is still rare; the South American oilbird (*Steatornis caripensis*) is one of the very few tropical species to rely entirely on a fruit diet, even for feeding its chicks.

Leaves

Along with other vertebrates, birds lack cellulases, the enzymes required to digest cellulose, so few rely exclusively on a diet of leaves and grasses (folivory). Many mammalian herbivores have evolved a symbiotic relationship with gut microbes (bacteria and protists) that break down plant material via fermentation in the foregut before food enters the intestine for absorption. However, only one bird species – the hoatzin (*Ophisthocomus hoazin* see page 145), from the Amazon – has evolved foregut fermentation. All other birds use hindgut fermentation, with populations of microbes housed in large, paired caecae located at the start of the large intestine, although this is less efficient for digesting plant material than foregut fermentation. As a consequence, grazing birds that consume mainly grass, such as geese and swans, have to eat continuously, relying on high rates of intake and fast

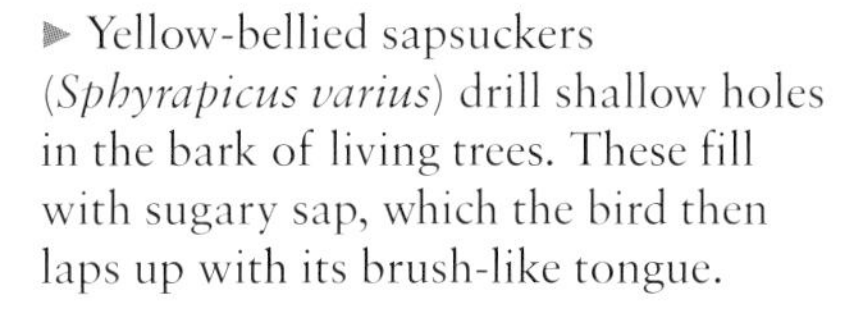

► Yellow-bellied sapsuckers (*Sphyrapicus varius*) drill shallow holes in the bark of living trees. These fill with sugary sap, which the bird then laps up with its brush-like tongue.

throughput to obtain sufficient energy. These birds also supplement their diet with the starchy roots and tubers of water plants. Other folivores such as grouse prefer eating buds or new leaves, which are more nutritious and easier to digest.

Nectar and sap

Nectar is a sugar-rich fluid produced by many plants as an attractant or reward for the pollinating animals – insects, bats, birds and other vertebrates – that visit their flowers. It is produced by special glands inside the base of flowers. When birds try to obtain nectar from flowers, they brush against the stamens (the male reproductive structures) and in doing so pick up pollen, which they then transfer to the next flower they visit, resulting in pollination. Many birds rely primarily on a diet of nectar, including New World hummingbirds, African and Asian sunbirds, honeycreepers in Hawaii and honeyeaters in Australia. All of these species show strong convergence of bill structure, with elongated bills allowing the birds to access nectar deep in tubular flowers, and most are adapted to utilise a particular flower type or structure (an example of coevolution).

As with fruit digestion (see above), digesting sugar-rich nectar is difficult for birds. For this reason, hummingbirds feed for less than a minute, storing the nectar in their crop. In any given hour they make ten to 15 foraging bouts, and then sit on a perch for the remainder of the time digesting the nectar. A few short-billed species cheat a plant's pollination system. For example, flowerpiercers (*Diglossa* spp.), members of the South American tanager family, make a hole in the base of a flower, accessing its nectar without touching the stamens, and therefore do not contribute to pollination. Other birds tap into the plant's internal vascular system, obtaining sugar-rich sap or phloem. Sapsuckers, a group of North American woodpeckers, drill neat rows of shallow holes – called sap wells – in the bark of living trees. Sugary sap or phloem accumulates in these wells and the birds then lap this up with their short, brush-like tongues. Hummingbirds are not averse to visiting these sap wells to get a free meal!

▼ Sword-billed hummingbirds (*Ensifera ensifera*) have incredibly long bills, allowing them to reach deep into *Passiflora mixta* flowers to obtain nectar.

◄ The hoatzin (*Ophisthocomus hoazin*) relies on a symbiosis with gut microbiota and foregut fermentation to digest the cellulose in its predominantly leafy diet.

Special diets

Most birds are omnivores and likely obtain a balanced diet in terms of carbohydrates, lipids and proteins simply by eating a variety of prey. There is little direct evidence that a lack of specific nutrients in birds' diets routinely causes ill health or reproductive failure, with the possible exception of calcium. Nevertheless, many birds do target dietary items that presumably provide them with specific nutrients. A few birds have also evolved very specialised - and strange - diets, including bones, dirt, blood and even 'milk'.

Obtaining calcium

Calcium is perhaps the most frequently limiting nutrient in birds, particularly for females during egg production, when it is needed for eggshell formation. Small birds obtain the mineral by foraging selectively on snails and other calcium-rich invertebrates. However, these prey items are less available or smaller in habitats with acidified soils, and this can limit egg production. Acid rain, associated with industrial development and pollution, can also decrease the availability of calcium-rich invertebrates, leading to eggshell defects and irregular laying.

Old World bearded vultures (*Gypaetus barbatus*) take consumption of calcium to extremes, with a diet consisting almost exclusively of large animal bones – although the birds are after the nutrient-rich bone marrow. The vultures can swallow bones whole, but they have also learned to fly up to 100 m above ground with large bones, which they then drop onto specially selected rocks below to break them into smaller pieces. The birds' highly acidic stomach acid then rapidly digests the bones.

▲ Bearded vultures or lammergeiers (*Gypaetus barbatus*) swallow and digest large bones – which they sometimes break by dropping them onto hard areas of rock.

Drawing blood

Few birds specialise in a diet of blood, but in the Galapagos the vampire ground finch (*Geospiza septentrionalis*) will feed on blood when insects are scarce. These birds peck at the soft base of the feathers of other birds, especially incubating Nazca boobies (*Sula granti*) or their chicks, consuming protein-rich blood in the process. Other birds will take a meal of blood when it is available. On subantarctic islands, sheathbills will peck at the flippers of southern elephant seals (*Mirounga*

CROP MILK

Pigeons and emperor penguins are the only birds to feed their chicks a form of 'milk', although this is quite different to mammalian milk. Crop milk is a secretion from the thickened lining of the crop or oesophagus that is regurgitated to chicks soon after they hatch. Like mammalian milk, pigeon milk contains antioxidants and immune-enhancing proteins important for the growth and development of the young. Male emperor penguins fast while they incubate their eggs for up to 110 days over the Antarctic winter, and provide this protein-rich 'curd' to their newly hatched chicks until the female returns to the nest with freshly caught food.

leonina) until they bleed and then feed on this – up to the point that the infuriated seal escapes back into the ocean! And in Africa, oxpeckers similarly feed on the blood oozing from small wounds on the skin of rhinos, zebras and giraffes, in between removing ticks and other pests.

Eating dirt

Geophagy is the consumption of soil, ground-up rock, termite-mound earth, clay or dirt. This is a relatively widespread but rare habit, occurring in 2 per cent of all birds – mainly those with primarily plant- or fruit-based diets that lack specific nutrients found in animal food. Many tropical birds, such as macaws, frequent outcrops of mineral-rich soil, salt or clay 'licks' to obtain essential minerals otherwise lacking from their diet. There is evidence that some birds also eat clay to 'self-medicate': clay neutralises toxins found in many tropical fruits and seeds and might also protect the birds from pathogens.

▲ Galapagos vampire ground finches (*Geospiza septentrionalis*) peck at the base of the feathers of Nazca boobies (*Sula granti*) to get a meal of blood.

▼▼ Red-and-green macaws (*Ara chloropterus*) eating mineral-rich clay from a clay lick.

Obtaining food: morphological and sensory adaptations

Bill, wing and feet adaptations allow birds to specialise in particular types of foraging, particular prey or even particular types of flowers. However, bird bills are equipped for feeding in many other ways – for example, by flexing to allow prey capture, or serving as sensory organs, allowing birds to detect prey they cannot see. In addition to sight and sound, birds use smell, 'remote touch' and even echolocation to find and capture prey.

◄ Many shorebirds like this common snipe (*Gallinago gallinago*) can flex their upper mandible, opening the bill tip while the rest of the bill remains closed. This is called rhynchokinesis and might help birds capture prey while probing deep within mud or soil.

Variation in bill structure

Many birds have elongated bills of varying shapes, allowing them to reach further into flowers for nectar (e.g. hummingbirds), or deeper in soil or mud to obtain prey (e.g. shorebirds). Crossbills and the New Zealand wrybill (*Anarhynchus frontalis*) have twisted mandibles for, respectively, extracting seeds from pine cones or flipping over stones to reveal hidden prey. But bills and tongues can be modified in many other ways as specialised feeding structures. Sandpipers and snipes are long-billed shorebirds that feed by probing in soil or mud, and can open the tip of their bill by flexing their upper mandible while the rest of the bill remains closed, an adaptation known as rhynchokinesis. This may reduce the force needed to open the bill underground, as well as improving the bird's grip on prey deep within the substrate. Rhynchokinesis also occurs in other birds (e.g. cranes, swifts, hummingbirds and some passerines), where it is associated with different feeding methods.

Birds can extend their tongue beyond the bill tip, sometimes up to four times the beak length, to reach hidden prey. In addition to the taste buds and salivary glands, the tongue may be armed with brush-like structures, be covered in mucus, or have backward-pointing barbs to capture prey and handle food. Birds that feed by probing cannot see their prey and so rely on non-visual sensory systems such as touch and smell (olfaction) to locate it. They also use a specialised sensory system called remote touch. In this, numerous pits in the bone of the bill tip are packed with sensory cells called Herbst corpuscles, which detect vibrations and small pressure waves from invertebrates burrowing through the soil or mud (see page 104).

Some shorebirds use their bill for suction feeding or grazing, slurping up biofilm from the surface of mudflats. Biofilm is a mat of microbes, organic detritus and sediment particles in a matrix of amino acids and lipids, secreted by diatoms and bacteria, and can form up to 60 per cent of the diet of small calidrid sandpipers during migration. The tongue of American western sandpipers (*Calidris mauri*) is specially adapted for a biofilm diet, with a very complex microstructure comprising long, dense papillae and keratinised lateral spines along its edges and at the tips.

Some aerial insectivores, which rely on a diet of flying insects, feed at night or in low-light conditions at dusk. Australian tawny frogmouths (*Podargus*

Ultrastructural details of the bill and tongue for feeding on biofilm

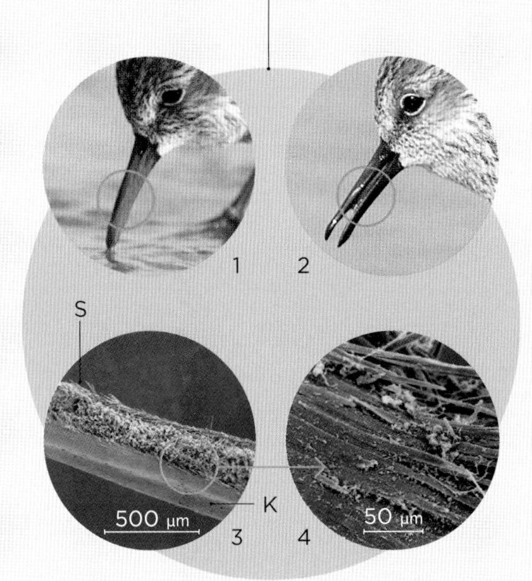

1–2. Mudflat biofilm adhering to the bill and in position for ingestion.

3. The numerous tongue spines (S) are used to gather the superficial biofilm; the spines arise from the keratinous ventral tongue surface (K).

4. Detail showing the origin of the tongue spines.

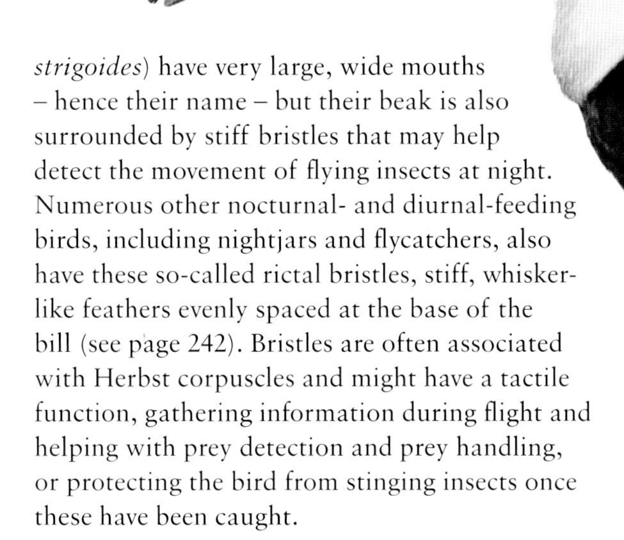

strigoides) have very large, wide mouths – hence their name – but their beak is also surrounded by stiff bristles that may help detect the movement of flying insects at night. Numerous other nocturnal- and diurnal-feeding birds, including nightjars and flycatchers, also have these so-called rictal bristles, stiff, whisker-like feathers evenly spaced at the base of the bill (see page 242). Bristles are often associated with Herbst corpuscles and might have a tactile function, gathering information during flight and helping with prey detection and prey handling, or protecting the bird from stinging insects once these have been caught.

◄ Western sandpipers (*Calidris mauri*) use suction feeding to suck up biofilm from the surface of mudflats.

▲ The toco toucan (*Ramphastos toco*) uses its massive bill to crush large nuts, but it can also dexterously manipulate fruits.

SENSING PREY

Kiwi, flightless ground-dwelling birds from New Zealand, were traditionally thought to detect soil-dwelling prey using smell since the olfactory bulb in their brain is ten times larger than that in other species. In addition, they are the only birds with nostrils at the tip of the bill – good for sniffing out earthworms in leaf litter. However, kiwi also have a shorebird-like sensory organ with an arrangement of pressure-sensitive cells within pits at the bill tip, allowing them to detect movements of buried prey – a perfect example of convergent evolution.

Sniffing out prey

Many birds use their acute vision, including UV and colour vision (e.g. vultures, falcons), or highly sensitive hearing to detect and catch prey (see pages 112–15 and 108–9). However, they have evolved adaptations that use the full range of senses for foraging. For a long time birds were considered to have a poor sense of smell, but we now know that many species use olfactory cues to locate prey. For example, turkey vultures can detect the chemical ethyl-mercaptan, emitted from decomposing carrion, from a distance of 1–2 km.

Albatrosses and petrels forage over large areas of the ocean for very patchily distributed prey – mainly fish, squid and krill. They do so using natural odour cues as guideposts in an olfactory landscape, allowing them to recognise potentially productive foraging areas. Early studies suggested that petrels could home in on smells from up to 10–20 km away, and it is now known that many seabirds can detect a chemical called dimethyl sulfide (DMS), which is produced when zooplankton graze on marine phytoplankton. Since concentrations of plankton often support high concentrations of fish and krill, DMS might provide seabirds with information on productive foraging areas. In experiments carried out by Gabrielle Nevitt, in which DMS was presented to birds at sea, several species responded by tracking the odour to its source using a zigzagging upwind search behaviour. Recent studies have shown that some insectivorous forest birds can detect the chemicals released by plants in response to herbivore attacks, allowing them to locate herbivore-infested trees that harbour concentrations of insect prey.

Echolocation

Bats are renowned for using echolocation to catch insects at night, but several birds have also evolved this adaptation. The South American oilbird, a species that feeds on fruits at night, and diurnal

▼ The tawny frogmouth (*Podargus strigoides*) has stiff rictal bristles around its beak that may help it detect and catch flying insects at night.

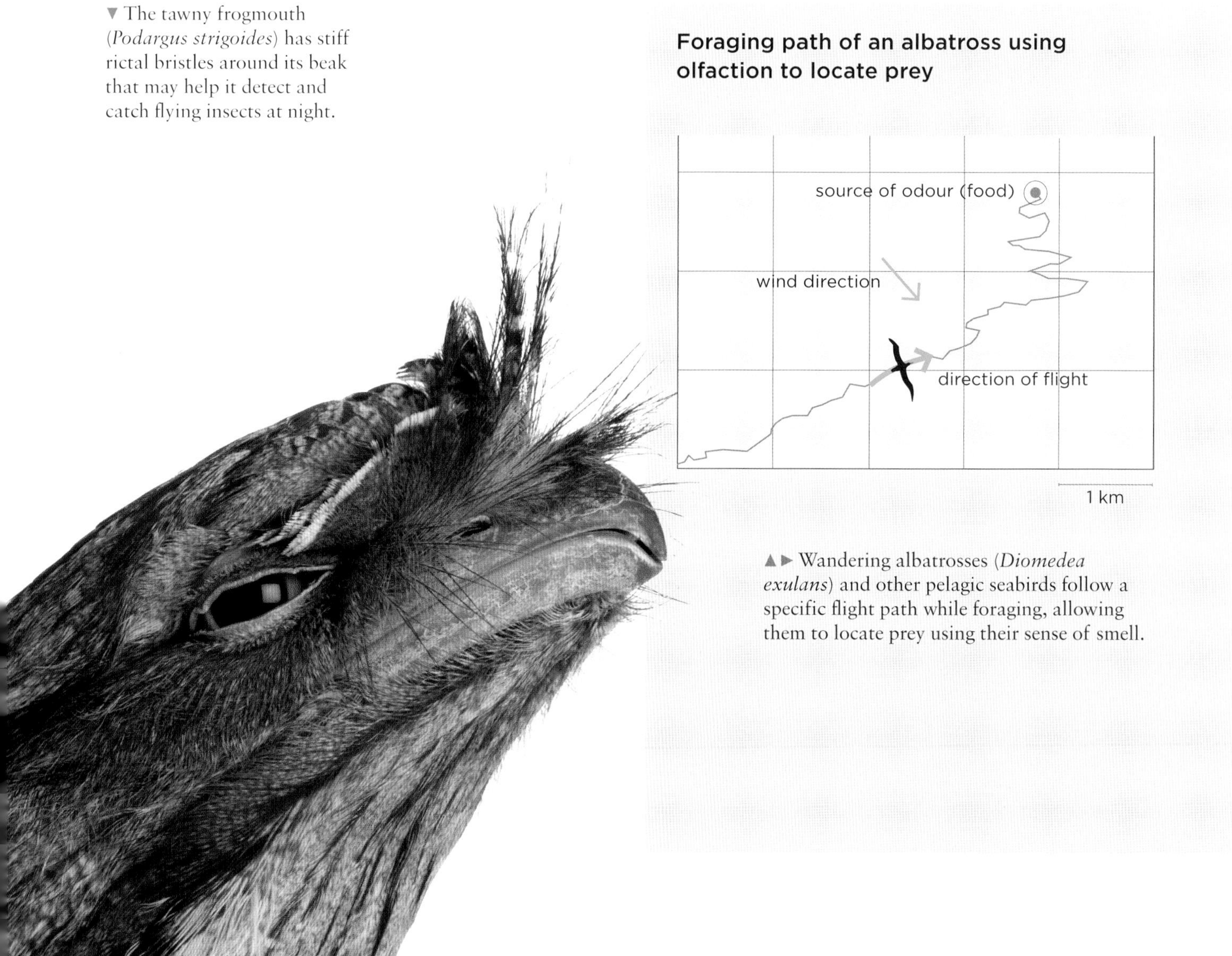

▲► Wandering albatrosses (*Diomedea exulans*) and other pelagic seabirds follow a specific flight path while foraging, allowing them to locate prey using their sense of smell.

insect-eating swiftlets both use what is considered to be a relatively crude form of echolocation. This involves lower-frequency clicks than those emitted by bats and might not be suited to the detection of insects smaller than 2–3 cm. Bird echolocation is therefore likely to be more important in allowing these species to nest in dark caves, where they face lower predation risk. That said, three-toed swiftlets (*Aerodramus papuensis*) in New Guinea click not only in their caves but also outside at night while hunting insect prey, so the use of echolocation in bird foraging remains an enticing possibility.

◄ Tahiti swiftlets (*Aerodramus leucophaeus*) use echolocation to navigate while nesting in dark caves, and perhaps also for foraging.

Finding food: behavioural and social adaptations

In addition to the morphological and physiological adaptations associated with foraging (see pages 240–3), birds use an amazing array of behaviours to increase their chances of catching prey. Rather than forage for themselves, some birds steal food from others, and some use other animals to disturb prey that they can then catch more easily. Many complex foraging behaviours are learned, and knowledge of novel foraging techniques can spread socially – and apparently rapidly – through populations.

Learned behaviours

Some gulls and crows fly up into the air and drop mussels and clams onto hard surfaces such as roads to smash the shells and extract the edible contents. The birds select medium-sized prey that will break relatively easily but that contain enough energy to make the behaviour worthwhile. Individual birds appear to get better at gauging the right size of prey and the correct dropping height as they age, suggesting that this is a learned behaviour. As discussed earlier (see page 236), bearded vultures use the same basic technique to break open large bones, dropping these onto rock areas to get at the rich marrow inside. Young vultures will initially drop the bones from an insufficient height, or onto soft ground, and as a result they do not smash. Eventually, however, they improve their skills through trial and error. In California and Japan, crows have been filmed using moving cars as nutcrackers, either dropping walnuts onto a road, or even waiting for a red light and placing the nuts in front of stopped cars!

Tool use

Once thought to be the preserve of humans, tool use is now recognised in several free-living animals – including birds. Woodpecker finches (*Camarhynchus pallidus*), from the Galapagos, use a cactus spine held in their beak to reach insect grubs inside tree bark. The birds reshape the spines (e.g. shortening a long spine) to form more manageable tools, and will reuse tools between foraging bouts.

New Caledonian crows (*Corvus moneduloides*) are the best-studied avian tool users and use a variety of tools made from plant material to extract insect prey from dead wood, bark and vegetation. The birds use at least three tool types. Their non-hooked stick tools are simple, unmodified twigs or leaf stalks, but the birds also make hooked stick tools from forked plant stems. For these, they carefully select and detach a small branch, and then 'sculpt' a hook using their bill. A third type of tool is made by stripping off the barbed leaf edges of screw pines (*Pandanus* spp.). When presented with a section of plant material that is too short to use for retrieving insects, New Caledonian crows are even able to combine different pieces to make a longer, functional tool.

▼ Several corvids have been reported dropping nuts onto roads and waiting for cars to drive over them to break the nut open.

Social feeding

Social foraging, or flocking, is widespread in birds, with single species or, more commonly, mixed-species feeding flocks occurring in waterfowl, wading birds, and granivorous and insectivorous passerines. Mixed-species feeding flocks are characteristic of forest bird communities outside the breeding season, and in tropical forests the majority of birds observed are in such flocks rather than foraging alone.

Flocking involves more than the simple aggregation of birds at food sources: there are complex social interactions between the members of a flock. In mixed-species flocks of passerines, birds call incessantly and signal to one another through posturing and wing fluttering. The size, stability and membership of feeding flocks is very variable, ranging from casual, transient feeding associations among seabirds, to the long-term associations seen in lowland Neotropical antbird flocks. The social dynamics within feeding flocks can be very complex, with some individuals or species being leaders and others satellites or followers, depending on their role in flock cohesion and their relative position within the flock.

Birds in mixed-species feeding flocks are thought to benefit from increased foraging efficiency because more individuals are searching, collectively, for good patches of food. Individuals may also benefit through social learning, using foraging sites where they observe that their flock mates have been successful. Mixed-species flocks facilitate foraging through the 'beater effect', where other birds in a flock flush out more insects from vegetation. Flocking can also provide benefits through a reduction in predation due to safety in numbers: individuals can take advantage of the vigilance and alarm calls of their many flock mates and have more time for foraging.

Feeding in a flock does, however, have its costs. There can be increased competition for food, and some individuals or species may dominate food patches within social groups. Many flocks have dominance hierarchies – in flocks of tits in Europe, younger, more subordinate birds are forced to move to peripheral foraging areas by older, experienced, dominant birds, where they might be more vulnerable to predation. That said, young, inexperienced birds likely benefit overall if they can follow adults to good foraging sites. In some flocks, individual birds will steal food items from their flock mates (called kleptoparasitism), and the trade-off between foraging and vigilance opens up the possibility of birds 'cheating'. In the Kalahari Desert, fork-tailed drongos (*Dicrurus adsimilis*) flock with ground-foraging southern pied babblers (*Turdoides bicolor*) and will act as sentinels, making alarm calls to warn of approaching predators. However, a drongo will also occasionally fake an alarm call – a rare example of deception in free-living birds – then swoop down to take the prey item the startled babbler drops. Drongos even know which birds are easier to trick, focusing their fake alarms on inexperienced, naive juveniles more often than on adults!

▼ Woodpecker finches (*Camarhynchus pallidus*) use cactus spines as tools to extract grubs from tree bark.

▼ Frigatebirds (*Fregata* spp.) kleptoparasitise other seabirds, chasing and attacking them in an attempt to steal food.

Hoarding food

In temperate and boreal regions, some species avoid the problem of variation in food availability by storing food when it is plentiful, especially in autumn, and then using this food in winter when natural food is less available. Food storing, or caching, is common in crows, jays, nutcrackers, tits and nuthatches. Retrieval of cached food requires highly developed spatial memory, and there is some evidence that this might be associated with variation in size of some brain regions such as the hippocampus. A few species use stored food to get an early start on breeding.

The size of the hoard

Most birds that store food do so in autumn, taking advantage of an abundance of ripening tree seeds at this time of year, and then use their caches or hoards to survive periods of low food availability over winter. In North America, Clark's nutcrackers (*Nucifraga columbiana*) each cache between 20,000 and 35,000 seeds of whitebark pine (*Pinus albicaulis*) and piñon pines from August to December. The birds are considered 'scatter hoarders', carrying the seeds in a pouch under their tongue for up to 10–20 km, and then burying between one and 15 seeds in each cache, in forest litter, soil, or cracks and holes in tree bark. Florida scrub jays (*Aphelocoma coerulescens*) and Eurasian jays (*Garrulus glandarius*) both store between 4,000 and 11,000 *Quercus* oak acorns in autumn. The scrub jays spend a lot of time digging up and recaching their acorns, perhaps to monitor the quality of their stored food. In California, acorn woodpeckers (*Melanerpes formicivorus*) store thousands of acorns in granaries in tree bark, drilling storage holes for the seeds. The birds constantly check the acorns individually as they dry out and are at risk of falling out of the granary, repositioning and resizing them to ensure they are stored securely.

▼ Eurasian jays (*Garrulus glandarius*) can store up to 11,000 acorns during autumn, which they then retrieve and eat over winter.

Retrieving a cache

Food-storing birds have highly developed spatial memory – some species can retrieve up to 80–90% of stored food for up to nine months after it has been cached. Stored seeds that are not recovered can germinate, and so caching also contributes to seed dispersal for plants. The caching of food is not random, with the birds placing seeds at sites that have specific environmental characteristics. They then recover their stores using these spatial relationships and visual cues of the landscape, such as rocks, logs or trees. Some studies have suggested that brain regions associated with memory (e.g. the hippocampus) are larger in caching species, or vary seasonally with the occurrence of caching behaviour (see pages 122–3), but other studies have found no clear association between brain morphology and caching behaviour. There is likely also a learning component to caching – for example, mountain chickadee (*Poecile gambeli*) adults perform better than juveniles in spatial memory tasks.

WHY STORE FOOD?

Stored food can form a substantial proportion of the diet of caching species, especially in winter. The importance of food caches for survival depends on environmental factors over winter, with harsher conditions associated with greater dependence on caches. A few birds also use food they store to support reproduction. Canada jays (*Perisoreus canadensis*), for instance, use stored food to nest much earlier than other birds, laying eggs between late February and March, when there is little natural food, and often incubate eggs in temperatures below –25°C. In western North America, Clark's nutcrackers also breed in February and March, raising their chicks partly on stored pine seeds.

▲ Acorn woodpeckers (*Melanerpes formicivorus*) store acorns in holes drilled in the bark of a tree used as a 'granary'. They constantly check the acorns, moving them to smaller holes to prevent them from falling out as they dry.

CHAPTER 8

Social Behaviour and Communication

Imagine walking through the winter woods in eastern North America. All around you are flashes of black and white as a flock of black-capped chickadees (*Poecile atricapillus*) flits by, giving their characteristic *chick-a-dee* call. By springtime, the flock dissolves into several mated pairs, each defending its own breeding territory. Magenta-throated woodstars (*Calliphlox bryantae*) in Costa Rica, by contrast, remain solitary and fiercely territorial nearly their whole lives. Australian splendid fairywrens (*Malurus splendens*) spend their lives in family groups of a mated pair and one or more adult offspring. The social environment is even more complex in African grey parrots (*Psittacus erithacus*) and their relatives, with ever-changing alliances and rivalries.

Birds provide examples of almost every possible social system, ranging from solitary to gregarious and cooperative. Together with this diversity in social behaviour, birds have evolved elaborate songs, dances and other displays to communicate with mates, rivals, and even members of other species.

◂ Great crested grebe (*Podiceps cristatus*).

Most birds are monogamous

Monogamy, where one male and one female share a territory and cooperate to raise their offspring is generally rare among animals but the majority of birds of all species show this behaviour. Why is this? It has been suggested that chicks grow so rapidly and are so vulnerable to predators that they often require care from both parents if they are to survive. Another possibility is that because birds do not rely on internal gestation or lactation, male birds are better able to care for their young than are male mammals, making two-parent care more likely to succeed in birds.

Both male and female birds often attempt to enforce monogamy and prevent their partner from acquiring a second mate. For example, female house sparrows (*Passer domesticus*) are aggressive toward other females that their partner attempts to court. Should a male house sparrow that is already paired to one female manage to attract a second, the first female may destroy any eggs or chicks the second female produces. For their part, males in many species follow their partners closely during the early breeding season to ensure they are not 'cuckolded', a behaviour termed mate guarding.

◀ Wandering albatrosses (*Diomedea exulans*), seen here at a nest on South Georgia in Antarctica, generally mate for life. Their monogamous attachments take years to form but can last for decades.

▶ Male (left) and female (right) zebra finches (*Taeniopygia guttata*) readily form monogamous pairs.

▼ Cape sugarbirds (*Promerops cafer*) are socially monogamous, but rarely genetically monogamous. A large proportion of offspring result from extra-pair dalliances.

Until death do us part?

Although most birds are monogamous based on the definition above, monogamy can take many different forms. Such a partnership may last for a single nesting attempt, for the length of the breeding season, for several years, or for life. Long-term partnerships are often seen in long-lived species such as wandering albatrosses (*Diomedea exulans*), scarlet macaws (*Ara macao*) and mute swans (*Cygnus olor*), although if one member of the pair dies or is evicted from the territory, the remaining individual will often mate again. Short-lived species like most songbirds generally have more temporary partnerships, and may leave one mate to pursue another, particularly after a failed breeding attempt. But there are exceptions to this general pattern. Australian zebra finches (*Taeniopygia guttata*) are small, short-lived songbirds with long-term social partnerships, perhaps to allow mated pairs to begin breeding immediately when conditions become favourable, such as when seasonal rains begin. For other birds, including herons and storks, long-term monogamous partnerships may have evolved because individuals return repeatedly to the same breeding locations.

Degrees of fidelity

Monogamous birds also vary widely in their degree of sexual fidelity. Parentage analysis shows that in most bird species that appear to be monogamous, many individuals also copulate with birds other than their social mate (extra-pair copulations). This finding led to the distinction between social monogamy and genetic monogamy. Some birds, including long-eared owls (*Asio otus*) and other raptors, are both genetically and socially monogamous, meaning that neither parent produces offspring with individuals other than their mate. But in other species – even those with lifelong pair bonds such as wandering albatrosses – individuals occasionally engage in copulations with birds other than their social mate. In South African Cape sugarbirds (*Promerops cafer*), more than 60 per cent of chicks in broods may have been fathered by a bird other than the attending male or social partner. In fact, whereas social monogamy is widespread in birds, genetic monogamy seems to be the exception rather than the rule.

Other mating systems

After monogamy, the second most common mating system in birds is polygyny, whereby some males attract and raise offspring with two or more females. Polygyny can take several different forms. In some polygynous species, males compete for the best territories or other resources, and then use these resources to attract females. In other species, males may monopolise harems of breeding females, or cluster together in multi-male groups, each displaying to attract the attention of females like a bustling marketplace. Some birds show polyandry, the opposite of polygyny, where a single female may raise offspring with more than one male.

Types of polygyny

In most polygynous birds, including Japanese bush warblers (*Horornis diphone*) and Eurasian great reed warblers (*Acrocephalus arundinaceus*), males compete for territories containing food, nest sites or other important resources that attract females. This pattern often emerges in patchy environments such as marshes and grasslands, where valuable resources are distributed unevenly and a single territory can support multiple breeding females. In other species, males attempt to control access to the females themselves rather than defending a territory in order to attract them. Although this mating system is widespread in mammals, in birds it is rarer than resource-defence polygyny. However, it may emerge in birds where females breed in social groups. In Costa Rica, for example, female Montezuman oropendolas (*Psarocolius montezuma*) establish breeding colonies with many nests hanging from a single tree. Male oropendolas visit these colonies and then compete for access to females.

Regardless of whether males compete for mates indirectly (through monopolising a territory or other important resources, as in Japanese bush warblers) or directly (through male–male competition, as in oropendolas), or through courtship displays aimed at attracting females (as in lekking species – see box), polygyny typically occurs in environments, populations and species where male care is

◄ Male Montezuman oropendolas (*Psarocolius montezuma*) do not defend a territory to which females are attracted. Instead, they visit colonies of females, which construct woven nests that hang from tree branches. A single tree can accommodate many such nests (below).

LEKS

Sage grouse
(Centrocercus urophasianus)

The term 'lek', from the Swedish 'to play', describes a form of polygyny in which dozens of males aggregate in a single location, displaying and competing for the attention of females. Lekking males provide no resources to females and no parental care to their offspring, instead focusing their efforts on courting females and intimidating neighbouring males through vigorous and conspicuous displays. In North America, lekking male sage grouse (*Centrocercus urophasianus*) inflate their gular (throat) sacs, strut and boom. In New Guinea, groups of male Raggiana birds-of-paradise (*Paradisaea raggiana*) clap their wings and shake their heads to impress females. Another distinctive characteristic of leks is strong variation in male mating success: of the matings that occur at a lek, about three-quarters involve just 10 per cent of males.

▼ In red phalaropes (*Phalaropus fulicarius*), shown here copulating, females (bottom) are brightly coloured and attempt to mate with many males. Most parental care is provided by the duller-coloured male (top).

not crucial to the survival of the offspring. Sexual selection, a special form of natural selection favouring traits that help to acquire mates (e.g. weapons or attractive ornaments and displays; see page 40), is particularly intense in polygynous birds.

Polyandry

Only around 1 per cent of bird species are polyandrous, whereby some females form social partnerships with two or more males, but males mate with only a single female and often provide the majority of parental care. In red phalaropes (*Phalaropus fulicarius*, known as grey phalaropes in Europe) polyandry is sequential: the territorial female produces a clutch of eggs with one male, then leaves him to care for the nest while she initiates a new clutch with a second male. Polyandry can also be simultaneous, as in Galapagos hawks (*Buteo galapagoensis*), where multiple males jointly defend a territory with a single breeding female.

Attracting a mate

Birds are conspicuous creatures, especially during courtship, when individuals – usually males, but occasionally females as well – pull out all the stops to impress the opposite sex. When we think of how birds attract mates, bright plumage and beautiful songs probably come to mind first. But courtship displays vary tremendously, including anything from vigorous athletic displays in the form of flights or dances, to architectural creations such as bowerbird courts, sounds such as songs or feather snaps, attractive odours and interactive displays in which both partners participate.

▲ Mate choice in great bowerbirds (*Chlamydera nuchalis*) centres around elaborate bowers. Males attract females by constructing a symmetrical avenue that opens into a decorated court.

Impressive displays

Research is increasingly showing that birds' mating signals are carefully orchestrated and have many interacting parts. In the southwestern USA, for example, male Costa's hummingbirds (*Calypte costae*) perform side-to-side 'shuttle dance' flights in front of a potential mate. The dancer places himself at a certain angle relative to the position of the sun and the watching female, and simultaneously puffs out his bright purple throat feathers with perfect timing so as to appear to best advantage. In northern Australia, male great bowerbirds (*Chlamydera nuchalis*) build and decorate elaborate bowers, positioning themselves just so while they display to visiting females. Decorations lining the avenue of the bower are carefully arranged according to size, resulting in a fixed-perspective illusion that may make the displaying male appear larger than he really is.

Serenading a partner

Birds are capable of many vocalisations and other sounds (see pages 28, 69, 102–3 and 172–3), but songs are particularly noticeable, often being relatively complex, loud and persistent. These vocalisations frequently serve the dual function of attracting mates and repelling rivals of the same sex. Although it may appear effortless to a human observer, birdsong represents a challenging feat of neural and muscular control. In Southeast Asia, for example, the rapidly trilled song of plain prinias (*Prinia inornata*) requires the singer to modulate both pitch and amplitude with a high degree of precision and speed. In many species of bird, females prefer males with better vocal performance. Birdsong can also advertise motor skills

▲ Courtship in bald eagles (*Haliaeetus leucocephalus*) includes a series of aerial cartwheel displays, dramatically described as a death spiral.

and athleticism when it is produced during flight. Anyone who has heard the warbling song produced by male Eurasian skylarks (*Alauda arvensis*) during their climbing, hovering and downward-parachuting flight display will appreciate the impressiveness of this accomplishment.

Showing commitment

In most bird species, males as well as females provide substantial care to offspring, so it is perhaps not surprising that both sexes are often particular in their choice of mates (see pages 256–9 and 291). Mutual mate choice is widespread in birds, at least in the majority of species that are socially monogamous. A striking example of gauging a potential mate's intentions and level of commitment is the cartwheel courtship display of bald eagles (*Haliaeetus leucocephalus*). Both the male and female fly high into the air, then the pair lock their talons. They fall toward the Earth in what has been described as a death spiral, before disengaging at the last possible instant. Less death-defying but also spectacular are the mutual dance displays that characterise the courtship of Japanese red-crowned cranes (*Grus japonensis*).

Why be choosy?

As in most animals, females in the majority of bird species are the choosier sex (except in polyandrous species; see page 253). Most female birds do not mate with the first available male that they encounter, and instead visit and assess several, even dozens, of potential mates before making their choice. This may seem surprising, as visiting many males before making a choice is costly in terms of time and energy. Why, then, are female birds so selective when it comes to choosing a mate?

The answer to this apparent paradox is that in birds, as in other animals, some potential mates are better than others. Thus, the more candidates a female screens or checks out, the greater her chance of choosing the best male available. Despite the costs of exercising mate choice, choosy females may benefit in several ways, such as gaining resources, or securing a high-quality male with 'good genes'.

Material benefits

In many birds, males defend territories containing nest sites, food, shelter or other important resources that females need. By selecting a male with the best territory, choosy females can gain access to the best or most plentiful resources. In addition to territorial resources, most male birds also assist their mates to some degree with the care to their offspring – for example, by contributing to incubating the eggs or feeding the female while she does so, feeding the offspring, or helping to defend the nest and offspring from predators.

If males with the most attractive displays are also capable of providing the best paternal care, then choosy females should benefit by exercising mate choice. In South American red-crested cardinals (*Paroaria coronata*), males with brighter red head plumage defend the nest more vigorously against predators. Similarly, male common nightingales (*Luscinia megarhynchos*) in Eurasia that produce more complex song also feed their offspring at higher levels. Tangible resources such as breeding territories and care of offspring, termed direct benefits of mate choice, provide one compelling reason why mate choice persists despite the associated risks and costs.

◄ Although female preferences for brightly coloured males may at first seem arbitrary, studies have shown that male house finches (*Haemorhous mexicanus*) with redder plumage provide more food for their offspring.

► A male red-crested cardinal (*Paroaria coronata*) displays his bright head plumage. This trait may advertise his ability to protect his offspring.

Good genes

Even in species where males provide no material support to females or their offspring, female birds are still very choosy. This is particularly evident in lekking species such as sage grouse (see box on page 253), where females visit several males and assess their displays before choosing one, despite the males contributing nothing but their genetic material. Similarly, females of many socially monogamous species sometimes copulate with males other than their social partner, even though these 'extra-pair' males provide no tangible benefits to the female or her offspring. Instead, attractive male ornaments and displays may provide reliable information about the individual's 'good genes', defined as genetic advantages that, when passed down to offspring, improve their ability to grow, survive or resist disease (see box on page 281).

SEXY SONS

Beyond the tangible benefits associated with male resources and improved survival of offspring associated with good genes, females that select a male carefully also benefit by producing 'sexy sons'. Assuming that highly ornamented and attractive males tend to produce sons that are also ornamented and attractive, females that mate with attractive males can therefore expect to produce attractive sons, which in turn are likely to provide the choosy female with many grand-offspring.

▼ A female satin bowerbird (*Ptilonorhynchus violaceus*), left, inspects the workmanship of a male, right.

▲ White-throated sparrows (*Zonotrichia albicollis*) have two distinct plumage types: white-striped (left) and tan-striped (right). The two types differ not only in plumage but also in behaviour, and individuals of one type consistently mate with the opposite type.

Compatible genes

In some species, females (or both sexes) may be choosy but their preferences are individualistic rather than uniform: each bird has a preference, but not all preferences are identical. This pattern suggests that the best mate for an individual bird depends on that bird's own genetic make-up. Mates may thus be chosen not just on the basis of their good genes or attractive genes, but because the genes they offer are compatible with those of the chooser. For example, European storm petrels (*Hydrobates pelagicus*) prefer the odour of unrelated individuals over that of their close relatives. These odour preferences may reduce the risk of mating with relatives (inbreeding), which is associated with high levels of genetic disorders in the resultant offspring.

A particularly striking example of individualistic mate choice comes from white-throated sparrows (*Zonotrichia albicollis*) in North America. In these birds, both males and females may have either white-striped or tan-striped crown plumage, and the two plumage forms are associated with very different behaviours. In both males and females, white-striped individuals sing more, are more aggressive and territorial, and provide less parental care than do tan-striped individuals. Individuals nearly always mate with the opposite colour and behavioural form. Thus, breeding pairs of white-throated sparrows may involve either a tan-striped male and a white-striped female, or a white-striped male and a tan-striped female, but virtually never two tan-striped or two white-striped individuals. This fascinating system is maintained in part by chromosomal differences between the two forms, and should two white-striped individuals produce offspring together, about 25 per cent of these would be inviable.

Social behaviour during breeding

In addition to diversity in their mating systems and patterns of mate choice, birds vary considerably in their social behaviour and how they use space during breeding. Many birds such as goldcrests (*Regulus regulus*) are solitary breeders, where a single individual or a mated pair defends a territory, from which they exclude other birds of the same species. Other species such as great black-backed gulls (*Larus marinus*) are sometimes solitary breeders but may also aggregate in breeding colonies. Yet other species, including the African red-billed quelea (*Quelea quelea*), can breed in large colonies that sometimes contain hundreds or thousands of pairs.

Colonial breeding – the pros

About 13 per cent of bird species are colonial breeders, meaning that many individuals nest and breed in close proximity. Breeding colonies can consist of hundreds or even thousands of nests, often within inches of one another. Although colonial breeders are generally socially monogamous and each mated pair provisions its own offspring, the group participates jointly in watching for predators. Safety in numbers therefore probably represents an important benefit of colonial breeding: with thousands of birds in a breeding colony, there are many eyes to notice potential predators and individuals to drive them away.

Another advantage of colonial breeding is increased opportunities for finding new or replacement mates, since there are many individuals to choose from. Colonial living may also facilitate access to information about the location of food. For birds that rely on unpredictable, unevenly distributed food sources – such as Guanay cormorants (*Leucocarbo bougainvillii*), which feed on schools of anchovy off the coast of Peru – search costs can be reduced by watching the bearing from which successful foragers return to the colony. Finally, some species likely form breeding colonies because safe nesting sites or other critical resources are distributed unevenly over the environment. Many seabird colonies, for instance, are located on cliffs or islands that are inaccessible to terrestrial predators.

Colonial breeding – the cons

While colonial breeding offers the many advantages discussed above, it also has many associated costs and complications. Although the high numbers of individuals at a breeding colony provide better opportunities to detect and drive away predators, the conspicuous sounds and smells emanating from the mass of birds make them easily noticeable: colonies may thus attract more predators than a single breeding site. Colonial breeding also increases birds' susceptibility to parasites (organisms that live on or inside a host, on which they feed), such as feather mites, as well as bacterial, viral and other infectious diseases, which can spread rapidly through a crowded colony.

The high density of birds within a breeding colony increases competition: for food resources, which may become locally depleted; for the best nest site; and for mating opportunities. Breeding in a large colony also introduces another challenge: parents returning to the colony after foraging must be able to recognise their nest site, burrow or offspring, and offspring must recognise their parents from an early age. In common murres (common guillemots, *Uria aalge*), offspring begin learning their parents' calls even before they hatch from the egg!

► Adult and juvenile king penguins (*Aptenodytes patagonicus*) at a breeding colony on South Georgia. Breeding in such large, densely packed groups offers advantages, but it also carries some costs.

Group living outside the breeding season

Just as bird species show tremendous variation in their breeding behaviour, so social behaviour outside of the breeding season is also diverse. As their name suggests, South American solitary caciques (*Cacicus solitarius*) are largely solitary for most of their lives, while other birds such as Australian zebra finches are social and gregarious. Even within the same species, social tendencies can change drastically between the breeding and non-breeding seasons.

Seasonal patterns in social behaviour

Some birds become less social outside of the breeding season. Atlantic puffins (*Fratercula arctica*), for example, nest in colonies during spring and summer, but defend individual territories and live alone in autumn and winter. More common, however, is the reverse pattern, in which birds become more social outside the breeding season and live in groups of various sizes. Black-capped chickadees are one example (see page 249), but many other species – from waterfowl to shorebirds and raptors – are territorial during breeding but form social groups during migration or over winter. Aggregations can even include members of other bird species: Australian buff-rumped thornbills (*Acanthiza reguloides*) regularly flock together with various species of *Malurus* fairywrens outside of the breeding season.

▼ Groups of American white pelicans (*Pelecanus erythrorhynchos*) cooperate to herd fish into shallow water, where they can be more readily captured.

◄ For small-bodied birds such as ashy woodswallows (*Artamus fuscus*), huddling together can conserve heat and energy.

▼ Viruses of the Flavidae family are the agents of West Nile and several other diseases, affecting birds as well as humans.

Group living – the pros

Group living outside the breeding season can bring both benefits and costs. Advantages include safety in numbers and often a reduced need for individual vigilance in watching for predators. Group-living birds also benefit from opportunities for cooperative hunting (or cooperative fishing, as in many pelicans), or information about the location of food or other resources. Huddling together in winter can conserve body heat in small birds (e.g. Australian white-browed babblers, *Pomatostomus superciliosus*) and large birds alike (e.g. emperor penguins, *Aptenodytes forsteri*). When groups are composed of relatives, opportunities can be provided to cooperate with kin. Another interesting feature of group living is seen in the North American brown-headed cowbird (*Molothrus ater*), females of which lay their eggs in the nest of another bird species. Young cowbirds are thus not raised by members of their own species, but shortly after reaching independence they form social flocks with other brown-headed cowbirds. The social interactions that occur in these flocks provide important cues for the young birds, allowing them to learn the species-appropriate vocalisations and other behaviours.

Group living – the cons

One disadvantage of group living is an increased risk of infectious disease. West Nile virus is widespread in some populations of American crow (*Corvus brachyrhynchos*) and is thought to be transmitted at winter roosts when infected birds defecate on their lower-roosting flock mates. Human activities such as attracting birds to garden feeders can also inadvertently contribute to the spread of diseases by increasing the density of birds in a particular area. In addition, living in groups increases competition for food and shelter, a cost that may be particularly acute for individuals that are already weakened by illness and less able to compete.

Finally, whereas all birds face daily challenges in their physical environment, those living in groups must also contend with social information and adjust their behaviour in response to that of other group members. In fact, group-living species of bird perform better than solitary-living species on cognitive tests, and within species, individuals living in large groups outperform those living in smaller groups.

Social dynamics outside the breeding season

Competition for food, shelter and other resources is a hallmark of group living. Most birds living in groups establish some form of dominance hierarchy, meaning that some individuals are higher in the 'pecking order' than others. Anyone who has spent time watching birds interact at garden feeders will be aware that some individuals easily displace others of the same species. Dominant birds thus have ready access to feeders or other resources, while more subordinate individuals may have to wait their turn.

What predicts dominance status?

An individual's position in the dominance hierarchy can be predicted somewhat by its sex and age. Older birds are usually socially dominant over juveniles, and males are often dominant over females. However, there are exceptions to this generalisation: females rather than males are socially dominant in raptors (presumably because of their larger body size), in polyandrous species such as sandpipers, and in wintering flocks of house finches. Body condition (health) is another predictor of dominance status. Although social dominance confers improved access to resources, dominant individuals may also have to spend more time and energy fighting off challengers, and evidence is mixed as to whether they are more likely to survive and reproduce than subordinates.

Dominance across different species

Dominance hierarchies can be described not only between individuals belonging to the same species, but also between different species of bird. In a FeederWatch study monitoring interactions at more than 1,500 garden feeders across North America, 136 species were ranked according to dominance. Not surprisingly, this ranking shows a major effect of body size: at the top of the hierarchy are wild turkeys (*Meleagris gallopavo*), which are never displaced from feeding sites by other species. Eurasian tree sparrows (*Passer montanus*), small songbirds closely related to house sparrows, are the least dominant of the species investigated. Doves and buntings are less dominant to other species than expected based on their body size, and crows and jays more dominant for their size than expected. The same study also described an interesting non-linear hierarchy, similar to the 'rock paper scissors'

dynamic, between common starlings, red-headed woodpeckers (*Melanerpes erythrocephalus*) and red-bellied woodpeckers (*Melanerpes carolinus*). Starlings are dominant to red-headed woodpeckers, which in turn are dominant to red-bellied woodpeckers, but red-bellied woodpeckers are dominant to starlings.

Social networks

An exciting new development in the study of the social lives of birds comes from applying principles of social network analysis. This approach, which identifies patterns of social ties by mapping interactions between pairwise combinations of individuals, has revealed the rich and complex social lives of species such as monk parakeets (*Myiopsitta monachus*). Network analysis of wild parakeets in Argentina and captive flocks in Florida shows that although pairs (not necessarily mates) are the fundamental unit of social structure, additional layers of social relationships – including complex dominance relationships and shifting alliances – occur within the flock. The social structure revealed by network analysis can predict the speed at which information flows between individuals, or how fast a disease will spread within a flock.

▼ A red-bellied woodpecker (*Melanerpes carolinus*) faces off against common starlings (*Sturnus vulgaris*) and blue jays (*Cyanocitta cristata*) at a feeding platform. Such squabbles can reveal species differences in dominance.

Personalities

Although we might think of personality as something uniquely human, many animals also show consistent behavioural differences between individuals. Birds are no exception. Different species of birds, but also individuals within a species, vary in many behavioural traits, including their activity level, their willingness to explore new environments, and their degree of risk-taking when approaching predators or unfamiliar objects. When this behavioural variation is consistent across time or across different contexts, it can be described as personality.

Defining avian personality

Eurasian blue tit (*Cyanistes caeruleus*) individuals that are highly active during the breeding season are also more active than their sedentary counterparts in winter. Variation in behaviour can also be consistent across different situations, such as when individuals that take more risks during foraging also take more risks in social interactions. In addition, behavioural traits are often correlated with one another, as seen in Eurasian great tits (*Parus major*), where individuals that are quicker to explore an unfamiliar environment are also less fearful of approaching new objects. The individuals are displaying different personalities as defined in the introduction above. Animal personalities appear to be at least partially inherited and are likely also influenced by early experience.

Variation in personality within a species is likely maintained because different personalities succeed in different contexts and different environmental conditions. Great tits show heritable variation in boldness, but in the wild, fitness as measured by offspring production appears to be similar for both bold and shy males. This is because although bold males are more likely to pursue copulations with females other than their mate, and thus father more offspring in the nests of other males, shy males are less likely to lose paternity within their own nests. Despite their different mating strategies, bold and shy male great tits end up having similar mating success on average, and in turn this likely explains why both behavioural types persist.

▼ In song sparrows (*Melospiza melodia*) and other birds, individuals living in cities show different behavioural tendencies than individuals of the same species living in less urbanised environments.

Human influence

Urbanised areas are expanding, and in the process are changing many features of birds' physical and social environments. Urbanised environments may thus favour different personality types than rural environments. Supporting this, song sparrows living in urbanised environments in North America are less fearful of humans and more aggressive in defending their territories than are song sparrows living in more rural areas. Similarly, aggression in territory defence is higher among urban than rural great tits, and Asian common mynas (*Acridotheres tristis*) from urban populations are quicker to approach unfamiliar food items than those from rural areas. Eurasian common blackbirds (*Turdus merula*) show a different pattern, however. When blackbirds are captured shortly after hatching and raised under identical conditions, individuals from forest populations approach novel objects like plastic cups more readily than birds from urban environments. Human-associated habitat alteration thus appears to be changing not only the distribution and abundance of bird species, but also the types of personality traits that are favoured.

▲ Despite their small size, Eurasian blue tits (*Cyanistes caeruleus*) can be bold and aggressive. This is particularly true during territorial disputes with rivals.

► City living provides opportunities to encounter new food sources. Here, a common myna (*Acridotheres tristis*) on the streets of Bangkok feeds on discarded bread.

Birds at play

Bird behaviours often have clear functions related to survival and reproduction. However, many birds also show behaviours that seem to serve no function at all, puzzling both scientists and birders. Swallows chase feathers, then toss them back into the air after capturing them. Crows sneak up behind cats to pull their tails, then dart away. Kea (*Nestor notabilis*) in New Zealand seem to be irresistibly drawn to unfamiliar objects and sounds. Ravens slide or roll down snowbanks, then fly back up to repeat the process again and again. Are some of these behaviours just for fun? Do birds play?

Child's play

Birds show many types of behaviour that seem to serve no obvious purpose, and that strike human observers as being playful. These include manipulating non-food objects with their bills or talons; collecting and hiding shiny items; approaching and investigating novel objects or sounds; chasing and pouncing on inanimate objects such as leaves; appearing to taunt potential predators by coming close, then flying away; balancing on thin twigs; and movements such as sliding, swooping or bouncing that seem reminiscent of play in human children. Indeed, play behaviour appears to be more common in juvenile birds than in adults of the same species. In European herring gulls (*Larus argentatus*), for example, the 'drop-catch' game with clams occurs more often in younger birds than in adults. Adult gulls also drop clams from the air, but rather than catching them, they allow the shells to smash on the ground so that they can then eat their contents. Play, especially social play, is more widespread in species where breeding is delayed and long-term associations exist between parents and their offspring, as in cooperative breeders such as Arabian babblers (*Argya squamiceps*). Species with complex social systems, such as parrots and corvids (crows, magpies and jays), are particularly likely to show play behaviour.

▼ Juvenile carrion crows (*Corvus corone*) engage in play-fighting. Corvids and other species with complex social systems are more likely to engage in playing.

▲ Many gulls intentionally drop shellfish from the air. Rather than allowing the shellfish to smash on the rocks below, as adults do, juvenile gulls swoop down to catch them – a behaviour called the 'drop-catch' game.

▼ Kea (*Nestor notabilis*) may be among the most playful of all birds. Play-fighting in this species persists well into adulthood.

Social games

One of the most striking examples of play in birds comes from kea, the alpine parrot of New Zealand. Both adults and juveniles jump spontaneously into aerial updrafts and shoot skyward, for no apparent reason. They toss and catch rocks, sticks, seedpods, bones and even snowballs. This game of catch can be conducted solo, but often attracts one or more partners: during play, kea produce an infectious, warbling 'play call' that invites others in the area to join in on the action.

Social play, especially play-fighting, can also be invited by specific postures and movements. To initiate a wrestling match, a kea may either roll over and expose its belly until another individual takes the bait and pounces, or grab the foot of a nearby playmate. Although such play behaviour may serve no immediate function, it can be a crucial part of developing physical skills such as coordination, balance, muscular strength and agility. Play in birds can also sharpen social skills and other mental capacities, including behavioural flexibility, reaction time, and learning to anticipate potential responses of predators, prey or social rivals.

Spreading information: learning to sing

Birds have wide repertoires of vocalisations, including specialised calls for raising alarms, mobbing predators, and maintaining contact with mates, family members, or other individuals in their social group. But it is their songs – vocalisations aimed at attracting mates, defending a territory, or both, often (but not always) complex in structure, and performed mostly during the breeding season – that are most conspicuous. In temperate-breeding birds, song is produced mostly by males, but in many tropical species both sexes sing, performing elaborate duets to maintain contact while jointly defending their territory.

Natural talent or learned skill?

In many birds, song development is innate: individuals develop the vocalisations typical of their species even if they never hear these sounds during early life. However, vocal learning has evolved in at least three groups of birds – parrots, hummingbirds and oscine songbirds. Just as humans learn to speak by listening and mimicking vocalisations, vocal-learning species of birds begin learning their songs or other vocalisations as juveniles. This means, for example, that in Australian zebra finches and Asian white-rumped munias (*Lonchura striata*) a male that is prevented from hearing the song of conspecifics during his first few months of life will not be able to sing normally as an adult. Zebra finches, munias and many other songbirds are closed-ended learners: song learning is restricted to a sensitive period during early life, after which they are unlikely to change or add to their vocal repertoire. In other birds, including common starlings and many parrots, learning is open-ended: individuals can continue learning new songs and adding to their vocal repertoire throughout life.

► For many birds, like this white-rumped munia (*Lonchura striata*), song learning is usually possible only during early life.

MIMICRY

While each species of bird has its own distinct sound, some species also learn to mimic the sounds of other species. For example, parrots are well known for their ability to mimic the sounds of human speech – indeed, 'to parrot' even means to mimic. Mimicry of other bird species is particularly common in open-ended learners with large vocal repertoires, such as the brown thrasher (*Toxostoma rufum*) and its close relative the northern mockingbird (*Mimus polyglottos*), both of which produce creditable impressions of many of the birds living around them in their North American habitats. In Australia, superb lyrebirds (*Menura novaehollandiae*) are even more accomplished at mimicry, accurately imitating not only the sounds of other birds, but also babies crying, dogs barking, car alarms and camera shutters. Although mimicry is striking, its function remains uncertain. One theory is that when females prefer large and varied repertoires, mimicry provides a way for males to acquire new 'songs'.

Superb lyrebird
(*Menura novaehollandiae*)

Repertoire variations

Bird species vary not only in the timing of song learning, but also in the number of songs that an individual can learn. In South America, male rufous-collared sparrows (*Zonotrichia capensis*) typically produce only a single song type; by contrast, each individual common nightingale (*Luscinia megarhynchos*) can have a repertoire of hundreds of different songs.

Just because young birds can learn songs does not mean that they are blank slates capable of learning the vocalisations of any bird species. Most birds are selective, and are much more likely to learn the song of their own species than that of other species singing in the local area, even in noisy environments. Young males may copy songs primarily from their father, as in Darwin's finches (*Geospiza* spp.), or from other same-species males in the local neighbourhood, as in common chaffinches (*Fringilla coelebs*).

◀ Rufous-collared sparrows (*Zonotrichia capensis*) are among the world's best-studied birds with respect to song learning. Most males produce just a single song type.

Spreading information: what's in a song?

The party game 'broken telephone' involves players whispering a brief story from one person to the next; over the course of many retellings, the message invariably changes in entertaining ways. This pattern illustrates an important feature of vocal learning. Birds, like humans, make the occasional mistake in learning and copying vocalisations from other members of their species. Over time, these mistakes generate geographic variation in vocalisations. Just as humans show regional accents and dialects, often providing clues as to the speaker's place of origin, birds often produce songs that are noticeably different from those in a different geographical area.

Variation within species

Learned birdsong can show gradual and continuous variation over a species' range, as in the red-faced cisticola (*Cisticola erythrops*). In these birds, syllable features change gradually across a distance of more than 6,500 km across sub-Saharan Africa. In other species, geographic variants in song are separated by distinct boundaries. Nuttall's white-crowned sparrows (*Zonotrichia leucophrys nuttalli*) breeding near Point Reyes, California, provide a striking example of this type of variation, termed song dialects. Male sparrows have stronger territorial responses to the songs of males from their own dialect, even when compared to songs from a neighbouring dialect just a few metres away.

In many species, female birds prefer males singing the local dialect. This finding could suggest that males of local origin might be better adapted to the local conditions and are preferred by females because they are better candidates as mates. Female preferences for the bird equivalent of 'local accents' illustrate an interesting tension between the importance of having a mate that is well adapted to the local environment, and the need to avoid inbreeding. Alternatively, females might interpret non-local dialects as poorly learned versions of the local dialect. If so, males that sing a dialect other than the local variant may be discriminated against for this reason.

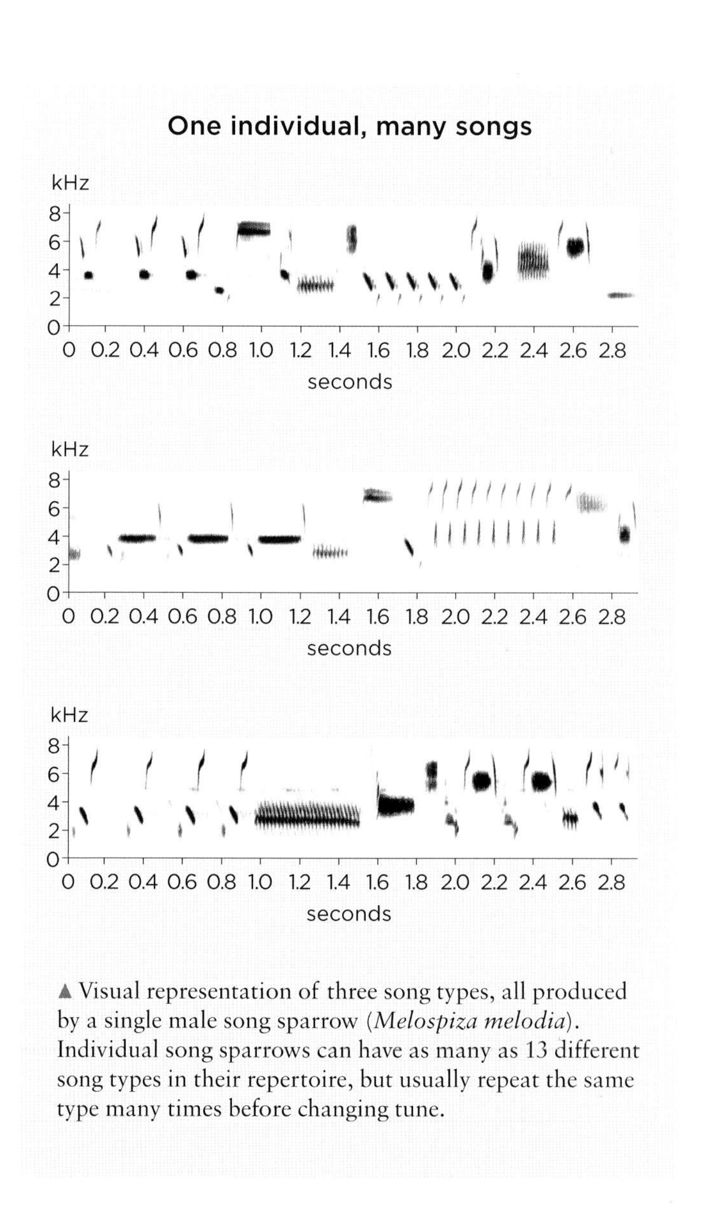

▲ Visual representation of three song types, all produced by a single male song sparrow (*Melospiza melodia*). Individual song sparrows can have as many as 13 different song types in their repertoire, but usually repeat the same type many times before changing tune.

Nuttall's white-crowned sparrow (*Zonotrichia leucophrys nuttalli*)

▲ Nuttall's white-crowned sparrows (*Zonotrichia leucophrys nuttalli*) provide a striking example of song dialects over small geographic scales. Regional differences in song can affect its effectiveness as a territorial and courtship signal, and are readily apparent to human listeners.

Variation among males

The ability to learn different song or syllable types, and learn them well, varies not just between species but also among males within a species. In Old World sedge warblers (*Acrocephalus schoenobaenus*), for example, an individual male may have as few as 55 syllables in his repertoire or as many as 80. Female birds pay attention to this variation, generally preferring males with many songs in their repertoire over those with fewer songs, and males that perform their songs consistently and accurately. Even in North American common grackles (*Quiscalus quiscula*), a species in which each male sings only a single song type, females prefer repertoires that are artificially or experimentally altered to contain multiple songs.

By choosing a mate that is a good song learner, a female bird may benefit by gaining access to high-quality territories or other resources, or by producing high-quality offspring. Exactly how females benefit remains an active area of research, but it is clear that such preferences exist.

Spreading information: innovations and problem solving

Early in the twentieth century, open-topped milk bottles were delivered each morning outside the doors of British households. The cream floated to the top of the bottles, and Eurasian blue tits and European robins (*Erithacus rubecula*) were quick to take advantage of this new food source. After the First World War, delivered milk bottles were capped with waxboard or foil, but homeowners soon noticed that the seals were pierced with small holes: blue tits had learned to pierce them to drink the cream underneath. By the 1950s, this behaviour had spread throughout the UK, suggesting that nearly all British blue tits had learned this innovation.

Learning a skill

In social birds like blue tits, individuals can learn new techniques by watching their flock mates. By contrast, European robins live mostly solitary lives, and although a few individual robins stumbled upon the same innovation to open foil seals on milk bottles, the behaviour did not spread through their population as it did in the more social blue tits.

▼ Eurasian blue tits (*Cyanistes caeruleus*) and other birds have learned to pierce the foil tops of milk bottles, allowing them access to the cream inside.

▲ A great tit (*Parus major*) at an experimental feeding station that requires solving a puzzle for a food reward. Individuals may learn faster from more experienced group members.

An experimental version of the classic milk bottle story has been conducted on great tits in Wytham Woods near Oxford in the UK. Researchers trained a handful of individuals with puzzle boxes: the birds learned to retrieve food rewards from the box by sliding open either a red or a blue door. The trained birds were then released back into the woods, and the population monitored by microchips to identify which individuals interacted with one another. In flocks of birds with trained demonstrators, naive birds were more successful in solving the puzzle compared to flocks without demonstrators. Presumably, this is because copying the tactics of a successful individual is quicker than working out a solution on one's own. When the social networks of the great tits were reconstructed based on the microchip data, the researchers noted that naive individuals that interacted with a trained demonstrator were 12 times more likely to solve the food puzzle than those that did not. These findings demonstrate how rapidly information can spread, particularly in highly connected social groups where each individual is separated from all others by only a few degrees of separation.

Clever crows

New Caledonian crows (*Corvus moneduloides*) provide another striking example of how information travels within a social group. Birds of this species rip up leaves of *Pandanus* plants into distinct shapes and use them as tools to dig for grubs (see also pages 244 and 352). The shape of the leaf tools that the crows construct varies geographically, suggesting that birds may pay attention to other individuals' tool-building techniques and copy them.

Social learning can transmit other information beyond how to get food. In an ingenious experiment in Seattle, USA, researchers wearing caveman masks captured, handled and released a number of American crows (*Corvus brachyrhynchos*). For months afterward, people wearing the caveman masks were harassed by crows, including crows that had not been trapped or handled in the first stage of the experiment. Presumably, the birds learned from their parents or other individuals which humans were best avoided, a cultural tradition that could allow grudges to be passed down through the generations.

◄▲ New Caledonian crows (*Corvus moneduloides*) use leaves and sticks as tools to dig for grubs. Individuals may observe others in their social group to learn which tool designs are most effective.

Birds and their predators

As well as interacting with members of their own species, birds interact every day with members of other species. One interaction of particular importance is predation, when one animal kills and consumes another. Being eaten is a risk for birds at every stage of life, from eggs, through early life and into adulthood. Well known, and major, predators of birds include many mammals; snakes, turtles and other reptiles; and larger avian species. However, some bird species are at risk of being killed and eaten by fish, frogs, and even (in the case of small birds like hummingbirds) large spiders.

Anti-predator defences

Birds have evolved a dazzling array of features that reduce their likelihood of being captured and eaten. Flight offers obvious advantages in escaping predators, but most birds have many other lines of protection in their appearance and behaviour. These include cryptic coloration to assist with blending into the background, as in the winter white plumage of rock ptarmigans (*Lagopus muta*) in the Arctic and subarctic, and the mottled brown feathers of Central and South American common potoos (*Nyctiobius griseus*). Even in birds where males are brightly coloured and conspicuous during the breeding season, females and offspring are generally more muted and camouflaged in their appearance. In songbirds, interestingly, smaller-bodied species are less likely to have bright plumage, perhaps because smaller birds are more vulnerable to predators and being camouflaged is especially critical.

Freezing behaviour is another defence reducing the risk that a bird (or its nest) will be noticed by predators. American bitterns (*Botaurus lentiginosus*)

▼ Ptarmigans (genus *Lagopus*) are well concealed from predators year-round: their plumage is mottled brown in the summer months and snowy white during winter.

▼ The cryptic brown plumage of common potoos (*Nyctiobius griseus*), combined with their tendency to freeze, makes them difficult to distinguish from the stumps or branches on which they sit.

▶ Outdoor cats – both pets and feral – kill billions of birds worldwide each year.

remain absolutely still in the presence of humans or other potential predators, keeping their heads and bills in an upright position that helps them blend unseen into the background of dense reeds. Similarly, in many ground-nesting birds the attending parent remains silent and still on the nest as predators approach, rather than flying away and revealing the vulnerable eggs or nestlings. Ground-nesting shorebirds and ducks may even alter the way they smell during incubation, producing less volatile feather oils to make themselves and their nests less conspicuous to mammalian predators that hunt using their sense of smell. Distracting predators from vulnerable eggs through the use of deceptive broken-wing displays, as in Australian masked lapwings (*Vanellus miles*), or positioning nests in inaccessible sites, as in the Scandinavian cliff nests of thick-billed murres (Brünnich's guillemot, *Uria lomvia*) and other seabirds, are other important defences.

Introduced predators

The anti-predator behaviours described above have been refined over many, many generations and reflect a long shared evolutionary history between birds and their predators. Birds are far more vulnerable to new or introduced predators with which they do not have a shared evolutionary history. This is keenly demonstrated in the devastation of native island birds after human-associated predators such as rats and cats are introduced (see box on page 343). Happily, eradicating such non-native predators from islands can allow native birds to recover. On the British island of Lundy, populations of Atlantic puffins and Manx shearwaters (*Puffinus puffinus*) have dramatically increased since invasive rats were eradicated in 2006. Domestic cats, both feral and pets that are permitted to roam out of doors, are the main predatory threat to most of the world's birds (see pages 344–5).

▼ A female common nighthawk (*Chordeiles minor*) feigns injury to distract a predator and lead it away from her nest.

▼ Bitterns (genus *Botaurus*) have streaky plumage that helps them remain concealed in their reedbed habitat, especially when they freeze in a bill-up posture.

Mutualistic relationships

While being eaten by another animal is clearly harmful, birds also interact with other organisms (including humans) in ways that benefit both parties. These interactions, called mutualisms, can include birds leading a larger animal to a food source that would otherwise be inaccessible to the bird; birds cooperating with members of other species in defending a territory, or in detecting and driving away predators (another key mutualism, between birds and plants, is covered on page 205).

Northern raven
(*Corvus corax*)

Feeding with wolves

In western North America, northern ravens (*Corvus corax*) often feed alongside wolves (*Canis lupus*) at the carcasses of deer (*Odocoileus* sp.) and elk (*Cervus* sp.). At Yellowstone National Park, for example, feeding wolf packs are nearly always accompanied by ravens. This may seem surprising, given that foraging alongside wolves should increase ravens' competition for food or even put them at risk of being eaten. In fact, ravens and wolves appear to have formed a mutualistic relationship. Ravens that come across an unattended carcass produce loud yells that attract the wolves' attention. On their own, ravens are unable to open large carcasses, especially in winter when they may have frozen, but once the wolves begin tearing in, chunks of meat become accessible to the ravens. Ravens may also be taking advantage of wolves' keen sense of smell and ability to identify carcasses that have been poisoned as bait by humans. For their part, wolves likely benefit from the vigilant eyes and ears of ravens as they consume their kill. Ravens have even been observed appearing to play with wolf pups, gently tugging on their tails and encouraging them to pounce.

◄▲ Northern ravens (*Corvus corax*) join a wolf (*Canis lupus*) feeding on a deer carcass. The two species have developed a unique mutualistic relationship, whereby the birds indicate the presence of a carcass to the wolves by calling, and in turn the wolves tear apart the frozen carcass, making the meat inside available to the birds.

AN ANCIENT PARTNERSHIP

Honeyguides, which feed on wild bee colonies, have established a remarkable mutualism with hunter-gatherer people in tropical Africa and Asia. African greater honeyguides (*Indicator indicator*) attract human attention by making loud chattering calls and conspicuous bounding flights, eventually leading the human hunters to the beehive. The honey-hunting people subdue the bees with smoke, open the hive, and remove part of the honeycomb. The honeyguide then feasts on the remaining eggs, larvae and beeswax. Human honey-hunters have also developed specific whistles or calls to invite the birds to join the hunt. This partnership between birds and humans is almost certainly ancient. It was first described in the 1500s, but some experts believe it was practised by our evolutionary ancestors, *Homo erectus*, more than a million years ago!

Greater honeyguide
(*Indicator indicator*)

Birds, parasites and disease

Birds occupy almost every corner of the Earth and many travel long distances during migration, encountering countless different environments. As a consequence, they are exposed to a wide diversity of parasites and other agents of infectious disease. Bird parasites include microscopic organisms like bacteria and other single-celled organisms, fungal spores and viruses, but also multicellular animals like intestinal worms, feather mites and fly larvae. Birds can become infected with parasites through social contact with another infected individual, by eating food or drinking water contaminated by droppings, through airborne spores, or from the bites of insects.

Avian epidemics

Bird parasites, and the infectious diseases they cause, are receiving growing attention both from the perspective of wildlife conservation and from that of human health. Many bird species have undergone severe declines or extinctions due to infectious disease outbreaks, from the devastating effects of malaria on Hawaiian birds, to the impact of avian poxvirus in Darwin's finches.

Disease outbreaks in domestic birds can spread to wild populations with deadly consequences, as when the bacterium *Mycoplasma gallisepticum* (MG) spread from chicken-processing plants in the US in the 1990s to infect house finches. This spillover from domestic to wild birds sparked a catastrophic

▲ The larvae of *Philornus* bot flies parasitise newly hatched birds. They lay their eggs in bird nests, and the larvae that hatch burrow under the nestlings' skin.

◄ A male house finch (*Haemorhous mexicanus*) infected with *Mycoplasma gallisepticum*. Cleaning bird feeders on a regular basis can reduce the spread of this condition.

GENETIC RESISTANCE

The risk of infectious disease has almost certainly had a huge effect in shaping mate choice in birds. Costly displays such as acrobatic dances, energetically demanding flight songs, brightly coloured plumage and skin patches, and well-learned songs are being increasingly studied through the lens of host–parasite interactions. Males that are healthy enough to perform an elaborate display presumably possess good genes that confer resistance to parasites. Such genes, when passed down to offspring, should improve their resistance to disease.

Lack of genetic resistance to avian malaria contributed to drastic reductions in the population size of i'iwi (*Drepanis coccinea*) and other Hawaiian birds.

outbreak of debilitating eye disease, resulting in the loss of an estimated 40 million house finches in just four years. Citizen science efforts were key to tracking the spread of MG across North America, as concerned birdwatchers monitored the numbers of house finches at their garden feeders and how many were affected by MG. Other diseases, such as the H5N1 influenza virus, are concerning not only for their deadly effects on wild waterfowl – including bar-headed geese (*Anser indicus*), brown-headed gulls (*Chroicocephalus brunnicephalus*) and Pallas's gulls (great black-headed gulls, *Ichthyaetus ichthyaetus*) – but also for their ability to spill over to infect domestic birds and humans.

Healthy behaviours

In addition to their immune system (see page 136), birds have evolved many behaviours that may help reduce the risk or severity of being infested by parasites. Behaviours that have been proposed to have an antiparasitic function (largely, although not exclusively, representing defences against external parasites) include scratching; feather preening (either one's own feathers or those of other group members); crushing and smearing ants into the plumage (anting) to take advantage of the formic acid ants contain (see box on pages 84–5); ruffling fine sand or dirt through the feathers (dusting); dust bathing; positioning oneself in direct sunlight (sunning); and sanitising the nest by physically removing ectoparasites or fumigating it with aromatic leaves.

▼ Sunbathing, as demonstrated by this female common blackbird (*Turdus merula*), can help to control ectoparasites such as lice and feather mites.

CHAPTER 9

Reproduction

Most free-living birds have seasonal breeding cycles, with egg-laying and rearing of young restricted to a particular time of year (often spring and early summer) when environmental conditions are optimal for successful reproduction. Among and even within species there is large variation in egg size, clutch size (from one to 15 eggs) and number of breeding attempts each year (from one to five). There is also considerable variation in parental care, including in patterns of incubation behaviour, development of chicks and post-fledging care. Many birds breed first at 12 months of age, the year after they hatch, but a few breed precociously, while others delay sexual maturation for up to ten years.

◀ Gentoo penguin (*Pygoscelis papua*).

Seasonal breeding

In most birds, breeding occurs seasonally and is restricted to a particular time of year when environmental conditions increasingly improve their chances of successfully rearing offspring. In temperate and boreal regions breeding occurs in spring and summer, but tropical regions are less seasonal and breeding can occur year-round. Most birds breed once a year, although a few species breed at shorter or longer intervals. Climate change is increasingly affecting the timing of breeding in some species, leading to a mismatch between the bird's breeding schedule and prey availability.

Cues for breeding

Why do most birds breed seasonally? The 'ultimate' or evolutionary explanation for this is that individuals that time their breeding so that their chicks are in the nest when food is most available will be more successful, producing more, larger and higher-quality offspring, which therefore have higher 'fitness'. However, before chicks can hatch, parents have to undergo a lengthy period of physiological and behavioural preparation for breeding, which can take many weeks. Thus, most birds must initiate the reproductive process well in advance of any seasonal increase in food availability and they therefore need to be able to predict the onset of their breeding season. They do this using information from environmental cues, or 'proximate' factors.

Outside the tropics, day length provides the most reliable proximate cue for seasonal timing, with increasing day length in spring providing long-term or 'initial predictive' information to birds. Increasing day length in spring initiates movements back toward the breeding grounds in migrants, and also initiates seasonal (re)development of the reproductive system. However, spring can be warm and early, or cold and late, in different years. So, females need to fine-tune the actual timing of egg-laying using information from a range of other cues, including temperature, food availability and social interactions. In the tropics, factors such as food availability (e.g. the appearance of fruits) or rainfall are likely involved in the timing of breeding.

The length of the breeding season

Some birds (e.g. quails) have symmetrical breeding seasons, where egg-laying can occur over a prolonged period as long as day length is above a minimum stimulatory level. However, in many species increasing day length not only initiates reproductive development but also initiates a slower process (called photo-refractoriness) that ultimately shuts down reproduction. This is important since, if

◀ Seabirds like this Atlantic puffin (*Fratercula arctica*) time their breeding season to coincide with an increased abundance of their marine prey (here sand eels, *Ammodytes* spp.) to feed their chicks.

laying were to continue until food availability began to decline, young would be in the nest at a time when there was insufficient food to rear them successfully. Many birds therefore have asymmetrical breeding seasons – for example, common starlings (*Sturnus vulgaris*) start breeding in April but stop laying even before the solstice on 20 June.

In general, breeding success declines through the season – in other words, earlier birds do better, producing larger clutches and more fledglings, and recruiting more young to the breeding population. This can be due to effects of time or date per se (the 'date' hypothesis), such as a seasonal decrease in habitat quality or food availability. Alternatively, it may reflect quality differences between individuals (the 'quality' hypothesis) – for example, younger or lower-quality birds with lower reproductive output might breed later than higher-quality birds.

▲ Female common starlings (*Sturnus vulgaris*) initiate laying over a single week, relying on a short seasonal increase in soil larvae to feed their chicks.

► In many tropical species, such as this spotted antbird (*Hylophylax naevioides*), birds can breed in any month of the year.

Breeding cycles

Most birds have annual breeding cycles, but there are some exceptions: sooty terns (*Onychoprion fuscatus*) breed subannually, with a mean period of 9.6 months, and king penguins (*Aptenodytes patagonicus*) breed two out of every three years on average. Breeding every other year (biennial breeding) is seen only in a few large, long-lived seabirds such as the wandering albatross (*Diomedea exulans*) and grey-headed albatross (*Thalassarche chrysostoma*) – although other long-lived species such as common eiders (*Somateria mollissima*) and macaroni penguins (*Eudyptes chrysolophus*) will simply skip breeding in some years.

◀ A male great hornbill (*Buceros bicornis*) feeds its female partner, which is sealed inside the nest.

Timing of breeding

In temperate, boreal and Arctic regions, birds typically breed in late spring and early summer when warmer temperatures support higher plant growth and greater insect abundance. Breeding can last two to three months, although in the Arctic the period of increased insect abundance might last only one month; western sandpipers (*Calidris mauri*) arrive on their Alaskan breeding grounds in mid-May and start migrating south again in late June. Some birds, such as American goldfinch (*Spinus tristis*) and European goldfinch (*Carduelis carduelis*), time their breeding with the late ripening of seeds in summer to feed their chicks. Similarly, Eleonora's falcon (*Falco eleonorae*) starts breeding in late July to take advantage of the abundance of passerines making their autumn migration across the Mediterranean. Emperor penguins (*Aptenodytes forsteri*) start breeding in the austral autumn, when sea-ice starts to form, taking seven months to complete reproduction so that chicks fledge the following summer.

Tropical variations

Some tropical birds can breed at any time of the year, but others are seasonal breeders and many individual birds still have approximate annual cycles of breeding and moult. Fruit production and insect availability are often seasonal in the tropics, but tied to rainfall regimes (wet and dry seasons), and provide proximate cues for timing of breeding. Many fruit eating tropical birds breed and moult during the wet season, coinciding with the highest abundance of fruits. Conversely, in Southeast Asia great hornbills (*Buceros bicornis*) rely on lipid-rich fruits from laurel trees (Lauraceae) and their breeding coincides with production of these fruits during the dry season. In arid regions of Australia, zebra finches (*Taeniopygia guttata*) are opportunistic breeders, initiating breeding only a few days after the onset of rainfall and growth of freshly sprouting grass seeds (see page 221).

Mismatched phenology

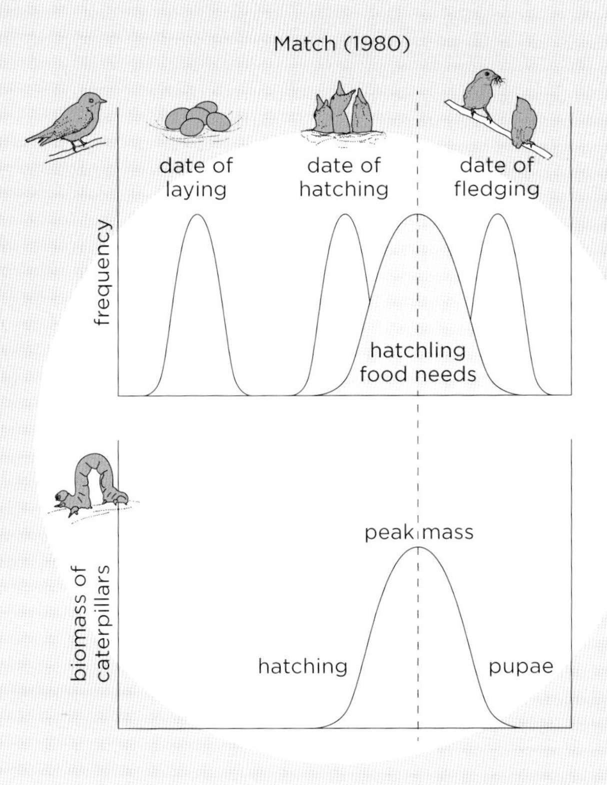

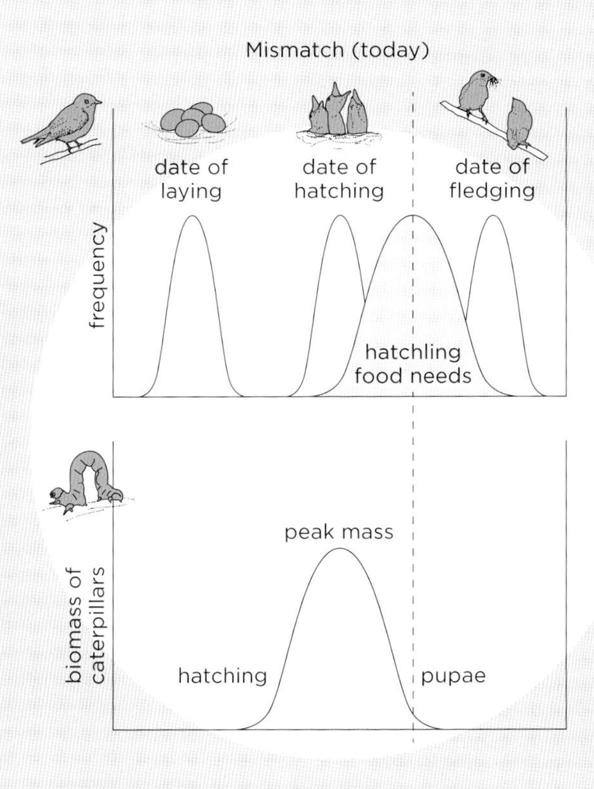

▲ Advances in the timing of spring leaf emergence and insect abundance may lead to mismatched timing between peak food availability and peak demand, when chicks are in the nest and rapidly growing.

THE EFFECTS OF A WARMING WORLD

There is abundant evidence that the majority of birds do successfully time their breeding to coincide with seasonal variation in food availability in most years. However, over the last few decades, at least in some populations, timing of breeding has become increasingly mismatched with prey availability due to climate change. In response to increasing spring temperatures, great tits (*Parus major*) in the Netherlands have advanced their breeding time less than the advance in peak availability of their caterpillar prey, and breeding success is reduced in years with an increased population-level mismatch. In contrast, great tits in the UK have adjusted their timing of breeding in synchrony with the earlier timing of peak caterpillar abundance. Climate change has not led to a uniform increase in temperatures over the entire year, and it seems that unequal increases in temperature can lead to differential timing shifts and different levels of mismatch.

Great tit (*Parus major*)

Territoriality, courtship and mating

In order to maximise reproductive success, males compete for mating opportunities with females, and females need to choose good-quality mates. Males therefore compete to hold breeding territories that can provide resources such as nesting sites or food. In other species, males dispense with territories and compete among themselves on communal arenas or leks, with females choosing to mate with the highest-quality males. Male–male competition is often ritualised, with extravagant plumage displays or complex songs. Ritualised behaviours, coordinated dances and even song duets between males and females help to establish and reinforce the pair bond for breeding.

Establishing a territory and finding a mate

Females need to choose high-quality males, but how do they differentiate between potential mates? Males compete with one another to obtain and defend breeding territories, which provide food resources or nest sites for females. But male–male competition can be dangerous when it involves direct physical conflict, so competition is often ritualised through the production of complex song or plumage displays. The dawn chorus, a medley of song from males of multiple species, is a beautiful result of territoriality. Song, and other displays, act as 'keep out' signs when directed at other males, but they also advertise a male's availability and the quality of his territory to females.

Territory size varies among birds. That of a male European pied flycatcher (*Ficedula hypoleuca*) consists of a very small area around the nest hole, and breeding murres (guillemots) simply defend the small piece of sea cliff on which their egg rests. In contrast, territories of European robins (*Erithacus rubecula*) average 5,000 m^2, with some lowland woodland supporting as many as 200–300 pairs per square kilometre, while larger birds such

▼ The Andean cock-of-the-rock (*Rupicola peruvianus*) forms leks of eight to 14 males, which display to one another while the females look on and then choose which to mate with.

▼ Male bare-throated bellbirds (*Procnias nudicollis*) are considered to have among the loudest songs in the world.

Chaffinch territories in a wood in southern England

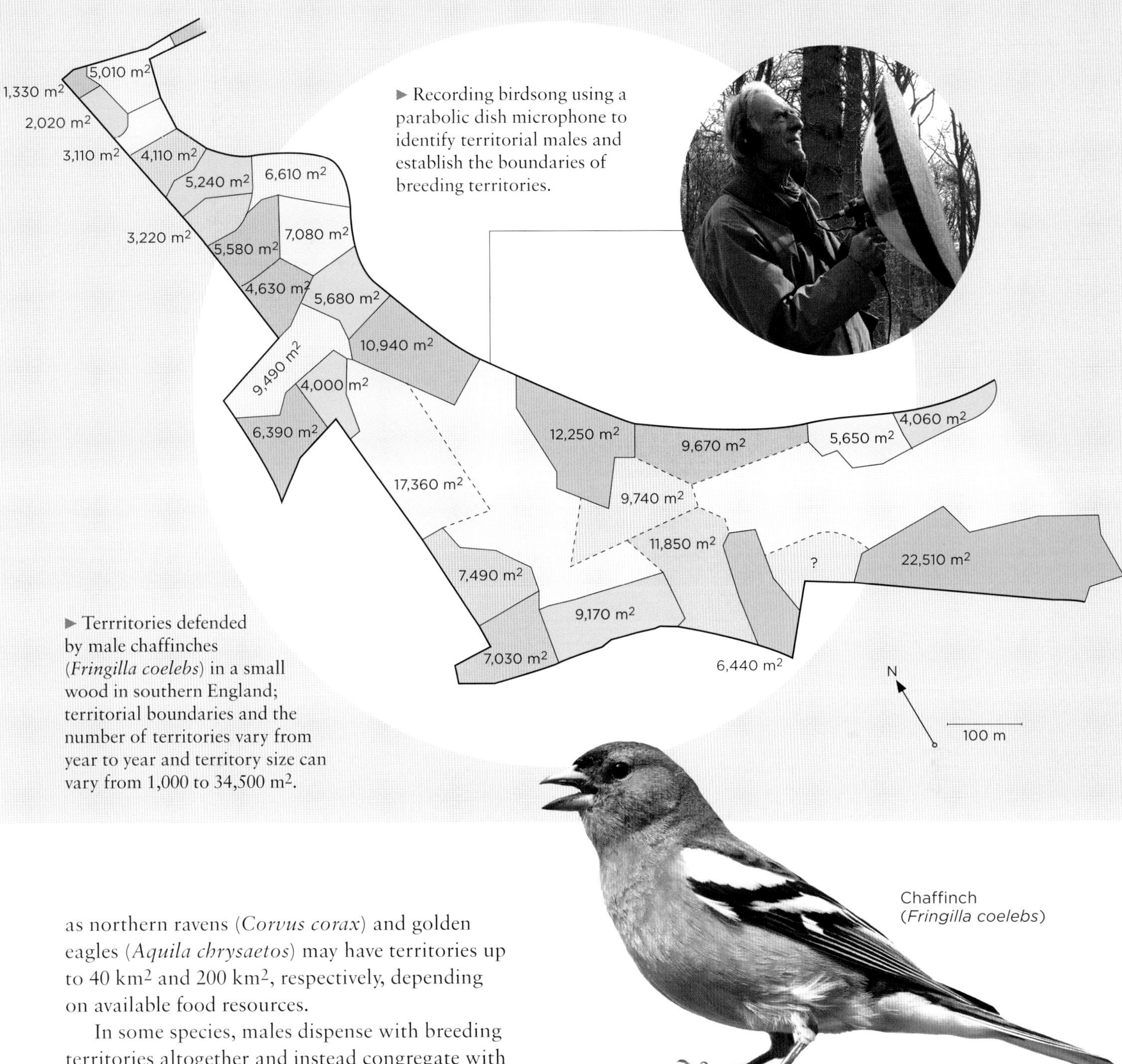

► Recording birdsong using a parabolic dish microphone to identify territorial males and establish the boundaries of breeding territories.

► Terrritories defended by male chaffinches (*Fringilla coelebs*) in a small wood in southern England; territorial boundaries and the number of territories vary from year to year and territory size can vary from 1,000 to 34,500 m².

Chaffinch (*Fringilla coelebs*)

as northern ravens (*Corvus corax*) and golden eagles (*Aquila chrysaetos*) may have territories up to 40 km² and 200 km², respectively, depending on available food resources.

In some species, males dispense with breeding territories altogether and instead congregate with other males on arenas, or leks, to compete for mating opportunities. Females visit leks and choose the best males to mate with. Lekking males provide no resources for females (other than good genes) and after copulation play no subsequent role in incubation or chick rearing. In the Andean cock-of-the-rock (*Rupicola peruvianus*), eight to 14 males form leks, often in steep, forested areas, and they display to each other while females look on. In North America, sage grouse (*Centrocercus urophasianus*) leks comprise 15–20 males on sparsely vegetated sagebrush prairie. Here, the males compete by puffing out their chests, inflating bulbous yellow air sacs to make booming vocalisations, fanning their tails and strutting up and down.

Defensive females

Territoriality, and even singing, is not the exclusive preserve of males: females can be aggressive toward other females, most likely to prevent their partner mating with a second female and the consequent reduction in paternal care. In North and Central America, female red-winged blackbirds (*Agelaius phoeniceus*) direct the same aggressive *cheer* call toward other females that their males use with other males. Female territoriality and singing is especially common in tropical passerines, such as the Central American white-bellied antbird (*Myrmeciza longipes*), where both sexes defend a territory year-round. Some species perform duets, with the male and female singing back and forth, and this also plays a role in mate defence and coordination of breeding activities, such as the timing of nest relief.

Courtship: strengthening the pair bond

Courtship behaviour serves to attract a receptive mate but can also help to coordinate reproduction, reduce aggression between males and females, and ensure compatibility of partners, thereby increasing the chance of successful breeding. Many birds have ritual greeting displays when they reunite on the nest – for example, penguin pairs bow to each other, use 'ecstatic displays', and sometimes offer small pebbles as 'gifts' during changeovers in care. In allopreening, or mutual preening, birds use their bill to preen their partner's feathers, especially on the head; this is more common in species where parents cooperate to rear offspring. In the North American western grebe (*Aechmophorus occidentalis*), pairs engage in elaborate dances with highly coordinated, synchronised movements, the male and female running across the water surface for up to 20 m in perfect unison. Finally, courtship feeding occurs in many species, where the male offers food to the female. This is taken to the extreme in southern yellow-billed hornbills (*Tockus leucomelas*) in southern Africa, where males feed their partner for up to a month before laying, but then the female cements herself inside a nest cavity using her own faeces. Her mate continues to feed her for up to 40 days while she incubates her eggs and broods the newly hatched chicks.

◄ Red-crowned cranes (*Grus japonensis*) engaged in a courtship dance.

► The decorated bower of a male Vogelkop bowerbird (*Amblyornis inornata*).

Monogamy and polygamy

Humans are familiar with marriage, long-term pair bonds and exclusive mating with one partner, but in birds things are much more complex (see Chapter 8)! Around 85 per cent of avian species are socially monogamous, rearing offspring with a single mate, and 15 per cent are polygamous, frequently having more than one mate. However, only 25 per cent of socially monogamous birds are genetically monogamous, where a single male fathers all offspring in a nest. So, mating with a bird other than the social partner (which might then not help in rearing offspring), called extra-pair copulation, is common, increasing male reproductive success while allowing females to shop around for good genes.

In short-lived monogamous species pair bonds can last for just a single breeding season, but in long-lived species many pairs reunite on the breeding grounds each year and in some vultures, eagles, albatrosses, swans and parrots the pair bond can last ten or more years. Nevertheless, separation of breeding pairs is common in longer-lived birds. Both males and females will mate with different partners in different years, especially after breeding failure. In a study of macaroni penguins, for example, 50 per cent of birds bred with the same partner in three successive years but, following mate change, some males failed to mate again for up to four years, even when they retained their nest site. So, few birds actually pair for life, and if a partner dies the remaining bird will try to find a new mate.

Avian architecture: nest-building

Many, although not all, birds build nests to hold eggs during incubation and chicks during their nestling phase. Some nests can be very complex, such as those of African weaver birds, but other birds use simple unlined scrapes or simple depressions in sand or gravel. Most birds use their own single nest, but sociable weavers build compound nests in which more than a hundred breeding pairs contribute to communal nest-building. A few species, including king and emperor penguins, dispense with nests entirely, resting their egg on their feet for incubation, while brood parasites lay their eggs in other birds' nests.

The function of nests

The nests of all birds have the same basic function: they provide a safe location where birds can lay and incubate their eggs and rear their chicks. Nests are often cryptic and well concealed by surrounding vegetation, and many birds place their nest in an inaccessible area such as a cliff ledge to provide protection from predators. However, nests also represent 'extended' components of an individual's phenotype, with complex nests indicating the 'quality' of the bird that built them. The specific construction of a nest, and the materials used, can also modify the microclimate experienced by incubating adults and affect chick growth. Some birds, such as common starlings, selectively place green leaves, flowers and even cigarette butts in their nests, and it has been suggested that these release volatile chemicals that can reduce nest parasites.

Variation in nest structure and size

Open, cup-shaped nests are the most common structure and may be placed on the ground, in trees and shrubs, or 'glued' to cliffs and barns. Open nests vary from rudimentary, unlined depressions or scrapes in gravel or pebbles, typical of many shorebirds, to the woven grassy cup nests typical

▲ Many shorebirds, including the pied oystercatcher (*Haematopus ostralegus*), lay eggs in a rudimentary nest consisting of a simple depression or 'scrape' in the ground.

▲ The nest of this chinstrap penguin (*Pygoscelis antarcticus*) might contain up to 1,700 pebbles, piled up to raise the eggs above any standing water.

of many passerines. They are relatively simple to construct but leave eggs and chick exposed to predators and the elements. Some penguins build nests with stones, occasionally lined with tussock grass, and one medium-sized nest can contain 1,700 small pebbles. American cliff swallows (*Petrochelidon pyrrhonota*) can make around 1,400 trips to collect the mud needed to construct their nests.

A smaller number of species build domed nests with roofs, which provide greater concealment and protection for chicks and eggs. For example, the baya weaver (*Ploceus philippinus*) from Asia uses grasses and palm fronds to weave an intricate hollow ball suspended from a tree branch, and then adds a long, downward-facing entrance tunnel to deter predators. These nests may incorporate more than a thousand strands of grass and take 18 days to complete. Also in Asia, common tailorbirds (*Orthotomus sutorius*) use their bill to stitch together a nest of leaves using grass 'thread' (see photo on p294). Females will inspect these complex nests as part of the process of choosing a mate. This is taken to the extreme in Australia and New Guinea where, rather than making nests, male bowerbirds build large structures called bowers, containing hundreds of often brightly coloured berries, flowers, shells, nuts or other decorative items (see page 291). Bowers serve only to attract females and advertise male quality, and are not subsequently used for incubation – the females go off and build their own nest after mating.

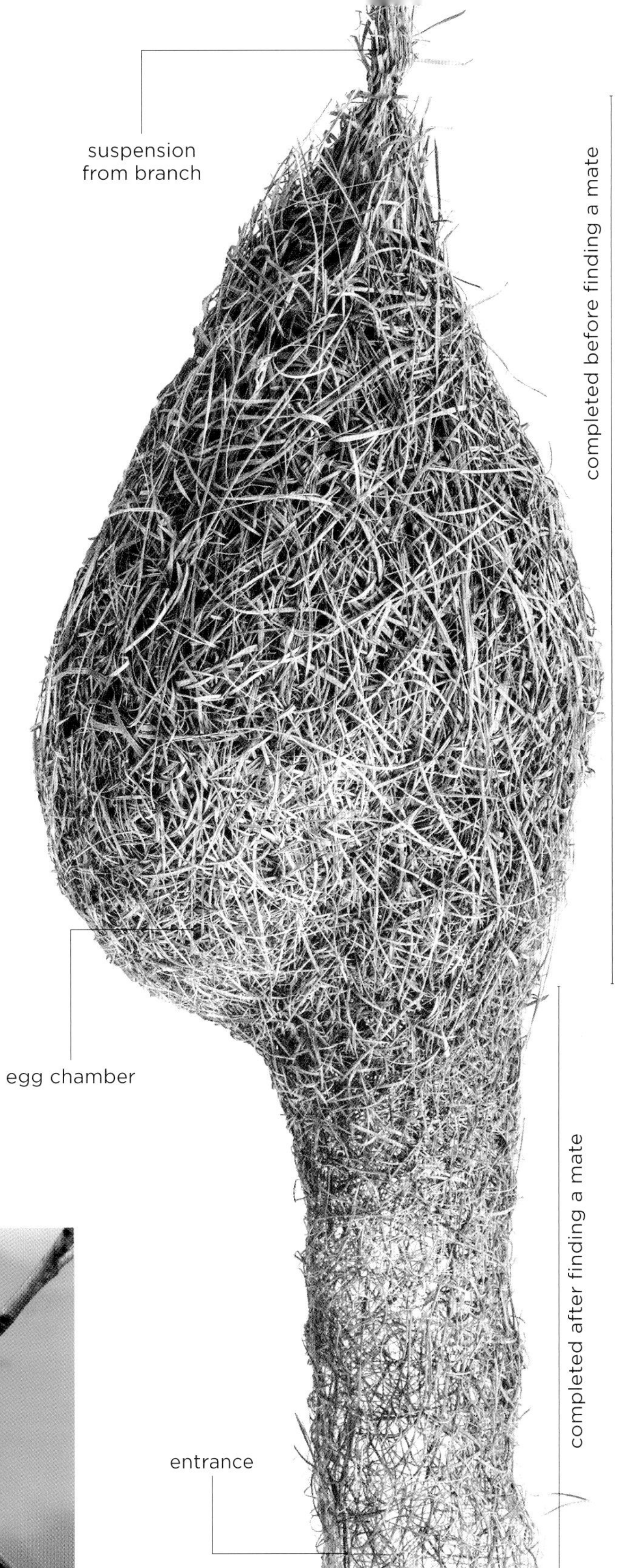

▲ An open, cup-shaped nest made of grass is typical of many passerines, including these black-naped monarchs (*Hypothymis azurea*), exposing chicks to the elements.

▲ Baya weavers (*Ploceus philippinus*) build complex domed nests with a roof and a long downward-facing entrance tunnel to deter predators.

◄ A common tailorbird (*Orthotomus sutorius*) collecting plant fibres, which it will use to stitch together leaves to make its nest.

▲ A black-cheeked woodpecker (*Melanerpes pucherani*) throwing out wood fragments as it excavates its cavity nest.

Cavity-nesters nest in hollows in trees, in underground burrows or in nest boxes provided by humans. Primary cavity-nesters (e.g. woodpeckers) excavate their own nest hole in a tree, whereas secondary cavity nesters, including some passerines, American tree swallows (*Tachycineta bicolor*) and wood ducks (*Aix sponsa*), and raptors (e.g. American kestrel, *Falco sparverius*), will use a cavity made by another bird. Other birds use existing, natural tree holes that they may then expand only a little (e.g. hornbills), excavate burrows in soil themselves, or take over burrows from other animals – for example, American burrowing owls (*Athene cunicularia*) rely on holes in the ground made by ground squirrels, badgers and foxes. Cavity nests are typically lined with grass or similar material.

Megapodes, such as the dusky megapode (*Megapodius freycinet*) in Indonesia, build a third nest type, constructing huge mounds of leaf litter more than 11 m across and 5 m tall. They bury their eggs in the mound, and these are then incubated by the heat generated from the rotting vegetation.

Nest-building

Most small birds build a new nest each year, although some (e.g. swallows) will reuse old nests. Some larger species reuse and refurbish nests over several years, eventually creating 'platform nests': one North American bald eagle (*Haliaeetus leucocephalus*) eyrie measured 3 m across and weighed an estimated 2 tonnes. In many species both males and females will contribute to nest-building, with females often completing the final lining of the nest, but solely female-built nests are less common. Where males primarily complete nest-building, the nest likely also has a social function, being used by males to attract females and by females to choose males. Males will often build multiple nests in a single territory, even though females will use only one, and males that build more nests have greater reproductive success, suggesting they are higher-quality individuals. In Eurasian penduline tits (*Remiz pendulinus*), males that construct larger nests are more successful in acquiring a female. In Africa, hamerkops (*Scopus umbretta*) build massive roofed nest structures in a tree fork, each of which takes eight weeks to complete and incorporates 10,000 twigs – not only that, but the pair build up to four nests a year, working year-round in the process.

ALTERNATIVE STRATEGIES

King and emperor penguins incubate their single egg, and brood their newly hatched chick, on their feet. Not being tied to a nest and hence being able to move around with the egg allows emperor penguins to huddle together – an essential survival mechanism while they incubate through the Antarctic winter. In the southwest Pacific, white terns (*Gygis alba*) also go without a nest, simply placing their egg precariously on a tree branch where it is safe from predators (although still at risk from being dislodged by high winds). Similarly, many cliff-nesting seabirds lay their eggs directly on exposed rock ledges.

A different strategy is employed by brood parasites, such as Old and New World cuckoos and New World cowbirds. These birds lay their eggs in the nests of others, often matching the colour and shape of their host's eggs, and so get the host parents to rear their chicks in place of the host's own (see pages 123 and 316).

Adult white tern (*Gygis alba*) incubating a single egg.

▼ King penguins (*Aptenodytes patagonicus*) dispense with nests and incubate the egg while it rests on their feet.

▲ Sociable weavers (*Philetairus socius*) build giant communal nests that can house up to 100 pairs of birds.

Solitary and colonial nesting

Many birds are solitary nesters, breeding some distance from others of the same species within a territory defended by the pair. This social structure has numerous advantages, such as ensuring access to plenty of food to rear chicks, and less 'interference' from other birds (e.g. through extra-pair mating attempts or intra-specific nest parasitism). In contrast, colonial nesters gather together and nest in colonies varying from a few hundred to many thousands of pairs. This is typical of many seabirds, flamingos, some herons, passerines such as weaver birds, sparrows and corvids. In Africa, sociable weavers (*Philetairus socius*) breed in giant hotel-like compound nests that can house more than a hundred pairs, but the largest penguin colony, on the subantarctic South Sandwich Islands, contains approximately 2 million breeding chinstrap penguins (*Pygoscelis antarcticus*), and up to 4 million sooty shearwaters (*Ardenna grisea*) nest on a single island off South America. Colonial nesting provides increased protection against predation, with more sets of eyes and beaks to detect and deter predators, and might increase efficiency of foraging since birds can follow more successful foragers to their feeding sites. However, this might come at a cost of more competition for food, as a large number of birds forage over a small area near the breeding colony, potentially decreasing food availability.

Keeping the enemy close

Some birds site their nest close to that of other species, including – perhaps surprisingly – potential predators. In the Arctic, snow geese (*Anser caerulescens*) will nest near gull nests, and in Eurasia breeding red-billed choughs (*Pyrrhocorax pyrrhocorax*) associate with the nests of lesser kestrels (*Falco naumanni*). In both cases the birds benefit from the extra vigilance and aggression of

the predator toward other potential nest predators. However, nesting birds have to strike a balance between protection and risk: Eurasian red-breasted geese (*Branta ruficollis*) suffer lower nest predation if they are closer to peregrine falcon (*Falco peregrinus*) nests, but the geese are also harassed by the falcons if they nest too close. In this trade-off, the geese have most success if they site their nest around 40–50 m away from the falcons. A few birds nest in association with insects, likely also obtaining some protection from potential predators. In Southeast Asia, rufous woodpeckers (*Micropternus brachyurus*) build their nest within the nest of acrobat ants (*Crematogaster* spp.), and in Central America rufous-backed wrens (*Campylorhynchus capistratus*) nest close to wasps' nests.

◄ Rufous woodpeckers (*Micropternus brachyurus*) build their nests within the nests of tree ants (*Crematogaster* spp.), which might afford protection against predators.

▲ Many seabirds, such as these northern gannets (*Morus bassanus*), nest in large colonies; colonial nesting might reduce predation risk and increase foraging efficiency.

Eggs and egg-laying

All bird eggs have the same function: they protect the developing embryo and provide all the resources needed for growth to hatching. Despite this common function, bird eggs vary greatly in size, colour and shape across different species. Larger birds lay eggs that are absolutely larger, but relatively smaller as a proportion of body mass, and in general species that lay larger eggs produce fewer eggs – they have smaller clutch sizes. Clutch size varies from a single egg to 15 or more, and birds can lay between one and five clutches each breeding season.

Egg size

Egg size varies among species, within species and even systematically within clutches of the same female. Among living birds, ostriches lay the largest eggs at 1.4 kg (1.4 per cent of their body mass) and the North American Anna's hummingbird (*Calypte anna*) lays the smallest at 0.5 g (or 12.4 per cent of body mass). In the same species, or population, the largest eggs laid by individual females are sometimes twice as large as the smallest eggs laid by other individuals – for example, egg mass in snow geese females range from 86 g to 166 g, a 93 per cent difference.

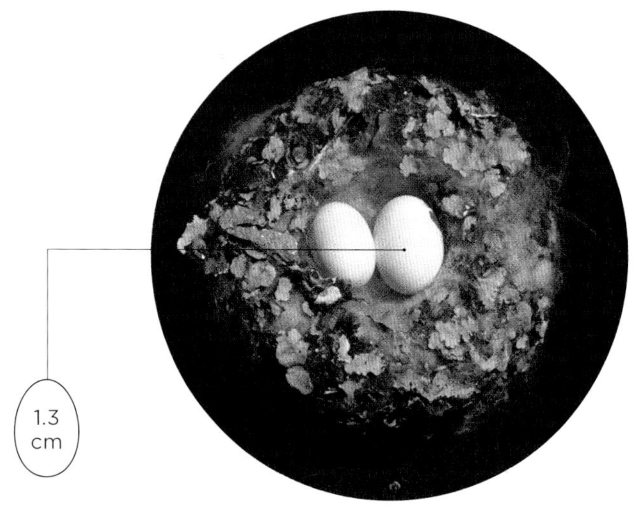

Anna's hummingbird (*Calypte anna*) egg (actual size)

► Egg size varies with body size in birds, but larger birds lay relatively smaller eggs for their body mass.

15cm long

Common ostrich (*Struthio camelus*) egg (actual size)

Variation in egg size within the same clutch is much less (1–3 per cent) but is systematic. Decreased egg size in eggs laid later (e.g. in raptors and egrets) is thought to reflect a brood-reduction strategy, facilitating the rapid sacrifice of late-hatched chicks if food becomes scarce. Conversely, where egg size increases within a clutch this promotes the survival of smaller, later-hatched chicks (brood survival). Egg size varies with age, experience, body mass, temperature and food, but these factors explain surprisingly little overall variation (around 10 per cent). Despite the large variation, the fitness consequences of egg size surprisingly remain poorly understood. For example, even though bigger eggs might be expected to produce 'better' chicks, there is relatively little evidence for strong, positive and especially long-lasting effects of egg size on offspring growth and survival.

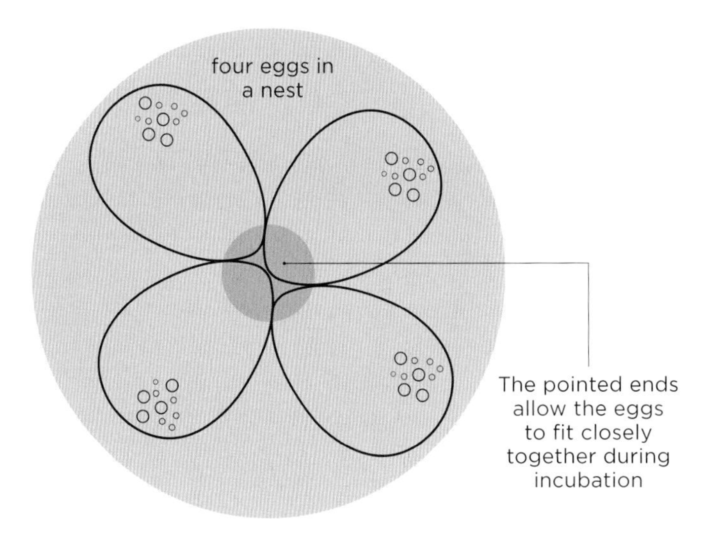

Egg shape

Egg shape varies from spherical (in owls), to elliptical (in megapodes), to pyriform or pointed (in seabirds and shorebirds). However, we still have a surprisingly poor understanding of why such differences in shape have evolved, and widely accepted explanations were not rigorously tested until recently. For example, murres (guillemots) have conical, pointed eggs and it was thought that these would rotate in place like a spinning top when disturbed rather than rolling off the rock ledges where these birds breed. But this does not seem to be the case; instead, this egg shape might confer mechanical strength or minimise the risk of contamination from dirt. Another idea is that the general egg shape, more rounded at one end and more pointed at the other, increases incubation efficiency in relation to clutch size, because eggs fit tightly together under the incubating bird's brood patch (see above). It has even been suggested that egg shape might be related to how much birds fly; good flyers, which have a more streamlined body shape (e.g. sandpipers, murres or guillemots), tend to lay more elongated eggs than species that fly less (e.g. tropical trogons), which lay more spherical eggs.

► Egg shape can vary from elliptical (owls), to pyriform or pointed (seabirds, shorebirds), to elongated (megapodes).

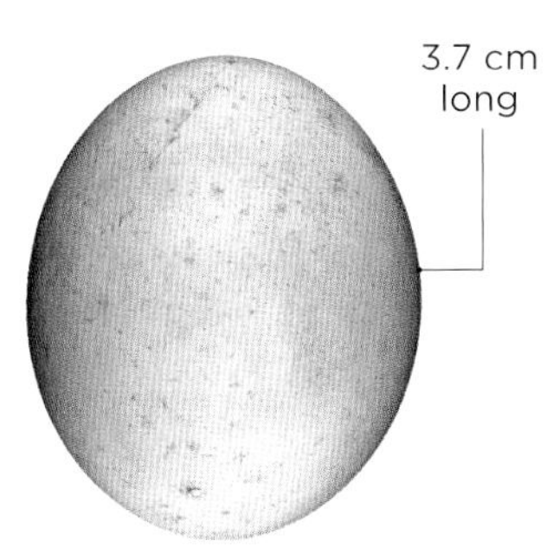

Eurasian pygmy owl (*Glaucidium passerinum*) egg (actual size)

Common murre (*Uria aalge*) egg (actual size)

Megapode (*Megapodius* sp.) egg (actual size)

Egg colour

Eggshell comprises mainly calcium carbonate, so many bird eggs are white, although both colour and patterning can vary enormously. Egg colour comes from two main pigments that are incorporated into the eggshell in the oviduct during the final stages of egg formation: biliverdin, which produces blue and green colours; and protoporphyrins, which produce reddish-brown colours. Cavity-nesting birds tend to have plain white or blue eggs, perhaps because colour is less important in the dark nest cup. In open-nesters, eggs often match the colour and pattern of the nesting surface so that

African jacana (*Actophilornis africanus*) egg (3× actual size)

▼ The huge diversity in bird egg colour as shown by these European species derives from two main pigments: biliverdin (producing blues and greens) and protoporphyrins (producing reddish browns).

Blue tit
size 1.6 cm

Wren
size 1.7 cm

Sedge warbler
size 1.8 cm

Dunnock
size 2.0 cm

Chaffinch
size 2.0 cm

Blackcap
size 2.0 cm

Yellowhammer
size 2.2 cm

House Sparrow
size 2.2 cm

Song thrush
size 2.8 cm

Blackbird
size 2.9 cm

Starling
size 3.0 cm

Jackdaw
size 3.6 cm

Rook
size 3.7 cm

Common Tern
size 3.9 cm

Kestrel
size 4.0 cm

Carrion crow
size 4.3 cm

Lapwing
size 4.6 cm

Shelduck
size 6.6 cm

Herring gull
size 7.0 cm

Great black-backed
gull size 7.7 cm

egg sizes are shown at 80% actual size

they blend into the background and are camouflaged against predators. Eggs of other species can be spotted or maculated (e.g. many passerines and shorebirds), or heavily, and crazily, streaked (e.g. the Australian spotted bowerbird, *Chlamydera maculata*, and the common murre or common guillemot, *Uria aalge*). The functions of this diversity of egg colour remain poorly understood, but pigmentation might signal female 'quality', allow individuals to identify their own eggs in crowded colonies or differentiate them from a brood parasite egg laid in their nest, or even act as a sunblock, protecting the embryo from harmful ultraviolet rays from the sun.

Egg-laying

In many species, egg-laying (oviposition) occurs at a specific time of day soon after sunrise and follows a daily, or circadian, rhythm, with one egg being laid every 24 hours until the clutch is complete. Female American yellow warblers (*Setophaga petechia*) return to their nest within ten minutes of sunrise and complete egg-laying within 30 minutes, whereas common starlings lay most eggs between 08.00 and 10.00. Larger birds, including many ducks, also lay one egg each day, but common eiders lay most of their eggs between 12.00 and 15.00. Other non-passerines lay an egg every two to four days, and North Pacific ancient murrelets (*Synthliboramphus antiquus*) lay a two-egg clutch with a seven-day laying interval. Laying skips or gaps – days in the laying sequence when no egg is laid – are common and occur more frequently toward the end of a laying sequence and in species with large clutches. The significance of such gaps is not understood.

first-laid A-egg

second-laid B-egg

EGG-SIZE DIMORPHISM IN CRESTED PENGUINS

Crested penguins (*Eudyptes* spp.) have extreme egg-size dimorphism, in which the first-laid egg (A-egg) is much smaller than the second-laid egg (B-egg). This is unique among birds and is associated with obligate brood reduction, whereby most birds rear only one chick despite laying a two-egg clutch. In macaroni penguins up to 95 per cent of A-eggs are lost before hatching, so the function of this egg-laying pattern has remained an enigma for decades. Recent work suggests that it might be related to trade-offs between migration and reproduction, with birds having to form eggs while still at sea as they are returning to their breeding colonies.

◄▲ Extreme within-clutch egg-size dimorphism in macaroni penguins (*Eudyptes chrysolophus*); the smaller egg is only about 60 per cent the size of the larger egg.

Clutch size and number of breeding attempts

Some avian taxa have invariant clutch size: many gulls lay three-egg clutches, most shorebirds lay four-egg clutches, most penguins lay two-egg clutches, and many large, long-lived seabirds lay a single-egg clutch. Other species lay larger and more variable clutches. For example, clutch size varies from two to five eggs in North American song sparrows (*Melospiza melodia*), from four to 17 eggs in Eurasian blue tits (*Cyanistes caeruleus*), and from seven to 12 eggs in mallards (*Anas platyrhynchos*). In species where clutch size is fixed, individual females will typically not lay additional eggs when any are lost or removed from the nest during laying, a strategy known as determinant laying. In contrast, indeterminant layers will lay additional eggs if any are removed as they are laid. Zebra finches have a normal clutch size of five eggs, but will lay up to 22 if eggs are removed from the nest daily as they are laid.

Clutch size is thought to have evolved in relation to the number of chicks the parents can successfully raise (called Lack's principle, after ornithologist David Lack, who proposed the original hypothesis in 1954), levels of nest predation, and perhaps incubation patterns and embryo viability. In general, tropical species tend to lay smaller clutches than temperate species, cavity-nesters tend to lay larger clutches than open-nesters, and young birds lay smaller clutches than older, experienced birds. There are also strong seasonal patterns of variation in clutch size in most species. Single-brood species and some multiple-brooded species (e.g. ducks, geese, swans, gamebirds) all show a linear seasonal decline in clutch size – in other words, clutch size is larger earlier in the season. In contrast, the clutch size in other multiple-brood species first increases with laying date and then declines. These patterns of variation reflect trade-offs related to seasonal environmental changes (e.g. food availability, predation) and date-dependent reproductive value of offspring (in general, earlier offspring have higher fitness).

Many birds, especially in Arctic, boreal and temperate regions, produce a single clutch each year primarily because there is only a short breeding window in which seasonal increases in food availability allow chicks to be reared. However, many birds lay replacement clutches if they suffer early breeding failure. In multiple-brooded species there can be considerable variation in the number of breeding attempts made per year. Common starlings can be double-brooded in many parts of their range, with 40 per cent of individuals laying a second clutch, but around 40 per cent of females that initiate a second clutch experience total failure of their second brood. Eurasian black redstart (*Phoenicurus ochruros*) females can fledge up to three broods per season, but on average 40 per cent initiate only a single clutch each year. Song sparrows can lay up to five clutches in a season, but many of these are unsuccessful due to high rates of nest predation. Variation in the number of breeding attempts has been related to female age, laying date of the season's first egg and shorter inter-brood intervals.

▼ Many single-brooded species show linear declines in clutch size with laying date (left), while multiple-brooded species lay smaller clutches earlier and later in the breeding season and larger clutches at intermediate laying dates (right).

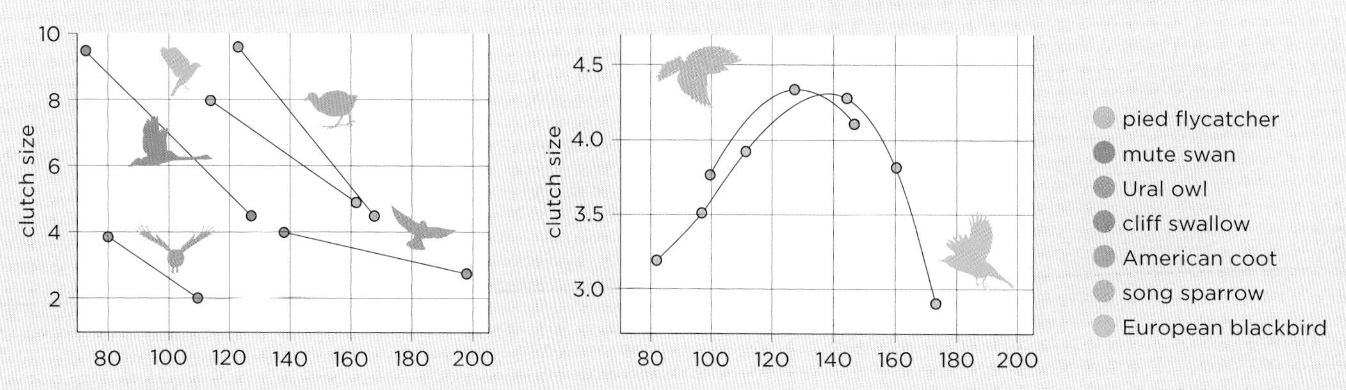

Clutch sizes

Chaffinch
(*Fringilla coelebs*)
typical clutch size 4–5

Eurasian blue tit
(*Cyanistes caeruleus*)
typical clutch size 8–10

Mallard
(*Anas platyrhynchos*)
typical clutch size 9–12

▲► Clutch sizes vary widely among different species (above). Many large seabirds, including this black-browed albatross (*Thalassarche melanophris*), lay only a single-egg clutch.

Parental care: incubation

Since all bird eggs develop outside the female's body, they need to be kept warm (above 35°C) for embryo development to occur. This requires almost constant attention from the parents, with birds sitting on the eggs and transferring heat to them from their own body via a highly vascularised, unfeathered abdominal area called a brood patch. Incubation takes around 11 days in small birds but up to 60–80 days in large seabirds. Males and females share incubation in many species, whereas in others females do most incubation and in a few the responsibility falls entirely on males.

Completing the clutch

After initiating egg-laying, most birds lay one egg a day until the clutch is complete (see page 301). Some birds start incubation before clutch completion, so that the embryos develop at different rates. This is associated with asynchronous hatching, in which some chicks hatch earlier and some later than their siblings. The resulting size hierarchy among chicks in the nest is thought to be a brood-reduction strategy to adjust the number of chicks to the prevailing food availability, with smaller, weaker and later-hatched chicks rapidly starving when there is insufficient food. This occurs in herons, pelicans, raptors and parrots. Other species delay the onset of incubation until the last egg is laid, so that all chicks develop synchronously and hatch at the same time. This is essential in precocial species such as the Nene or Hawaiian goose (*Branta sandvicensis*), where chicks hatch fully feathered and are able to thermoregulate, and where the whole brood leaves the nest together with their parents shortly after hatching.

Incubation periods last 11–14 days in many small and medium-sized passerines, around 35 days in most penguins (although as long as 60 days in the emperor penguin) and 80 days in wandering albatrosses. In general, larger eggs of larger species require a longer incubation period, but there are exceptions – for example, the very small eggs of hummingbirds take 14–23 days to incubate. Fork-tailed storm petrels (*Oceanodroma furcata*) have especially variable incubation periods, lasting from 37 to 68 days, and this is associated with periods of prolonged 'egg neglect', during which the parents do not incubate the eggs for several days at a time.

▼ In the Imperial shag (*Leucocarbo atriceps*), from the coasts of southern South America, incubation begins before laying is complete, and the smaller, later-hatched last chick usually starves to death during its first days of life.

Incubation schedules where both parents take turns

Macaroni penguin (*Eudyptes chrysolophus*): 36 days

11 days

Gentoo penguin (*Pygoscelis papua*): 35 days

2 days

Blue petrel (*Halobaena caerulea*): 49 days

12 days

Shy albatross (*Thalassarche cauta*): 72 days

3 days

female

male

◀ The variation in incubation schedules in two penguins (one with long and one with short incubation bouts) and two petrels (one with long and one with short incubation bouts; the time periods are not to scale).

Variation in incubation schedules

About 50 per cent of all birds undertake continuous incubation. In this, both parents contribute to incubation, coordinating and alternating bouts on the nest, which allows for a high level of nest attendance (greater than 90 per cent). Thus, eggs can be maintained more or less constantly at 36–38°C, above the temperature required for embryo development. Many smaller species alternate frequent, short incubation bouts, so parents can feed often, albeit intermittently. However, incubation bouts can range from every one to three days in some seabirds (e.g. gentoo penguin, *Pygoscelis papua*, and shy albatross, *Thalassarche cauta*) and 10–20 days in others. In macaroni penguins the male and female share a first incubation bout of 11 days, the female completes a second bout of 12 days while the male goes to sea (after a total fast of 35 days), and then the male returns for a third incubation bout of 12 days while the female goes to sea. Such prolonged incubation bouts involve prolonged fasting and cause significant loss of condition – male macaroni penguins lose 30–40 per cent of their initial body mass at this time.

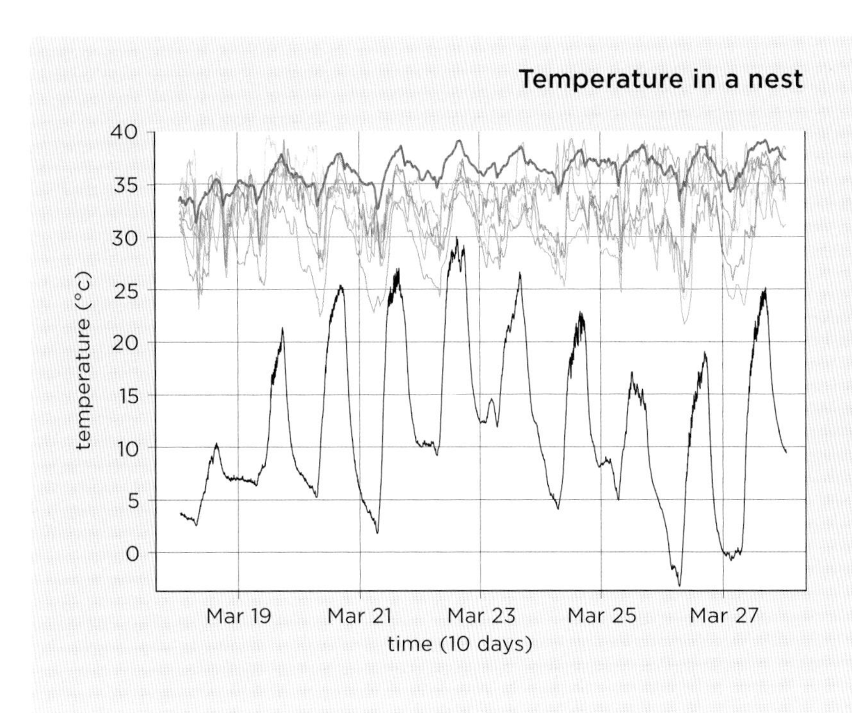

◀ Variation in nest temperature of a wood duck (*Aix sponsa*) nest over 10 days of incubation, recorded using a stationary temperature logger (thick red line) or free-moving artificial eggs containing temperature loggers (all other colours), compared with ambient temperature (black line).

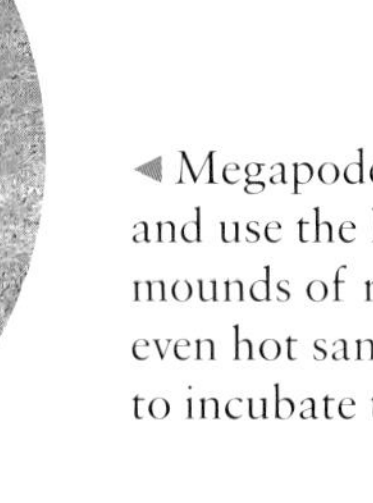

Malleefowl (*Leipoa ocellata*)

Two avian orders (passerines and swifts) have lower levels of nest attendance during incubation (less than 75 per cent), a behaviour commonly associated with female-only incubation. Females have to leave the nest to feed during the day and eggs therefore go through cycles of cooling and rewarming, called intermittent incubation, although incubation is typically continuous at night. There is little evidence that this natural variation in egg temperatures affects hatching success. European pied flycatcher (*Ficedula hypoleuca*) females spend 75 per cent of the day on the nest, in bouts averaging 16 minutes, and 25 per cent off the nest, with periods of feeding averaging five minutes. A few larger species of female-only incubators actually maintain very high levels of nest attendance (greater than 95 per cent), with only very short and infrequent periods off the nest and no male courtship feeding. Female common eiders have a 26-day incubation period and average only 11 minutes off the nest every 50-plus hours.

◀ Megapodes bury their eggs and use the heat generated from mounds of rotting vegetation or even hot sand (geothermal heat) to incubate them.

KEEPING EGGS COOL

A few species living in hot, arid environments face the alternative problem of trying to keep their eggs cool, so that they do not overheat and reach lethal temperatures such that the developing embryo dies. In these cases parents will often stand over their eggs rather than incubating them, providing shade from the intense sun. Egyptian plover (*Pluvianus aegyptius*) and the Namaqua sandgrouse (*Pterocles namaqua*) in Africa soak specialised feathers on their breast in water and carry this back to the nest to cool their eggs.

Namaqua sandgrouse (*Pterocles namaqua*)

Female Canada geese (*Branta canadensis*) spend an average of 98 per cent of their 28-day incubation period on their nest, and can lose up to 30–40 per cent of their body mass due to fasting. In contrast to female-only incubation, the small, Arctic-nesting phalaropes display sex-role reversal, where only males incubate.

▼ In Canada geese (*Branta canadensis*), although only the female incubates, she maintains a very high level of nest attendance and, consequently, incubation is associated with substantial loss of body mass.

Parental care: chick rearing

In precocial species chicks hatch fully feathered, able to thermoregulate and self-feed. Although they can't yet fly, they can leave the nest immediately, with parental care mainly involving defence against predators. In altricial species chicks are born naked, blind and unable to thermoregulate, so parents must initially brood them to keep them warm and provision them with food until they develop sufficiently to fledge or leave the nest. Post-fledging parental care is common and can last longer than the nestling phase. In a few species, including geese, family groups are maintained through migration to the wintering grounds.

Altricial and precocial development

Once chicks hatch from the egg they vary considerably in their development. Altricial chicks hatch at a relatively early stage of development, often blind and naked, and unable to maintain their own body temperature. These chicks must remain in the nest with their parents brooding them, to keep them warm, and feeding them, until they complete growth and development. Once they can fly, they leave the nest, or fledge. Parents of altricial species also defend the nest and young from predators. In contrast, the chicks of precocial species – including many ducks and geese, shorebirds and gamebirds – hatch at a more advanced stage of development, being fully feathered, and able to thermoregulate, feed themselves and walk (although not fly). These chicks leave the nest with their parents shortly after hatching and can move to foraging areas as a family group, guarded by the parents.

Feeding chicks

During chick rearing in altricial species, parents are central place foragers – they fly between foraging areas, where they obtain food, and the nest, where they feed their chicks. Eurasian blue tit parents can make up to 100 nest visits each hour with food, and house sparrows (*Passer domesticus*) make up to 45 visits per hour. However, the rate of feeding visits is highly variable among individuals: some birds make only 20 and five visits per hour, respectively, in these two species and surprisingly this is only weakly related to the number of chicks the parents are feeding. This might be because parents can adjust other components of their foraging effort, such as bringing back different amounts of food or prey of different types and sizes, or changing their foraging distance and travel time to and from the nest.

In larger species such as seabirds, where parents make long-distance foraging trips at sea, chicks are fed once a day or even less often – for example, wandering albatross parents return only every three or four days with food for their chick through the southern winter. When adult penguins return with food, they can recognise their own chicks among thousands in a crèche using unique calls. Feeding chases then often ensue: after initially approaching its chicks, the adult will run off and the chicks will follow in close pursuit. Feeding bouts will occur intermittently as the adult stops and attempts to restrain the chicks by pecking them away!

◄ American robin (*Turdus migratorius*) chicks remain in the nest after hatching and rely completely on their parents to feed them.

► Altricial chicks are born blind and sparsely feathered, but have large yellow beak flanges that help stimulate the parents to feed them.

Chick growth and development

In general, the duration of growth and development of chicks is related to adult body size, with larger species taking longer to mature, and mass or size at fledging predicts subsequent offspring survival and recruitment into the breeding population. Small-bodied passerine chicks stay in the nest for between ten and 21 days after hatching, chicks of larger eagles and parrots can take four to five months to fledge, and wandering albatross chicks take 11 months to fledge. In many species chicks reach adult size at or just before fledging (i.e. they are somatically mature), although recent evidence suggests that chicks are not physiologically mature at fledging. In some species chicks actually lose mass before fledging, perhaps to optimise flight costs in their first few days after leaving the nest. The chicks of passerines (e.g. Eurasian skylarks, *Alauda arvensis*) and some seabirds (e.g. ancient murrelets) depart the nest early, before they can fly, moving to 'safer' habitats in dense vegetation or the open ocean, respectively. There, they complete their development while still being fed by their parents.

Post-fledging parental care

Taking place before offspring become independent of their parents and often lasting longer than the nestling period, post-fledging care is a common behaviour among birds. In great tits parental care continues for 20 days after fledging, while dunnock (*Prunella modularis*) chicks fledge at 12–13 days of age but remain completely dependent on their parents for another 10–20 days. In Australian brown thornbills (*Acanthiza pusilla*) the relative duration of these phases of parental care is even more extreme: the nestling period lasts 14–18 days but post-fledging care continues for up to eight weeks. In large precocial species, especially geese and swans, family groups can be maintained for much longer, with parents and their offspring staying together through migration and on the wintering grounds.

Cooperative breeding

Most birds are socially monogamous, with a single male and female rearing their own chicks (see pages 50–1), but some parents get help raising their offspring through cooperative breeding. To date, this has been recorded in around 9 per cent of all avian species. Cooperative breeding systems are very diverse but typically involve extended family groups. In New Zealand, rifleman (*Acanthisitta chloris*) pairs can have one to three helpers, which may include adults of either sex that hatched in previous breeding seasons, juveniles from first broods that help with rearing in subsequent broods, or even successful breeders that simultaneously help at another nest. In contrast, African superb starlings (*Lamprotornis*

▼ In precocial species like these tundra swans (*Cygnus columbianus*), chicks leave the nest with their parents shortly after hatching and can move to foraging areas as a family group, guarded by the parents.

superbus) are 'plural' cooperative breeders, living in groups of up to 35 birds that defend a joint territory in which up to six individual pairs breed

In most species cooperative breeding is facultative, and the mated pair can successfully rear offspring without helpers. In a few species, however, such as the Australian white-winged chough (*Corcorax melanorhamphos*), cooperative breeding is obligate: successful reproduction is impossible without helpers. So why do helpers help? Helping involves costs (e.g. foraging for food for offspring), but helpers might gain indirect fitness benefits through increased productivity of related offspring (kin selection) and through reduced reproductive costs of their own parents. If nests or breeding territories are in short supply, helpers might have limited opportunities to breed themselves, but they might have a higher probability of inheriting the breeding territory in which they help if one of their parents dies – in other words, there can be a delayed pay-off. In longer-lived birds, where age and breeding experience are important for successful reproduction, helping parents can be one way to gain this experience until an opportunity arises for the helper itself to breed.

► Three superb starlings (*Lamprotornis superbus*) preying upon the nest of a solitary-nesting weaver finch (*Ploceus* sp.).

▼▼ A male vitelline masked weaver (*Ploceus vitellinus*) building its nest.

Alternative breeding strategies

In most species males and females work together to incubate their own eggs and rear their own chicks in their own nest. However, in some species females are nest or brood parasites, laying their eggs in the nests of other females – either from the same species or a different one. After the parasitic female lays her eggs, she plays no further role in parental care, instead getting the host parents to rear her offspring. This alternative female reproductive behaviour leads to an 'arms race', with hosts evolving adaptations to reduce parasitism and parasites evolving adaptations to increase its success.

Intraspecific nest parasitism

Intraspecific nest parasitism, or laying eggs in the nests of birds of the same species, has been reported in more than 230 species, including waterfowl, grouse, rails, estrildid finches, swallows and weaver birds. This behaviour is relatively straightforward because the eggs of the parasite and host female are similar in appearance, thereby decreasing the chance of parasitic eggs being detected and rejected. The main goal for the parasitic female is to add eggs to a clutch before the host female starts incubating, thus ensuring that the parasite chick hatches at the same time as its new nest mates. Parasitic females spend considerable time monitoring and checking potential host nests, quickly sneaking in to lay eggs in unattended nests, although successful parasitism can occur even when birds aggressively defend their nest. Typically, parasites will lay several eggs in the same host nest, but in the common starling parasites lay a single egg per host nest and will remove a host's egg to keep the number of eggs the same.

▼ Intraspecific nest parasitism is common in Eurasian coots (*Fulica atra*), and in some populations as many as half of all nests can be parasitised by other coots.

Females may lay parasitically if nesting opportunities are limited or as a 'salvage' strategy if they lose their own nest early in laying, and in some species (e.g. barnacle geese, *Branta leucopsis*, and common starlings) most parasitism appears to be by non-nesting females. In other species, females with their own nests lay parasitically before initiating their own clutch, which they then care for. Parasitism is therefore associated with an increase in total fecundity – in other words, parasites can potentially rear more chicks by laying additional eggs in other nests. Brood parasitism can reduce host fitness and should select for host defence to avoid parasitism, such as nest guarding and active repulsion of prospecting brood parasites. However, the ability to recognise, discriminate and eject parasitic eggs and chicks appears limited in hosts of intraspecific nest parasites.

▲▲ Canvasback (*Aythya valisineria*) nests are commonly parasitised by female redheads (*A. americana*), but canvasback females can, in turn, lay eggs in the nest of another canvasback.

▲ Nest parasitism in barnacle geese (*Branta leucopsis*) appears to be a facultative, 'best-of-a-bad-job' strategy, with parasitic females rarely having their own nests in the years when they parasitise.

Interspecific nest parasitism

About 100 species of birds are obligate interspecific brood parasites – females never have nests of their own, but instead lay eggs in the nests of individuals of other species, which then care for the parasite's eggs and young. Old World and New World cuckoos and American cowbirds are the most commonly studied avian brood parasites, but this behaviour is seen in honeyguides (Africa and Asia) and indigobirds (Africa), and some ducks. This complex female reproductive strategy has led to the coevolution of a suite of adaptations in both parasites (e.g. egg mimicry) and potential hosts (e.g. egg recognition and ejection), representing an evolutionary 'arms race'.

Parasitic females must find nests and lay their eggs without being detected by the host pair. Females search large areas for potential nests and monitor other birds' egg-laying so that they can lay their eggs at an appropriate time, often in the afternoon, when the host female is less likely to be in attendance. Some obligate brood parasites lay eggs that mimic the shape and colour of the host's eggs, deceiving the host into accepting them. This is an impressive feat even for those brood parasites that are specialists, perhaps targeting only a single host species. However, common cuckoos (*Cuculus canorus*) parasitise a wide range of hosts across their Eurasian breeding range, including warblers, wagtails, buntings and pipits, and can produce eggs that mimic a wide variety of host eggs. Brown-headed cowbirds (*Molothrus ater*) parasitise at least 220 species but do not mimic the eggs of most of their hosts.

Eggs of parasitic species have harder shells than host eggs, to reduce puncturing and rejection, and shorter incubation periods, ensuring the parasitic chick hatches first, gaining an immediate size advantage. In some species, the parasitic chick ejects all the host eggs or newly hatched chicks from the nest, so that it has sole access to food brought to the nest. Cowbird chicks do not oust host eggs and young but, by hatching earlier, the chick rapidly outcompetes its nest mates for food. In bronze cuckoo chicks, even the skin colour mimics that of host nestlings so the host parents cannot differentiate between chicks. Many parasitic chicks have colourful markings on the bill and mouth that mimic those of their nest mates providing a visual stimulus to parents to feed them. Common cuckoo chicks echo the begging calls of a whole brood of reed warblers (*Acrocephalus scirpaceus*) to get triple the amount of food.

Hosts are not completely helpless when it comes to parasitism and have coevolved behaviours to avoid being parasitised. Egg rejection based on egg recognition is a widespread and effective host defence against interspecific brood parasitism, and some pairs will abandon a parasitised nest, since the cost of rearing only a parasitic chick outweighs the cost of relaying. In Australian superb fairywrens (*Malurus cyaneus*), mothers teach their own chicks specific calls or 'passwords' by singing to them while they are still in the egg. If the parasitic chick does not make this call correctly, it is ejected!

► A fully grown, fledged common cuckoo (*Cuculus canorus*) chick still being fed by its much smaller host parent, a spotted flycatcher (*Muscicapa striata*).

Behaviour of a nestling common cuckoo

► A common cuckoo (*Cuculus canorus*) egg laid in the nest of a marsh warbler (*Acrocephalus palustris*).

◄ The parasitic cuckoo chick often hatches first and ejects the host warbler's eggs by pushing them out of the nest.

Age and reproduction

Since many birds have highly seasonal annual breeding cycles, it is common for breeding to commence at around one year of age, or 10–12 months after they hatched. A very small number of species can breed precocially at ages as young as three months. Conversely, some species – including many seabirds – delay breeding for up to 12 years, and generally larger species with lower reproductive rates breed at older ages. Age-dependent patterns of reproduction therefore reflect a trade-off against lifespan, forming a fast–slow continuum. Deferred breeding likely involves prolonged periods of both physiological and behavioural maturation.

Painted bunting, second-year male

Painted bunting, adult male

► Age-dependent variation in plumage in painted buntings (*Passerina ciris*), with an immature second-year male (top right) and a mature breeding male (right).

Most small birds breed for the first time in their first spring after hatching, at about a year old. Nestlings do not mature sexually in the year they are born because they are born in a 'juvenile photorefractory' state. Exposure to short days in autumn activates the hypothalamus–pituitary–gonadal axis, stimulating birds to respond to the appropriate environmental cues for timing of breeding in their first spring (see page 147). Australian zebra finches can show precocial sexual maturation, breeding as young as three months of age, likely an adaptation to arid environments where rainfall occurs sporadically and birds must be ready to breed even if they only recently hatched.

Other species delay breeding for one or more years after hatching, with females typically breeding at younger ages than males. In peregrine falcons females and males start breeding at two to three years and three to four years of age, respectively. Long-lived albatrosses and penguins show long-deferred sexual maturation, breeding for the first time between six and 17 years of age. Delayed onset of breeding has a physiological component (deferred sexual maturation, similar to puberty in mammals) and a subsequent social or behavioural component.

Wandering albatrosses first return to the breeding colony at two to five years of age and first breed between the ages of seven and 17 years. Testis size and reproductive hormone levels increase with age, but male albatrosses are likely physiologically capable of breeding at earlier ages than they do. Delayed breeding is likely due to young males being unable to compete with older, more experienced males. In contrast, female wandering albatrosses can first breed at seven years of age, although only 5 per cent of birds breed by 12 years of age. Young female albatrosses return to the colony after most birds have mated and do not show follicle development or elevated reproductive hormone levels. Even when physiologically mature, breeding occurs in females only after several seasons of group courtship displays and formation of short-duration pair bonds as birds 'practise' the complex behaviours required for breeding.

Most birds that show delayed sexual maturation also show deferred development of adult breeding plumage, displaying a distinctive subadult plumage for one or more years. This might benefit younger birds by minimising agonistic interactions when they cannot effectively compete with adults for breeding opportunities.

◄ Wandering albatrosses (*Diomedea exulans*) show long-deferred sexual maturation, breeding for the first time between the ages of six and 17 years. This individual is probably still several years away from breeding age.

Human Dimensions

Human lives are intertwined with the lives of birds. We eat them, keep them as pets, watch them for leisure, draw inspiration for technology and art from them, and hunt alongside them. Yet, we are also responsible for the most serious threats facing birds. Our relationships with birds are nuanced, and sometimes fraught. Birds capture the human imagination, perhaps more so than any other group of animals, with bird-like deities and symbolism found in almost all religions, and artistic representations of birds dating back to the Stone Age. This inspiration can be ascribed to our frequent encounters with birds in our day-to-day lives, as well as their extravagant displays, spectacular plumage and varied songs. Above all else, however, our inspiration is likely fuelled by our fascination with birds' defiance of gravity. The ability of birds to move between the terrestrial and aerial, and in some cases even aquatic, realms suggested a spiritual significance to many of our ancestors and earned birds a special place in human cultures.

◀ Mongolian huntress with a golden eagle (*Aquila chrysaetos*).

Symbolic and supernatural birds

Birds are prominent figures in the religions, folklore and traditions of cultures from around the globe. They are often symbols of creation, fertility, power, resurrection and more. In some cultures, gods are even depicted as having a bird-like form. This widespread and repeated representation of birds speaks to their spiritual significance in human traditions.

Bald eagle (*Haliaeetus leucocephalus*) on US Department of Justice logo

Creation, ascendance and resurrection

The creation stories of many cultures hinge on supernatural acts of birds or bird-like creatures. In ancient Egyptian mythology, the Bennu bird/god – usually depicted as a heron – created and populated the universe. Diving birds are featured in creation stories of some indigenous North Americans, which tell of birds that dove to the bottom of a primeval sea to bring up mud that formed the Earth.

Birds are also often associated with the ascendance of the spirit and resurrection, perhaps because of their ability to fly, and to shed and regrow their feathers, but also their reappearance in spring, foretelling the beginning of new life. The ancient Greek philosopher and polymath Aristotle thought swallows spent the winter hibernating underground, and then rose up, reborn, in spring. In some Christian writings and art, Indian peacocks (*Pavo cristatus*) are symbols of the resurrection, often depicted alongside images of Christ.

Fertility and birth

According to ancient European folklore, storks bring babies. Aphrodite, the Greek goddess of fertility, and Venus, her Roman counterpart, were attended by doves. Among the Hopi people of southwestern North America, a bluebird feather woven into a prayer stick with strips of yucca was used in a rite of passage to ensure later fertility.

Power

The association of eagles and other raptors with power is ancient. A dozen or more countries claim an eagle as their national symbol, including the golden eagle (*Aquila chrysaetos*) in Mexico and Germany, the African fish eagle (*Haliaeetus vocifer*) in Zambia and Zimbabwe, and the majestic harpy eagle (*Harpia harpyja*) in Panama. A golden eagle was companion to the Greek god Zeus, and eagle feathers are essential components of the traditional dress and customs of many indigenous peoples of the Americas.

Māori legend tells of hōkioi or pouākai, huge, monstrous birds that hunted humans across New Zealand in pre-European times. This folklore was probably inspired by reality; the now extinct Haast's eagle (*Hieraaetus moorei*) was a massive raptor that primarily fed on moa, large, flightless birds that are now also extinct, but likely occasionally took children as prey.

◄ Zeus, enthralled by the beautiful mortal Ganymede, took the form of an eagle and brought him to be cup-bearer to the gods at Mt Olympus (sculpture by Bertel Thorvaldsen, 1817).

BIRD GODS

Horus, a deity depicted as human-like but with the head of a falcon, was one of the most important gods of ancient Egypt. Horus is a sky god, associated with the sun and the moon, and thought to be an important protector of the people of Egypt. Some scholars think that the story of Horus and his parents, Isis and Osiris, influenced early Christian narratives.

The god Quetzalcoatl was likely inspired by a combination of the resplendent quetzal (*Pharomachrus mocinno*) and a large snake, possibly an anaconda. This feathered serpent god is common to the cultures of many of the indigenous peoples of the Americas, and may have originated with the Maya or their ancestors. Among many cultures, Quetzalcoatl is an essential part of creation stories, and a symbol of death and resurrection.

Quetzalcoatl

Resplendent quetzal (*Pharomachrus mocinno*)

Birds as muse

Birds have provided inspiration to artists for thousands of years; we find representations of the flying animals in some of the earliest known examples of art. More recently, birds have inspired, and are featured in, the work of artists across an array of media, from ballet to fashion, and from music to street art.

Birds in art

Some of the earliest known artistic representations of birds can be found in cave art, where they are drawn alongside other animals and components of the natural world. Many of these Palaeolithic engravings and drawings probably reflect the universal human drive to understand the world around us. From the long-eared owl in the cave at Chauvet in France, engraved into stone some 30,000 years ago, to the work of American ornithologist and painter John James Audubon in the early nineteenth century and contemporary works by Canadian naturalist and painter Robert Bateman, artists have long sought to capture the beauty and ecology of birds. More abstract works also capture the essence of birds, including famous works by Dutch graphic artist M. C. (Maurits Cornelis) Escher (e.g. *Bird Fish*, 1938) and Russian painter Wassily Kandinsky (e.g. *Exotic Birds*, 1915), and also amazing murals by street artists!

▼ *Defoliation*, a street mural in Dunedin, New Zealand, by artist DALeast, depicts the extinct Haast's eagle (*Hieraaetus moorei*).

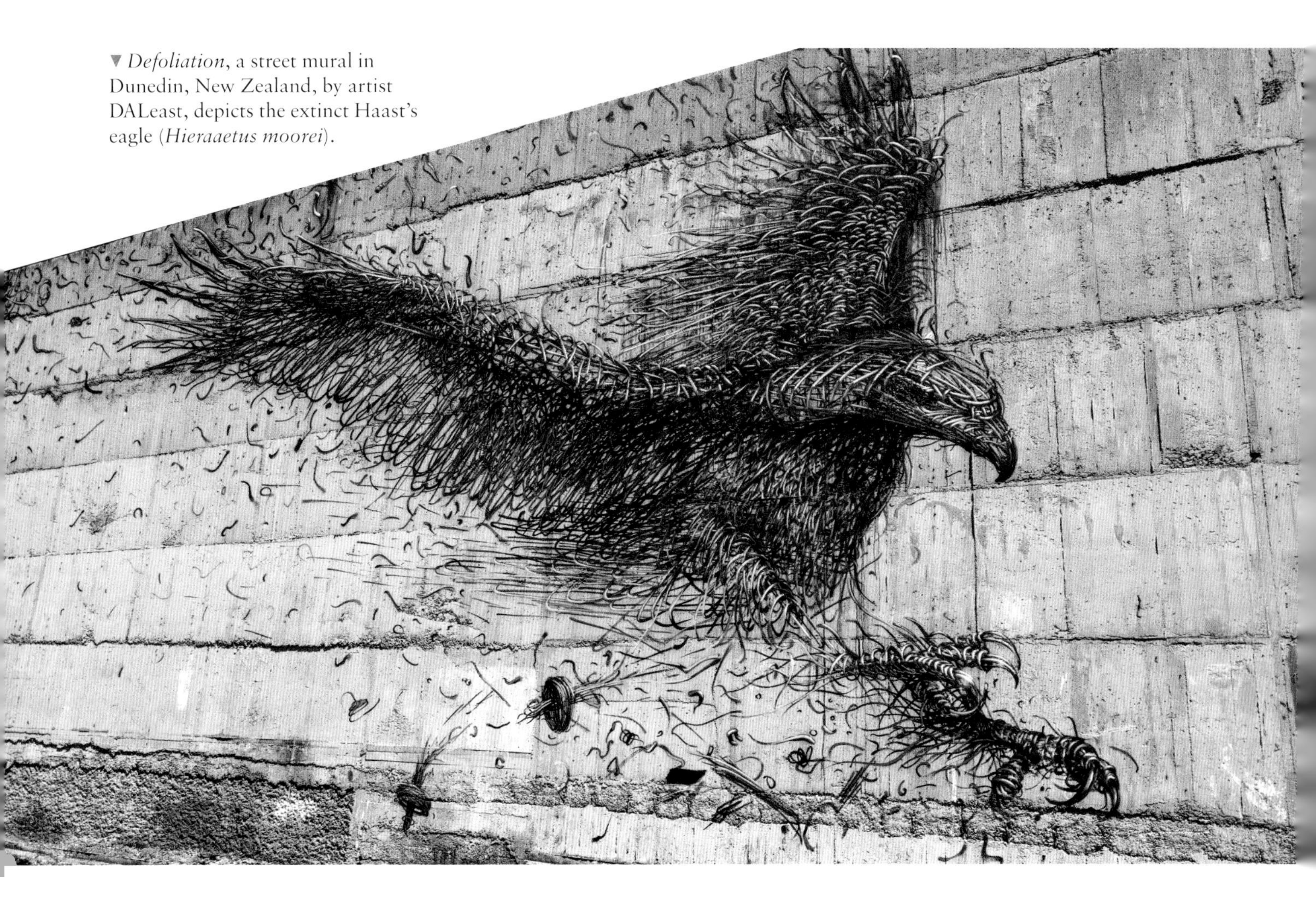

▲ In the early twentieth century, demand for feathers in fashion drove the annual harvest of millions of birds.

▲ Dancers performing the traditional White Heron Dance at the Yasaka Shrine in Kyoto, Japan.

Birds in fashion

The runways of Paris, Milan and New York are often graced with extravagant, bird-inspired designs. These exotic looks might not often make their way to department stores or casual Friday looks, but feathers have been components of high fashion for centuries: the feathered fan, a feather in a cap, feather boas and even full garments made of feathers. However, this focus on feathers has not been without its costs. The plumage of birds-of-paradise from Indonesia and Papua New Guinea have been used in fashion since as far back as the eighth century CE, but demand peaked in the USA and Europe in the early 1900s, leading to the decimation of some populations. Other birds have also been in high demand for fashion items, including the snowy egret (*Egretta thula*) in North America. Some estimate that more than 100,000 of these white birds were killed for their plumage in the name of fashion in a single year alone.

Birds in music and dance

The melodious songs and powerful symbolism of birds can be credited with inspiring musicians, from composer Wolfgang Amadeus Mozart in the eighteenth century, to the Beatles in the 1960s. Mozart had a pet common starling (*Sturnus vulgaris*) for three years, and even held a funeral for it when it died. The starling apparently learned to mimic some of Mozart's music and is thought to have inspired one of his later works, *A Musical Joke* (1787), which some experts think invokes the clumsy and repetitive way that a starling would mimic music.

Traditional dance rituals often invoke bird courtship, including the Shirasagi-no-mai, or White Heron Dance, of Japan, which is thought to have originated as a rite to banish plague and purify spirits as they pass from the world. More modern music and dance are also inspired by birds, from Russian composer Pyotr Tchaikovsky's *Swan Lake* (1875–1876) to the infamous 'Chicken Dance', which originated as a drinking song in Switzerland in the 1950s and was popularised by Canadian band the Emeralds in a 1982 polka adaptation. The version by Henry Hadaway and the Tweets, 'The Birdie Song', reached number two in the UK singles charts in 1981.

Birds and technology

Given the gracefulness of birds in flight, it is unsurprising that scientists and engineers often look to them when designing aircraft. But avian-inspired tech can also be found in some unexpected places, with designers turning to birds for solutions to technological challenges associated with robotics and train design.

Learning to fly

The Renaissance polymath Leonardo da Vinci was fascinated by avian flight, and in his ornithopter he envisioned a contraption that would allow humans to take to the sky like birds. He is quoted as writing, 'A bird is an instrument working according to mathematical law, an instrument which is within the capacity of man to reproduce with all its movements.' The ornithopter design included large, lightweight wings that flap under the power of a foot pedal and hand crank. Unfortunately, da Vinci never built and tested his ornithopter, but in the early 1500s, well after he sketched this flying machine, he wrote the *Codex on the Flight of Birds*, revealing insight into aerodynamics, lift and powered flight that were well ahead of his time.

Early in the design of planes, researchers looked to birds and bats for insight, just as da Vinci had. They soon recognised that a fixed wing with power generated by an engine, rather than flapping flight, would be more stable in the sky. So, planes resemble gliding birds rather than birds that fly by flapping their wings, and they fly just as smoothly. Much of the design of planes was bird-inspired, including the streamlined fuselage, the shape of the wings and even the small winglets found on the trailing edge of the wings.

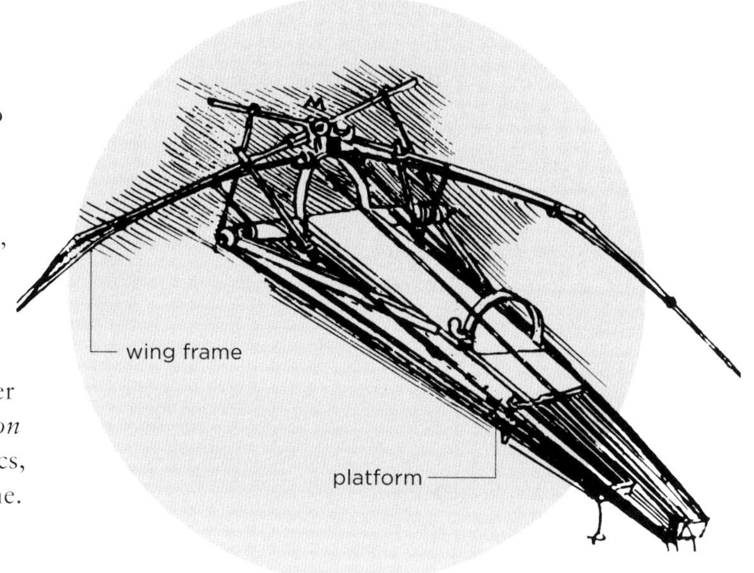

▲ A *c.* 1485 sketch of an ornithopter by Leonardo da Vinci. The wing frame would support a cloth wing that a person, strapped to the platform, would power by flapping.

▼ Winglets on planes (left) increase flight efficiency by reducing drag, and are inspired by the upturned flight feathers of soaring birds like the turkey vulture (*Cathartes aura*, right).

Avian-inspired robotics

Birds have inspired far more than the design of planes. Some birds that cannot fly, like the emu (*Dromaius novaehollandiae*) and common ostrich (*Struthio camelus*), provide guidance for the design of walking robots. These birds are excellent two-legged runners, a skill that is rare in the animal world. They can move quickly over uneven terrain and rarely trip over, whereas humanoid robots are often completely, and humorously, flummoxed by bumpy ground. By looking to running birds, engineers have more successfully designed robots that can handle a bump in the road.

▲ The efficient and well-balanced running of birds like the emu (*Dromaius novaehollandiae*, above) has inspired the design of robots (like the one shown at right), improving their ability to negotiate uneven terrain in comparison to more humanoid robots.

Lessons in aerodynamics

The designers of Japan's Shinkansen faced a problem: their super-fast bullet trains were too noisy. When the trains sped through a tunnel, they would make a loud booming noise because of a build-up of pressure in front of the train. Engineers looked to the streamlined shape of a diving common kingfisher (*Alcedo atthis*), which can plunge into water with barely a splash, to improve the trains' aerodynamics. Shinkansen began to sport a beak-like nose that eliminated the sonic boom problem. Another noise-producing problem was the sound of the pantographs – the cables that connect the train to power lines – which vibrate and make noise as the train moves at speeds in excess of 300 kph. Birds inspired the solution to this problem as well – engineers studied the serrated edges of owl flight feathers, which cut through wind noiselessly, and so added a comb-like structure to the leading edge of the pantographs to cut noise generation dramatically.

▼ The shape of the nose of the Japanese bullet train, or Shinkansen (left), was inspired by the streamlined head and beak of the common kingfisher (*Alcedo atthis*, right).

Birds as food

Birds and their eggs have been part of the hominid diet since before the origin of our species. Palaeontological evidence, and studies of primate teeth and jaws, suggest that humans and our ancestors have long been omnivorous, incorporating meats, fruits and vegetables into our diets for millions of years. Our earliest hominid ancestors likely ate young birds and bird eggs opportunistically, gathering them from nests when they were available – as we see today in several indigenous communities.

Archaeological sites provide evidence of active hunting of wild birds dating back to at least the late Stone Age, about 12000 BCE. Birds probably added much-needed calories and nutrients to diets in parts of the Middle East as early human populations expanded and became more sedentary. Specimens collected at the proto-city of Çatalhöyük in southern Turkey, dating to 7000 BCE, indicate the use of as many as 79 avian species from 23 different families, although waterbirds dominate. The artefacts at this site suggest multiple uses of birds, not just for food, but also for ceremonial costumes and spiritual traditions.

Wild birds in modern diets

The heritage of hunting wild birds continues in many modern societies, and hunting ducks and other waterfowl remains popular across much of the Western world. However, hunting is undergoing demographic changes, with fewer people engaging in this activity now than was once the case. For example, in Canada in 1975 an estimated 380,000 people went duck hunting; by 2017, that number had dropped to about 117,000. People also hunt smaller birds for sport and food, particularly in parts of Europe, where several small songbird species are considered delicacies and are hunted in huge numbers every year – with devastating effects (see page 342).

EDIBLE NESTS

Wild birds are usually hunted for meat and eggs, but some people also eat birds' nests. The edible-nest swiftlet (*Aerodramus fuciphagus*) is found across Southeast Asia and makes its unusual nest out of saliva. A soup made from these constructions is considered a delicacy in several Asian countries. Traditional harvesting occurs on sheer rock faces in the caves where the birds naturally nest, and harvesters often collect nests with eggs or young. More recently, the birds have taken to nesting on the walls of buildings and even inside houses, where bird-nest farmers protect them and are more careful to gather their harvest sustainably, after the nest occupants have fledged.

◄ Nestlings in a nest of the edible-nest swiftlet (*Aerodramus fuciphagus*) attached to a wooden wall.

► A Labrador retriever brings back a duck hunter's quarry – a male mallard (*Anas platyrhynchos*).

SÙLA SGEIR

In parts of Scotland the harvesting of northern gannet (*Morus bassanus*) chicks has been documented since the mid-sixteenth century, but likely has been practised since the Bronze Age or even earlier. This tradition continues today with people from the parish of Ness on the Scottish Isle of Lewis, who harvest 2,000 chicks annually on the island of Sùla Sgeir. The birds are salted and dried, and are then served boiled or roasted, often with a side dish of potatoes. The meat is so fatty that some modern recipes call for repeatedly draining off the oil during cooking – one can imagine how that extra fat probably helped a Bronze Age Gael make it through the winter.

▲ Northern gannet (*Morus bassanus*) adult with a fluffy nestling. These birds nest on North Atlantic islands off the coast of North America and Europe, and their young have been part of a traditional harvest in Scotland for centuries.

Domestication

Exceedingly few of the approximately 10,800 wild bird species have been domesticated. Galliformes (e.g. chicken, turkey, quail) and Anseriformes (ducks and geese) are the groups most of our domesticated birds derive from. These birds are notable for the large number of eggs they lay and also for the precociality of their offspring, which can walk, run and forage independently within hours of hatching. These traits undoubtedly made them appealing for domestication to human populations thousands of years ago.

Chicken

The red junglefowl (*Gallus gallus*), native to tropical Asia, is ancestor to domesticated chickens, but other wild birds likely also contributed to the modern chicken family tree, including grey junglefowl (*G. sonneratii*) and Sri Lankan junglefowl (*G. lafayettii*). Chickens might have been initially domesticated between 6000 BCE and 2000 BCE in the Indus Valley in modern-day Pakistan, and domestic chickens spread rapidly through Asia and Europe.

Bones from an archaeological site in Chile provide some of the earliest evidence for chickens in the Americas. DNA and radiocarbon dating suggest these bones date to around 1000 CE, and by the time the Spanish conquistador Francisco Pizarro arrived in the New World in the 1500s, chickens were already present and well integrated into Incan culture – possibly having been brought to the continent by Polynesians, who may have arrived more than a century before Europeans. Worldwide, people now consume more chickens than any other animal – about 65 billion birds per year – and chickens are by far the most abundant bird, wild or domesticated, on the planet.

▼ A female (left) and male (right) red junglefowl (*Gallus gallus*). Domesticated chickens descended from these tropical birds, which are native to South and Southeast Asia.

Turkey

Turkeys are one of the few domesticated animals that originated in the Americas. Their wild ancestor, the North American wild turkey (*Meleagris gallopavo*), was common, abundant and widespread from Canada to Mexico. The birds were probably domesticated in the southwestern US around 200–500 CE, with selection for their large pectoral muscles, but hunting of wild turkeys was common practice long before then. Turkey bones and feathers were used by indigenous peoples in costume and rituals, and as tools, and this association with people probably contributed to their early domestication.

Ducks and geese

Ducks and geese were domesticated in ancient Egypt by about 3000 BCE, but modern domestic ducks may trace their roots back to a later domestication in China. Modern domestic geese are the product of at least a couple of independent domestication events, of the wild swan goose (*Anser cygnoides*) in China and the greylag goose (*Anser anser*) in ancient Greece or Egypt. Domestic ducks are descendants of the common mallard (*Anas platyrhynchos*) and, like most domesticated animals, bear evidence of a history of intense selective breeding by humans. For example, the domestic Pekin duck is about three times larger than its mallard ancestor and is completely white – traits that are desirable for production of meat and feather down.

◄ A male wild turkey (*Meleagris gallopavo*) in full courtship display, with ruffled body feathers and fanned tail feathers. Domesticated turkeys descended from these North American birds.

UNCOMMON FOODS

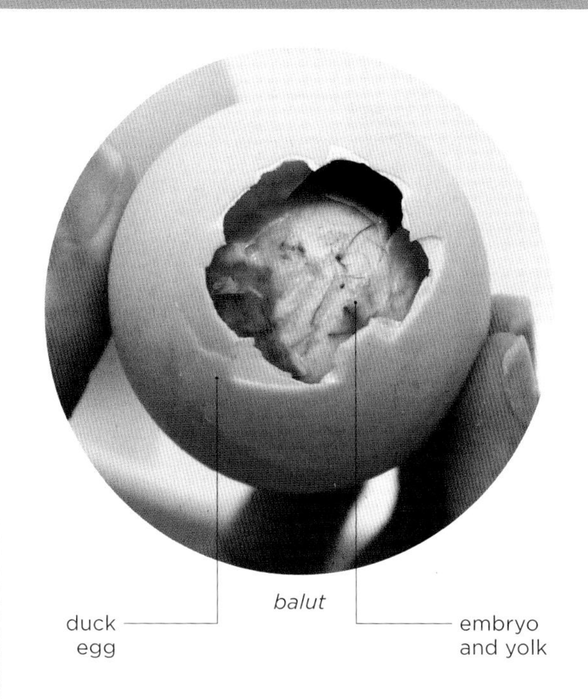

Poultry is a favourite food in many cultures, but *balut* is a poultry dish that many of us have not experienced. In some Southeast Asian cultures, particularly in the Philippines, this is a familiar and relished treat. *Balut* is a fertilised duck or, more rarely, chicken egg that has been allowed to develop long enough to produce a fairly large embryo, but not deplete the yolk. It is boiled and immediately eaten, yolk, embryo and all, directly out of the shell!

Birds as companions and tools

Human societies have long used birds for an array of reasons other than to eat them. People have kept birds as pets for as long as we have written records, and the tradition of using trained birds of prey as tools for hunting wild game, or falconry, likely originated thousands of years ago. Similarly, humans have fished with cormorants in parts of Asia for millennia. All of these ancient traditions continue today.

Among the millions of birds currently kept as pets, the Australian budgie or parakeet, mynas and several finch breeds are favourites. Pet birds are generally not truly domesticated, although several are bred in captivity. Through history, caged birds have also had practical uses – for example, miners famously kept a canary with them, which signalled the presence of dangerous gases when it ceased singing or appeared ill – the origin of the phrase 'canary in a coal mine'.

Hunting with birds

Falconry was practised in Mongolia in about 4000 BCE but it could have originated even earlier. The practice reached Europe by 500 CE, and falconry was a popular sport among the English nobility by the 1300s. Medieval English law even dictated which birds could be owned by the various social classes, from the gyrfalcon (*Falco rusticolus*), reserved for the king, to the common kestrel (*F. tinnunculus*), flown by servants and children. Falconry may also have been present in Aztec culture by 1500.

Today, falconry is practised in many parts of the world. Favoured species vary regionally, and include the gyrfalcon, peregrine falcon (*Falco peregrinus*), saker falcon (*F. cherrug*) and red-tailed hawk (*Buteo jamaicensis*). Highly social and intelligent Harris's hawks (*Parabuteo unicinctus*), native to southwestern North America, are increasingly popular with falconers from around the world. Birds used for falconry are often collected from the wild, usually subject to strict regulation. The captive breeding of birds for falconry began in the twentieth century, and

◄ The highly social Harris's hawk (*Parabuteo unicinctus*) of southwestern North America breeds and hunts cooperatively, contributing to its popularity with falconers.

▼ Atlantic canaries (*Serinus canarius*) and budgerigars (*Melopsittacus undulatus*; known as parakeets in North America) are both commonly kept as pets.

▲ *Ukai* is an ancient tradition and now a popular tourist attraction in Japan. Here, a master fisherman prepares his Japanese cormorants (*Phalacrocorax capillatus*) for an evening of fishing.

falconers produced thousands of birds for release to the wild in the 1970s and 1980s to help with the successful recovery of the peregrine falcon in North America after its populations were decimated (see page 347).

Fishing with birds

Cormorants are exceptional fishers, a fact noticed by people around the world, who have harnessed that skill for thousands of years. Cormorant fishing is most common in Japan, where it is called *ukai* and relies on Japanese cormorants (*Phalacrocorax capillatus*), but has also been practised in China, Europe, and possibly even ancient Egypt and Peru. In modern *ukai*, birds are fitted with a tether attached to a collar that prevents them from swallowing their catch. The cormorant master fishes at night from an open boat with a fire burning at the bow, to attract fish and light the waters ahead. The cormorants swim alongside, and when they come up from a dive with a fish, they are brought on deck so that their catch can be collected.

SPECIAL DELIVERY!

In fiction, birds are often used as messengers – owls in *Harry Potter* and ravens in *Game of Thrones*, for example. In reality, however, the only bird that has been used successfully for deliveries is the humble and familiar pigeon. Homing pigeons are descendants of rock doves (*Columba livia*) that have been bred over centuries and generations to enhance their strong innate drive to return home. This ability is generally limited to one-way delivery: homing pigeons brought to a location within their homing range return home when released. With strategic placement of nesting sites, pigeons have delivered important messages that have influenced world events.

Birds as threats to human health

We share our planet with billions of birds, so it is perhaps unsurprising that some aspects of our interactions with them could be harmful to our health. Some diseases can jump species from birds to humans, and the geographic ranges of some insect-vectored diseases, like Lyme disease, can be altered by migratory birds. On the whole, however, we pose a far greater threat to birds than they pose to us.

Relatively few pathogens infect both birds and people, and even fewer are transmitted directly from birds to people. However, a few diseases that are spread by insect vectors can be dispersed by migrating birds. For example, Lyme disease, caused by *Borellia* bacteria, is transmitted via tick bites, and can infect both humans and birds. If a wild bird contracts Lyme disease in one location and then migrates to another, it can pass *Borrelia* along to naive local ticks, allowing the disease to expand its geographic range.

▼ Working in close quarters with birds, like these chickens at a market in Indonesia, can put people at risk of transmission of diseases, including avian influenza.

Bird-to-human transmission

Diseases readily transmit among crowded poultry and, more rarely, from these birds to people. Avian influenza A (bird flu) presents one of the biggest threats of disease transmission from birds. Although this virus can infect wild birds, domesticated birds are almost exclusively responsible for cases of transmission to humans. These viruses are generally not capable of sustained human-to-human transmission, so outbreaks can be contained as long as the source of infection is identified and controlled. In parts of Asia, some strains of bird flu circulating in domesticated birds are fatal in as many as 60 per cent of human cases. Due to concerns that these virulent strains could evolve to better transmit among humans, they are closely monitored by international health agencies.

THE FLU PANDEMIC OF 1918

The influenza pandemic of 1918 killed more than 50 million people worldwide – far more than were killed by the world war going on at the same time. This deadly pandemic was caused by an H1N1 strain of influenza that owed some of its deadliness to avian influenza. Occasionally, two different strains of influenza infect the same host cell, where they can shuffle their genetic components, generating entirely new viral forms. Evidence suggests that some genetic elements of the 1918 flu were derived from avian strains, but others likely existed in mammals for a long time, and this novel combination set off one of the most lethal disease outbreaks in human history.

Emergency hospital at Camp Funston in Kansas, USA, during the 1918 influenza pandemic.

▲ American herring gulls (*Larus smithsonianus*) congregate on a rooftop in Massachusetts, USA. Their accumulating droppings can host thriving populations of microbes.

Health hazard

Bird droppings, like most excrement, can contain a dizzying array of microbes, several of which are potentially pathogenic. The most serious risk comes from cases where birds congregate in large numbers close to human dwellings. Their accumulated droppings can produce thriving populations of potential pathogens, like the fungi *Histoplasma* and *Cryptococcus*, as well as bacteria like *Salmonella* and *Escherichia coli*. Exposure to these organisms sometimes causes illness, especially in people with weakened immune systems, and can also transfer diseases to farm animals such as dairy cows in modern, intensive cattle lots.

Birds we have lost

Human actions threaten almost every extant avian species. In North America, avian populations decreased by an estimated 3 billion birds between 1970 and 2018 – a loss of almost 30 per cent of the continent's wild birds. The outlook across the globe is similarly bleak. At least 180 avian species have gone extinct since 1500 – species that will never return – and the rate of extinction has accelerated in recent decades.

Island vulnerabilities

Birds, and other flora and fauna, on islands are particularly vulnerable to human-generated, or anthropogenic, threats. These organisms often have limited experience of people and other predators, and so can easily be disturbed and exploited. For example, humans colonised many of the islands of the Pacific from about 3,500 to 700 years ago, and wherever they went, avian extinctions followed. This was likely mainly because of hunting, but also habitat destruction, the introduction of harmful invasive species like rats and the disruption of food chains. New Zealand's giant Haast's eagle may have gone extinct in part because humans drove its primary prey, the flightless moa, to extinction.

▼ The extinction of the New Zealand endemic Haast's eagle (*Hieraaetus moorei*) may have been caused by human overexploitation of its prey, including the South Island giant moa (*Dinornis robustus*).

▲ An illustration of the great auk (*Pinguinus impennis*) from John James Audubon's book *Birds of America*, published shortly before the species was last seen in 1844.

The great auk (*Pinguinus impennis*) was a large flightless seabird that once bred in huge numbers in coastal regions and on islands of the North Atlantic. It stood about 75 cm tall, weighed about 5 kg and was clumsy on land, and so was both appealing and vulnerable to human hunters. From the mid-sixteenth century, fishermen and sailors harvested the birds in enormous numbers for meat and oil. Large operations were set up in the mid-eighteenth century on Funk Island, off of Newfoundland, to harvest the birds for their feathers, which were used for stuffing mattresses and quilts. By 1844, the last known breeding pair was killed on an island off Iceland. Sadly, accounts of this event report that the fishermen who caught these last great auks also crushed their egg – literally stamping out the species.

A tragic demise

Passenger pigeons (*Ectopistes migratorius*) provide one of the most infamous tales of human excess and hubris – these birds were long considered far too prolific to ever be at risk. Indeed, they were once North America's most abundant landbird, numbering in their billions. Huge nomadic flocks of passenger pigeons darkened the skies of eastern and central North America as they travelled among deciduous forests, settling where trees were masting – events where whole forests of trees simultaneously produce a superabundance of nuts such as acorns and beech nuts. The birds nested in colonies of hundreds of thousands to millions of pairs, and their ecology was reliant on these large numbers.

As people intensified their hunting of the birds and disruption of their breeding colonies, which were considered a nuisance, the pigeon populations plummeted from abundant to extinct at a shocking pace. More than 100 million birds bred at a single colony in Wisconsin in 1871, and the species was extinct in the wild by the early 1900s. The last passenger pigeon, a female named Martha, died in the Cincinnati Zoo in 1914. She was more than 25 years old, meaning she had hatched at a time when there were still thousands of her species remaining.

Anthropogenic challenges to birds

A wide array of anthropogenic, or human-induced, challenges impact every ecosystem on our planet and often constitute significant threats to birds. We destroy natural habitat as we convert land for our uses; our activities often generate toxic pollutants and contribute to the ongoing climate crisis; we directly harvest birds, often at unsustainable levels; and our buildings, power lines and, especially, pet cats present significant mortality risks to birds.

A world in crisis

Conservation biologists identify loss of suitable habitat as the top threat facing most populations, not only of birds, but of organisms in general. Our demand for meat represents one of the most significant threats to nature: more than a quarter of the Earth's land area is currently used for raising livestock and the crops to feed them. That represents more than 4 billion hectares, an area as large as all of the Americas combined. In the Amazon, more than 17 million hectares of rainforest were lost between 2001 and 2012, and hundreds of thousands of hectares are cleared every year for livestock production. This is having devastating impacts on hundreds of bird species, including Spix's macaw (*Cyanospitta spixii*), made famous by the 2011 animated movie *Rio* and now thought to be extinct in the wild.

▼ Map illustrating the land area occupied by different natural and anthropogenic landscapes, aggregated by landscape category. Land used for raising livestock occupies an area as large as North and South America combined.

More than half of the world's human population now lives in cities, and over two-thirds of us are projected to be city dwellers by 2050. Cities and towns cover about 150 million hectares – less than 1 per cent of the Earth's land area, an order of magnitude less than is used for livestock production. However, this comparison likely underestimates the impacts of urbanisation. Urban areas can have a disproportionate effect on birds and other organisms, in part because our cities are clustered around areas of productivity, like coastal regions, which also support biodiverse ecosystems. Urbanisation is also expanding in the biodiverse tropics more rapidly than at temperate latitudes.

Most birds do not persist in cities, and, for those that do, urban populations are often what ecologists call 'sinks', failing to produce enough surviving offspring to be self-sustaining. Yet, a few species not only persist but also thrive in cities. These include the peregrine falcon, which has found city skyscrapers to be suitable substitutes for its natural cliff-nesting sites and a bountiful food source in the form of feral domestic pigeons.

▲ Two adult peregrine falcons (*Falco peregrinus*) tend to their nestlings. These birds thrive in cities, in part because buildings resemble their natural cliff-side nest sites.

▼ Spix's macaw (*Cyanospitta spixii*) was declared extinct in the wild in 2000, but several remain in captive breeding programmes, with hopes for their eventual reintroduction.

Escalator to extinction

Altitude (m)
1,400
1,300
1,200
1,100
1,000
900
800
700
600
500
400

1985

2017

Extirpated
Eight ridgetop specialists, including the variable antshrike (*Thamnophilus caerulescens*), were found in 1985 but not in 2017.

Range loss
The ranges of mountain birds like the versicolored barbet (*Eubucco versicolor*) have shrunk. When mountain birds move upslope because of climate change, less space is available and populations decline.

More space
Lower elevation species like the common scale-backed antbird (*Willisornis poecilinotus*) that move up the mountainside can expand their ranges. This species' range was enlarged by 17 per cent.

▲ Birds with mountaintop distributions are vulnerable to climate change because their ability to shift their ranges as temperatures warm is limited. This figure shows the results of a study by scientists from the University of British Columbia and Cornell University, who in 2017 repeated surveys conducted in 1985 on the slopes of Cerro de Pantiacolla, Peru. They found that most species had shifted upslope and several had disappeared altogether.

The climate crisis

The current climate crisis is caused by human activities, primarily our mass release of carbon dioxide and other greenhouse gases into the atmosphere. Climate change affects bird populations by contributing to habitat loss, but also through more complex impacts that can be devastating. Ocean warming has been blamed for the complete collapse of some seabird colonies. These birds rely on abundant schools of fish to feed their young, and those fish rely on tiny, temperature-sensitive plankton. Warming temperatures disrupt the entire food chain, with disastrous effects. On land, mountain birds in tropical regions shift their ranges upward in response to climate change. Often these birds are found in fragmented areas, surrounded by unsuitable habitat and lowland competitors, so they cannot easily disperse from the problems posed by warming temperatures. Their only escape is upward, which is, of course, a solution with limitations – some biologists have termed this response to climate change the 'escalator to extinction'.

Chemical pollutants

Humans are great at finding innovative ways to use chemicals to improve technology, grow crops and treat illnesses. Unfortunately, however, these chemicals can contaminate the air, water and soil, and are sometimes toxic to birds.

Dichloro-diphenyl-trichloroethane, better known as DDT, is one of the most notorious chemicals ever to poison birds. The insecticide had widespread agricultural, public health and household applications. It was first used in the 1940s, and harmful health effects on humans and wildlife were suspected almost from the start. In 1962, the effects of indiscriminate use of pesticides were the focus of Rachel Carson's influential book *Silent Spring*, which called attention to the devastating effects of DDT on bird populations. Among several harmful impacts, DDT caused thinning of eggshells, which was an important factor in the dramatic declines of several raptors, including bald eagles (*Haliaeetus leucocephalus*) and peregrine falcons. Most uses of DDT were banned in the US in 1972, in the UK in 1986 and worldwide in 2004, but the chemical is still produced in India, North Korea and China, and is endorsed by the World Health Organization for limited use to control insect vectors of serious diseases such as malaria. Some countries also continue the use of DDT in agriculture despite the global ban, and even in countries where DDT was banned, it can persist in soils as a 'legacy' pollutant.

One category of 'new' agricultural insecticide, the neonicotinoids, has been receiving increasing scrutiny due to suspected hazards for important insects, such as bees, as well as for birds and other vertebrates. Neonicotinoids are similar to nicotine, and can be similarly neurotoxic, so their uses in agriculture are tightly regulated and even banned in some countries. Research suggests important sublethal effects of exposure to these insecticides for migratory birds. In experiments in Canada, white-crowned sparrows (*Zonotrichia leucophrys*) lost weight, reduced foraging behaviour and altered their migratory behaviour after they were exposed to very small doses of neonicotinoids. Given how critical fat stores, foraging, navigation and timing are to successful migration, these sublethal effects could reduce survival and contribute to population declines.

VULTURES AT RISK

Eleven of the 16 species of Old World vultures found across Asia, Africa and Europe are at risk, with eight classified on the International Union for Conservation of Nature (IUCN) Red List as Critically Endangered, on the brink of extinction. The causes of these dramatic declines are unquestionably anthropogenic. Across Africa, vultures are often poisoned through lacing of carcasses, sometimes by poachers, who kill the birds to prevent them from alerting rangers to the locations of illegal kills, and also by people who target them for use of their body parts in traditional medicine. In South Asia, as much as 99 per cent of the subcontinent's vulture population was wiped out in the 1990s and early 2000s due to use of diclofenac, an anti-inflammatory drug, in livestock. After this cause of the precipitous population declines was finally identified in 2003, India, Pakistan and Nepal banned the use of diclofenac by 2006, although its continued illegal use still kills vultures today. Unfortunately, the drug has been approved for veterinary use in Europe since 1993, where it could threaten several populations of Old World vultures, including the magnificent Griffon vulture (*Gyps fulvus*).

▶ While populations of the Griffon vulture (*Gyps fulvus*) in western Europe have recently increased, it is declining in other regions, in part because of poisoning and persecution by people.

Pet trade

Every year, birds are taken from the wild in their millions for sale as pets. Many wild birds collected for the pet trade do not survive their transit to market, so for each bird sold as a pet, several are collected. Demand for African grey parrots (*Psittacus erithacus*) has driven wild populations of this intelligent, long-lived bird towards the brink of extinction. Similarly, fewer than 300 red-fronted macaws (*Ara rubrogenys*) remained in the wild by 2018, with illegal capture for the pet trade being a primary threat facing the tiny population hanging on in Bolivia. Some agencies and organisations are trying to combat this threat. In 2019, Interpol and the World Customs Organization seized more than 4,000 wild birds and made numerous arrests as part of Operation Thunderball, which targeted large networks that coordinate the illegal trade of exotic animals.

Overharvest

Humans have hunted birds for thousands of years, but as our populations increase and bird populations shrink, many hunts are no longer sustainable. As much as 20 per cent of the estimated 5 billion birds that migrate through Europe every year are killed in hunts fuelled by demand for wild birds on the menu – in Italy, for example, polenta with grilled songbirds is a favourite. In the 2010s, tens of millions of songbirds were killed in Cyprus alone, many to be served in the traditional dish *ambelopoulia*, comprising grilled, fried, pickled or boiled songbirds. Although these dishes and the hunts are banned, black-market demand remains high, as does the pressure on bird populations.

Introduced and invasive species

As human populations move around the globe, we also shuffle wildlife, introducing rats, snakes, birds and other animals to places they never previously existed. These invasive, introduced species successfully establish and can increase to the extent that they harm local wildlife. Invasive animals threaten birds around the world, with island populations being particularly vulnerable. One of the most notorious examples is the venomous brown tree snake (*Boigus irregularis*), which was accidentally introduced in the 1940s to Guam, an island that previously had only one blind, subterranean snake species. Birds and other animals on the island were naive to the threat posed by the tree snake, which has driven almost every native forest bird species on the island to extinction. We see similar devastating effects of invasive species on bird populations on other islands, including Hawaii and New Zealand, and these effects will spread as the rate of new species introductions increases.

▼ Palm cockatoos (*Probosciger aterrimus*) seized from smugglers in Indonesia, who would likely have sold them into the pet trade.

▲ A Eurasian blackcap (*Sylvia atricapilla*) caught illegally in Cyprus on a stick coated with adhesive, destined to be served in the traditional dish *ambelopoulia*.

▼ New Zealand's kākāpō (*Strigops habroptila*) persist only on small islands where the introduced mammals that drove them to near extinction have been eradicated.

NEW ZEALAND'S KĀKĀPŌ

Weighing up to 4 kg, the charismatic kākāpō (*Strigops habroptila*) is the heaviest parrot in the world. It is also flightless and nocturnal, and has an unusual life history. The birds don't breed until they are six to ten years old, and they can live into their 60s – and possibly much longer. This forest-dwelling species was once common throughout New Zealand, but introduced mammals such as cats and rats have decimated its populations. There were only about 50 kākāpō remaining in the 1990s, but through intensive recovery and management efforts, including the eradication of invasive mammals from some islands, the population had slowly recovered to more than 200 birds by 2019.

◀ Domestic cats kill more wild birds annually than any other human-related threat. Luckily, this threat has a simple solution – keeping our cats indoors.

Collisions

Billions of birds die every year due to collisions with our vehicles and human-made structures, including windows in buildings and homes, communication towers, power lines and wind turbines. Combined, collisions are estimated to be the second-biggest anthropogenic source of avian mortality, possibly killing a billion or more birds each year in the US and Canada alone, and collisions with buildings reportedly caused almost a third of deaths among resident birds in Singapore over a four-year period. During migration, millions of birds can die in a single night due to collisions with buildings, especially illuminated glass-covered skyscrapers in our cities. A lot can be done to reduce the number of birds killed in these collisions, such as the use of patterned or non-reflective glass in skyscrapers (see page 211), reducing night-time lighting around buildings and communication towers, altering the design of wind turbines and shutting them down on important nights during bird migration, and erecting fencing along roadways.

Enemy number one: cats

Pet and feral domestic cats are the biggest direct anthropogenic source of mortality for birds. More than 2.5 billion birds are killed by cats every year in the US and Canada alone, and a 2017 study estimated that a million native Australian birds are killed across the country by cats every day. A 2016 study estimated that cats have directly contributed to the extinction of 40 avian species since 1500 CE, and currently threaten another 186 species, which consequently have IUCN Red List classifications ranging from Vulnerable to Critically Endangered.

In one frequently reported example of feline predation, a lighthouse keeper's pet cat named Tibbles is credited with having driven a rare flightless songbird to extinction in the 1890s. Lyall's wren (*Traversia lyalli*) was once found on Stephens Island (Takapourewa), off the northeastern tip of New Zealand's South Island. The island was uninhabited until lighthouse keepers and their families arrived in the early 1890s. Every known specimen of the bird was brought to the lighthouse keeper, David Lyall, by his cat. Lyall prepared the gifts that Tibbles brought, when they weren't too badly damaged, to be preserved as natural history specimens – at least 15 of them – and he sent them to several ornithologists. Within about a year of the arrival of Tibbles, and right around the same time that ornithologists recognised the bird as a previously undescribed species, Lyall's wren no longer existed.

Birds worldwide continue to face the threat of cats, but the good news is that this enormous source of mortality has a relatively simple solution if people are willing to help: keeping our pet cats indoors and controlling feral populations.

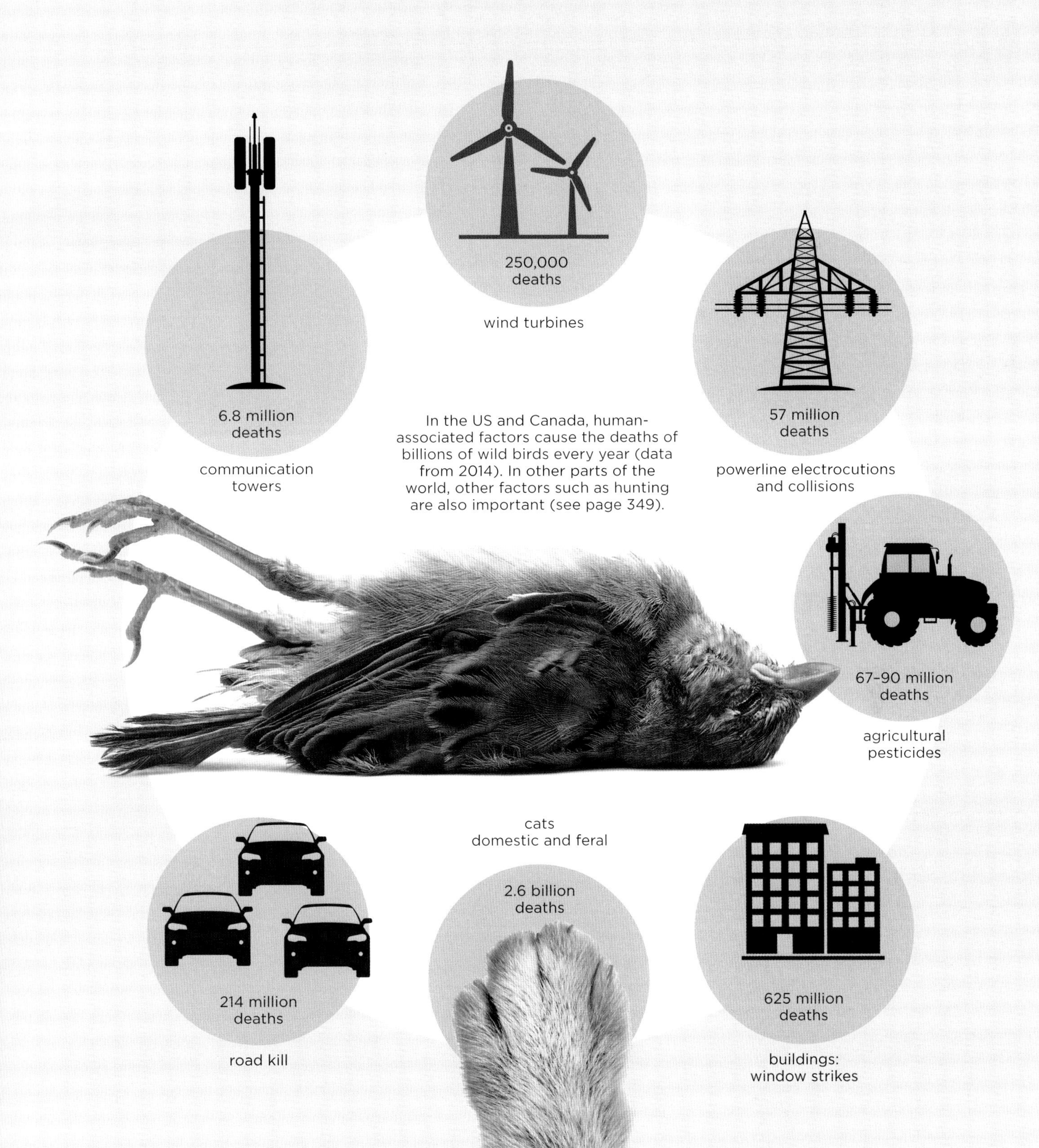
Anthropogenic causes of bird mortality annually in North America
250,000
deaths
wind turbines
6.8 million
deaths
communication
towers
In the US and Canada, human-associated factors cause the deaths of billions of wild birds every year (data from 2014). In other parts of the world, other factors such as hunting are also important (see page 349).
57 million
deaths
powerline electrocutions
and collisions
67–90 million
deaths
agricultural
pesticides
cats
domestic and feral
2.6 billion
deaths
214 million
deaths
road kill
625 million
deaths
buildings:
window strikes

Conservation of birds: necessary steps

Conservation requires information: we have to be able to detect populations in decline, determine why they are declining and devise strategies to mitigate the causes of decline. Fortunately, birds are among the best-studied organisms on the planet, so we are perhaps better equipped to protect their populations.

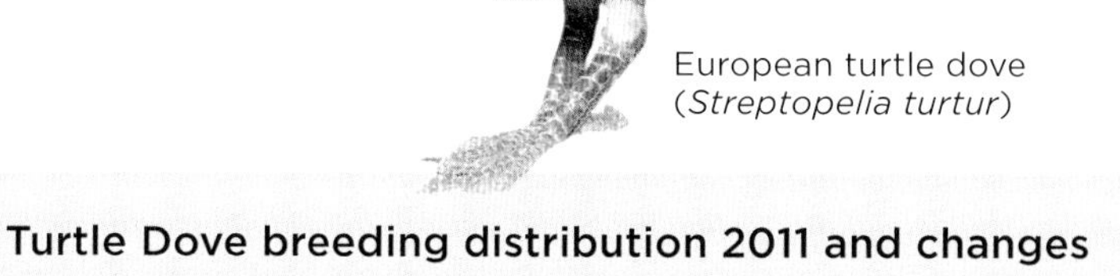

European turtle dove (*Streptopelia turtur*)

We need baseline data on the abundance of birds to be able to detect declines. Naturalists and birders have been recording their observations of birds for centuries, and several organised citizen science efforts also document bird abundance, like the North American Breeding Bird Survey and Christmas Bird Count, and the European Breeding Bird Atlas. More recently, similar efforts have expanded to many areas across the globe, including tropical regions.

Identifying decline

When a population is in decline, we know that either fewer birds are being born, more are leaving, fewer are entering and/or more are dying in the population, relative to historical rates. Determining which is actually happening can be surprisingly difficult, because it requires both pre-decline and current data on these demographic parameters – data that can be extremely hard to come by! Long-term studies of box-nesting birds have provided some of the most detailed datasets on population demography, and have been used to identify mechanisms of population decline in a variety of species, including the European pied flycatcher (*Ficedula hypoleuca*) and North American tree swallow (*Tachycineta bicolor*). These well-studied species can provide insight into mechanisms of decline in related birds, where similar data are lacking.

Turtle Dove breeding distribution 2011 and changes

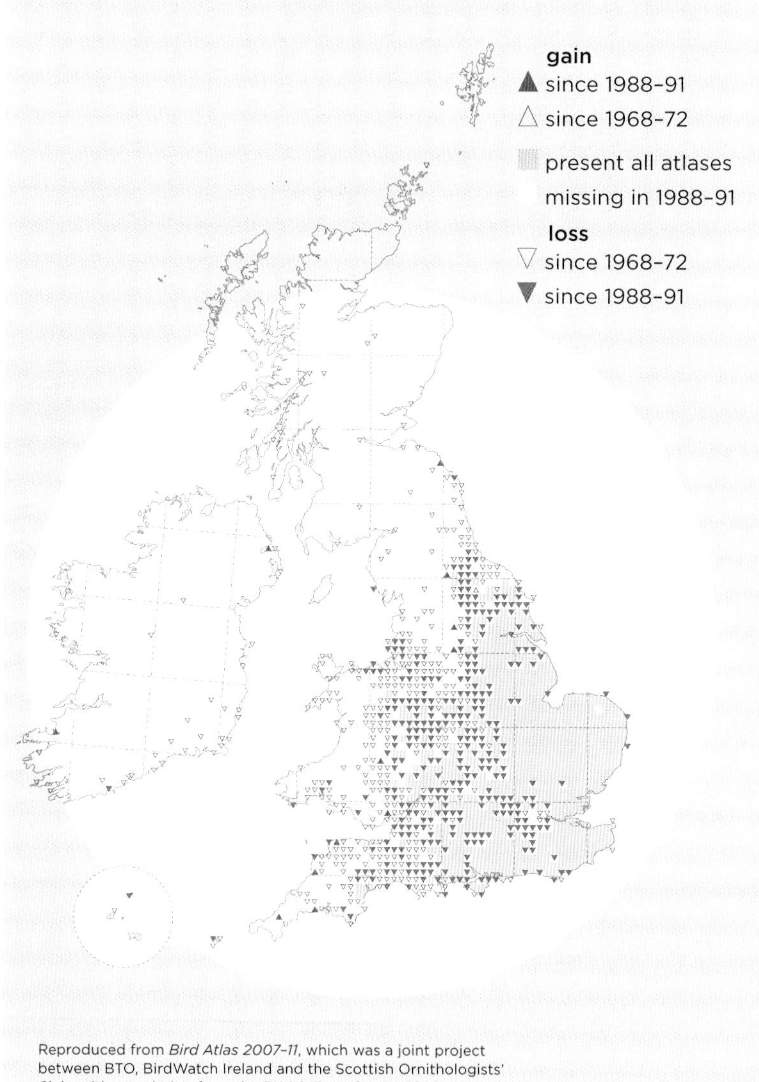

Reproduced from *Bird Atlas 2007–11*, which was a joint project between BTO, BirdWatch Ireland and the Scottish Ornithologists' Club, with permission from the British Trust for Ornithology.

▶ The decline in the breeding distribution of the European turtle dove (*Streptopelia turtur*) in Britain and Ireland, 1972–2011.

DDT AND NATURAL HISTORY MUSEUMS

Natural history museums house far more than specimens for display; they also curate extensive collections that can provide essential historical information about populations. In one classic example, more than 1,700 eggs from museum and private collections allowed biologists to discover that eggshells of several raptors had become thinner by the 1960s. Populations with eggs that had thinned the most were in severe decline, including bald eagles and peregrine falcons. This finding explained the demographic changes that had been noted in these populations - their eggs often weren't hatching. Other experimental work identified DDT toxicity as the culprit, with the subsequent ban of the chemical being critical for the successful recovery of these species.

Identifying the causes

Once we know a population is declining, and we know which demographic factor is causing the decline, the real mystery begins. We have to identify the factors that are causing the shifting demography: habitat loss? Invasive species? Pollutants? Climate change? Often, declines are driven by combinations of several factors. With ecological detective work, drawing from historical records, contemporary fieldwork and experiments, these mysteries can be solved.

If we can determine why a particular population is declining, we often have the tools to stop the decline. Some conservation mechanisms rely on legislation and enforcement, such as bans on hunting and trade. Others require active management, such as programmes eradicating invasive species. The most important and widespread threats facing birds – habitat loss and climate change – require large-scale shifts in how we use natural resources, build our cities and feed our populations.

▼ An example of the recovery of a raptor population after DDT was banned. The graph shows the number of known breeding sites for peregrine falcons (*Falco peregrinus*) in central Norway, 1976–2017.

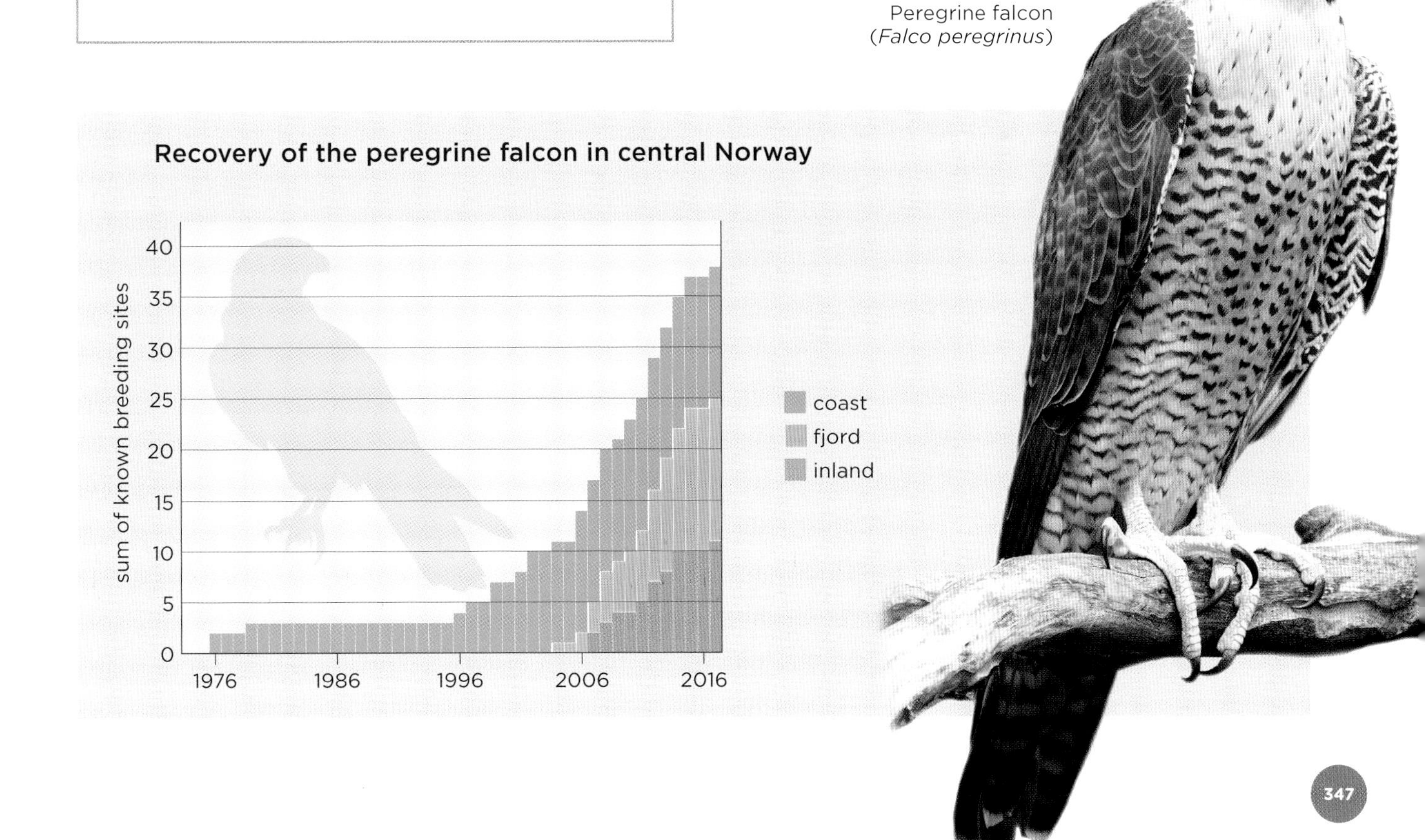

Conservation of birds: advocates

Protecting birds requires work across all scales. Enthusiastic ornithologists and birders have long been advocates for birds, and form the foundation of efforts to enumerate, understand and protect them. Local efforts are an invaluable component of conservation efforts, but with species distributions spanning countries and continents, and individual migratory birds travelling thousands of kilometres, protecting birds also requires cooperation at an international level.

Ornithologists and birders provide a wealth of information that is essential for the most fundamental aspect of bird conservation: detecting declines in populations. These grassroots advocates – hundreds of millions of people worldwide – support an array of programmes and citizen science efforts. Birders also support economies; a 2011 survey in the US estimated that birders spend US$41 billion a year on equipment and travel.

Ornithological societies

Birders, ornithologists and conservationists often form advocacy groups that promote the understanding and protection of birds. Ornithological societies can be found in most countries, including the German Ornithologists' Society (founded in 1850), the National Audubon Society in the US (1886), the Royal Society for the Protection of Birds in the UK (1889), the Wild Bird Society of Japan (1934) and many more. International societies are also important advocates, such as the International Ornithologists' Union and BirdLife International. In one example of their work, BirdLife scientists documented the enormous scope of illegal killing of birds in the Mediterranean in 2014–15 – an estimated 25 million birds annually in this region alone. In turn, this has brought increased attention to the plight of birds, and is putting pressure on governments to strengthen legislation and better enforce existing laws.

▼ Birders travel across the globe to watch wild birds such as the northern gannets (*Morus bassanus*) and great skua (*Stercorarius skua*) seen here.

International agreements

Governments at various levels began enacting legislation to protect bird populations in the nineteenth century. The earliest international agreement to protect birds was the Migratory Bird Treaty between the US and UK (on behalf of Canada) in 1916, which expanded to include Mexico in 1936, and Japan and the Soviet Union in the 1970s. The name of this treaty is a bit misleading – the Act initially restricted hunting and harvesting native birds deemed beneficial or harmless to humans and has since been revised to include most native birds, regardless of their migratory status. Such international cooperation is essential to modern bird conservation, so it is worrisome that the US government reinterpreted the Act in 2017, deciding that actions that incidentally harm birds – such as land development, uninsulated power lines or oil spills – would not be limited by this law, even though they had been under previous administrations. Advocates for birds are actively lobbying for new legislation to strengthen international protections.

▼ In 2014–15, biologists from BirdLife International and partner conservation organisations documented that millions of birds are killed illegally every year in the Mediterranean, largely to satisfy the demand for songbirds on the menu.

BIRD RINGING

The study of bird behaviour and survival was made much more effective through the innovation of fitting the animals with uniquely numbered leg bands, or bird ringing (bird banding in North America). The practice originated in 1890 with Danish ornithologist Hans Christian Mortensen, who fitted more than 5,000 birds with aluminium leg rings stamped with a unique number and his address. When rings were sent back, often by duck hunters, he gained the very first insights into where individual birds were travelling. Numerous ringing stations and trained individuals across the globe currently monitor bird populations, and much of what we now know of bird migration today has been derived from the millions of birds that have been ringed worldwide.

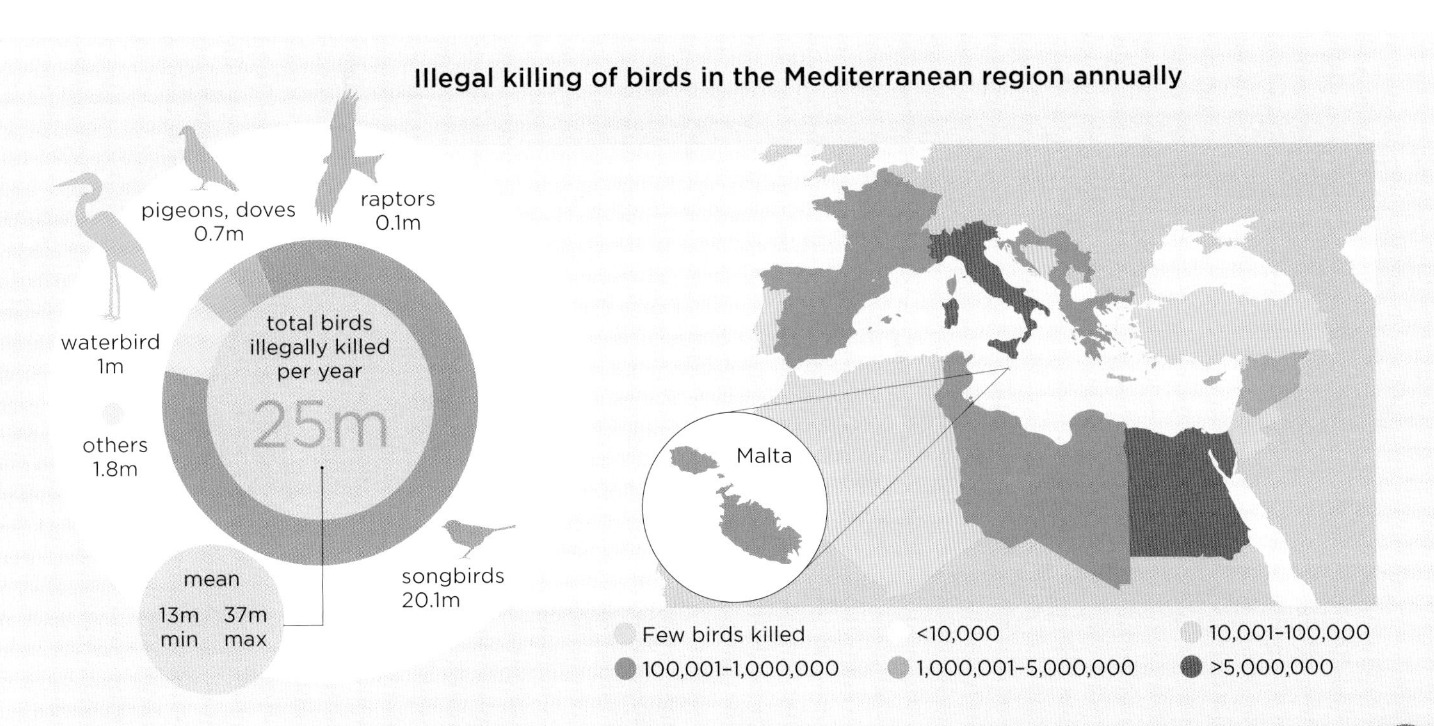

Citizen science

Citizen science – the generation of data by members of the public – is the engine of some of the most widespread, long-running efforts to enumerate bird populations and detect their declines. Data generated by citizen science have also been used to answer some fundamental research questions.

Northern cardinal (*Cardinalis cardinalis*), male

Breeding Bird Surveys and Christmas Bird Counts

Breeding Bird Surveys (BBSs) are conducted in several countries by volunteers familiar with local avifauna. These citizen scientists travel specific routes along a road, beginning 30 minutes before sunrise and stopping at set points to survey for birds by sight and sound. These standardised surveys generate invaluable data for detecting changes in bird populations.

Christmas Bird Counts (CBCs) are held in various countries across the Americas. They differ from BBSs, in that they occur over a 24-hour period during the northern hemisphere winter and involve intensive surveying of a set area – a count circle, rather than a roadside route. More than 2,500 count circles were surveyed in 2018, with participants reporting over 2,600 species; the record for one circle was 502 species, in Napo, Ecuador! Volunteers for these counts often aim to tally as many species as possible, rather than enumerating individuals, so CBC data are most helpful for detecting changes in species distributions rather than abundance.

Geographic coverage of the 119th Audubon Christmas Bird Count

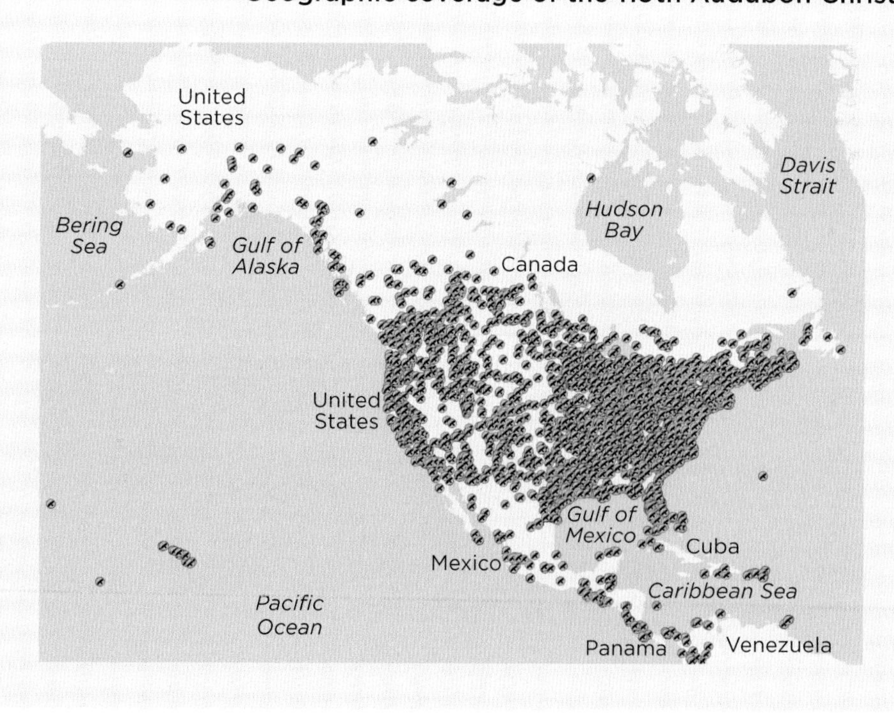

◄ Map showing the area covered by the Christmas Bird Count of 2019, from Alaska in the north to northern South America. During these counts, birders and naturalists spend a day documenting as many bird species as they can find inside a specified count circle.

Breeding Bird Atlases

Breeding Bird Atlases (BBAs), which are produced all over the world, take place less frequently than BBSs or CBCs but involve more intensive surveying. In these surveys, citizen scientists often receive training from BBA coordinators and are assigned one or more cells of a grid within an area to be surveyed (e.g. a city, state or nation). Participants categorise birds as possible, probable or confirmed breeders in their assigned grid cells. Observations are generally compiled over a period of several years, and then the survey may be repeated after a period of time – usually 20 years or more. By comparing sequential BBA results, researchers can detect changes in distributions and numbers of breeding birds, and identify important areas for conservation, such as regions with high species diversity or with many species that are at risk.

A 2018 citizen science project drawing on the expertise of birders and naturalists around the world revealed that some birds appear to be excluded from cities due to competition with close relatives (e.g. birds from the same genus). This study of factors that limit avian biodiversity in cities across the globe would be impossible without the contributions of knowledgeable citizen scientists.

CITIZEN SCIENCE AT CORNELL

The Lab of Ornithology at Cornell University in the US is unrivalled in the scope of its citizen science programmes aimed at understanding birds. Programmes include eBird, FeederWatch and the Great Backyard Bird Count. The most popular of these programmes, eBird (https://ebird.org), allows birders to report their sightings and keep track of their own garden, year and life lists, but also curates these reports for use by researchers. With hundreds of millions of bird sightings reported from around the world, the eBird database has been used in numerous published studies, on an array of subjects, including the effectiveness of conservation efforts and shifts in timing of bird migration.

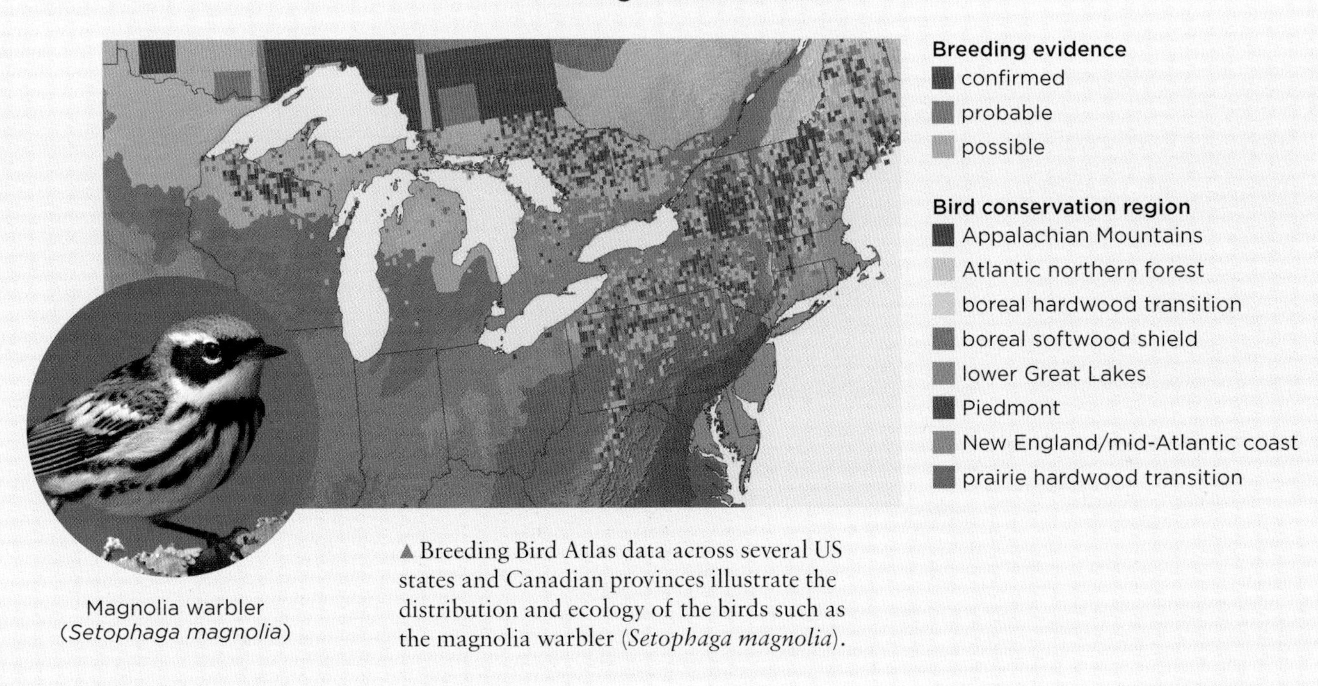

Magnolia warbler (*Setophaga magnolia*)

▲ Breeding Bird Atlas data across several US states and Canadian provinces illustrate the distribution and ecology of the birds such as the magnolia warbler (*Setophaga magnolia*).

Research advances

Birds have been studied for hundreds of years by ornithologists seeking to better understand their evolution, behaviour, populations and other factors, but also as model organisms to understand more about biology in general. The typically slow pace of research has been punctuated by important discoveries and innovations that offer amazing insights.

Bar-tailed godwit (*Limosa lapponica*)

Navigation skills

One of the biggest unsolved mysteries of bird behaviour was how migrants navigate enormous distances with apparent pinpoint accuracy. In one of many examples of migratory exceptionalism, some bar-tailed godwits (*Limosa lapponica*) fly more than 11,000 km non-stop from their breeding grounds in Alaska to winter in New Zealand. Ornithologists have tested numerous ideas about how birds find their way, with captive studies in the 1950s and 1960s revealing that some birds use the sun, magnetic fields or stars as a compass. Experimental studies in 2003 elegantly demonstrated that migrating thrushes, and likely many other birds, rely on a magnetic compass that can be calibrated using cues from the setting sun (see also page 200).

Problem-solving skills

At one time, experts in behaviour thought that self-awareness, tool use and complex problem solving set humans apart from other animals. One by one, these behaviours have been found in non-human animals, from chimpanzees to spiders. The ability to make complex plans also stood out as rare outside of our species. Enter the New Caledonian crow (*Corvus moneduloides*), a species known for its tool use and problem solving. A 2019 study revealed that these birds can form a mental image of as many as three steps in a planned strategy to attain a reward. Research on birds is challenging our ideas of what it means to have a bird brain!

Successive Doppler radar images of five dispersing tree swallow roosts

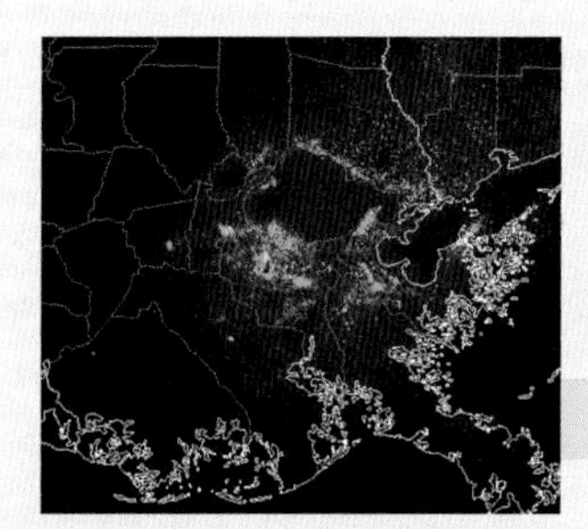

▲ A Doppler radar image from southeastern Louisiana, USA, captured early in the morning in November.

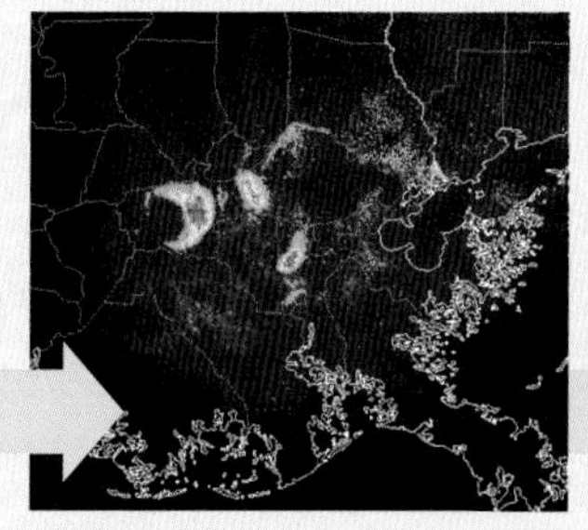

▲ As the morning progresses, the radar image detects the signature of tree swallows (*Tachycineta bicolor*) as they emerge from their overnight roosts, primarily in sugarcane fields.

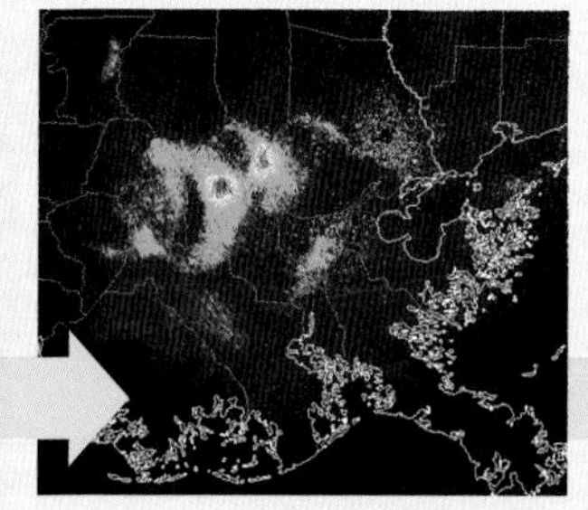

▲ The warm hues show hundreds of thousands of birds in wintering flocks, departing their roosts to forage during the day.

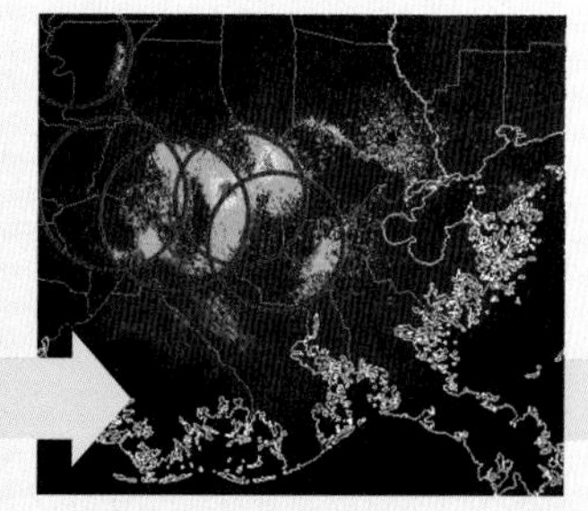

▲The circles identify roost sites, illustrating the value of using radar to study the behaviour of these birds.

ICARUS INITIATIVE

The ICARUS (International Cooperation for Animal Research Using Space) Initiative, launched in 2019, represents a remarkable innovation in tracking technology. Conceived by biologists at the Max Planck Institute in Germany, ICARUS uses small trackers to record the location of an animal, as well as its small-scale activity. Most other tracking devices require the researcher to gather data by following the animal or by removing the tag – not ideal for migratory birds that travel thousands of kilometres and suffer high rates of mortality. With ICARUS, an enormous receiver antenna on the International Space Station allows for remote transmission of data as the station orbits the Earth. Researchers hope to launch additional receivers on satellites, which could provide global coverage.

▲ A young common blackbird (*Turdus merula*) fitted with an ICARUS transmitter, which measures the location and movement of the bird.

The ICARUS Initiative could revolutionise our understanding of animals, including bird migration, social behaviour and responses to the environment. Biologists are also proposing ICARUS for use as an early-warning system, harnessing the instincts of animals to detect imminent natural disasters such as earthquakes and volcanic eruptions.

Technological innovations in ornithology

Several important breakthroughs in ornithology came about thanks to technological innovations. Bird banding provided insight into migration (see box on page 349), genetic tests revolutionised our understanding of mating systems, and an array of innovations in tracking technology have generated advances in our understanding of bird behaviour. Several of these were pre-existing technologies repurposed for bird research (see example opposite).

◄ An adult tree swallow (*Tachycineta bicolor*) in flight. These birds winter in enormous flocks along the coasts of southern USA and Central and South America.

Glossary

ADAPTATION The evolutionary process by which organisms become fitted to their environment; a phenotypic trait that improves an individual's fitness.

ALTRICIAL A species whose chicks hatch in an undeveloped state, typically naked, blind and unable to thermoregulate, requiring care and feeding by the parents.

ANTIOXIDANTS Endogenous or dietary compounds that decrease damage caused by reactive oxidising agents such as free radicals.

ARCHOSAURIA The group including all extinct dinosaurs, pterosaurs and extinct relatives of crocodilians, as well as birds and living crocodilians.

BARBS Keratin structures branching off the rachis or main shaft of a feather and forming the feather vane.

BARBULE Smaller keratin structures branching off feather barbs, with hooklets that 'zip' barbs tightly together.

BETA-KERATIN The protein making up integumentary structures (scales, claws, feathers) in birds.

BIODIVERSITY The number and diversity of species in a particular habitat or ecosystem.

BROOD The offspring (chicks) from a single clutch of eggs.

BROOD PARASITE A bird that lays its eggs in the nest of another female of the same or a different species, and does not help rear the chicks.

CARNIVORE An animal that feeds on other animals.

CAROTENOIDS A group of fat-soluble yellow, orange and red pigments that birds must obtain from their diets.

CITIZEN SCIENCE The collection and analysis of biological data by members of the general public, often in collaboration with professional scientists.

CLOACA The single external opening of a bird's digestive, urinary and reproductive tracts; used to expel faeces, for mating and to lay eggs.

COEVOLUTION The process by which two or more species evolve in tandem by exerting selection pressures on one another.

CREPUSCULAR Mainly active during twilight, at dawn or dusk.

CROP An expanded muscular part of the oesophagus in some birds used to store food temporarily.

CROP MILK A secretion from the thickened lining of the crop or oesophagus in pigeons and emperor penguins (*Aptenodytes forsteri*) that is regurgitated to chicks soon after they hatch.

DISPERSAL The movement of young birds away from the nest location where they were hatched, or movement of adults between breeding attempts.

DNA Deoxyribonucleic acid, the genetic material in cells making up chromosomes and genes.

DOWNSTROKE The movement of the wing downwards during flapping flight, pushing air down and creating lift.

DRUMMING A sound produced by snipe as part of their courtship display flights by vibrating the outer-tail feathers in the air while flying rapidly downwards.

ENDOCRINE SYSTEM The collection of ductless glands producing hormones that are secreted into the circulation and regulate metabolism, growth, development and reproduction.

ENDOGENOUS Internal to, or originating in, an organism.

ENDOTHERMY The maintenance of a high and stable body temperature using internal metabolic heat production.

ENZYMES Proteins that act as catalysts, regulating the rate at which chemical reactions proceed without being altered themselves.

FITNESS Of an individual, the relative contribution of its genotype to the next generation, determined by the total number of offspring produced.

FLEDGING When chicks leave the nest; in some species this occurs at the same time as chicks are able to fly.

FLOCKING When birds form large, socially interacting groups with others of the same, or different, species.

FLYWAY Continental-scale flight paths used by large numbers of birds migrating between their breeding grounds and their non-breeding overwintering areas.

FOLIVORE Feeding on a diet of plant leaves and grasses.

FRUGIVORE Feeding on a diet consisting primarily of plant berries and other fruits.

FURCULA The 'wishbone' of a bird skeleton, comprising two fused clavicles.

GENE A length of DNA on a chromosome that encodes and directs synthesis of a protein or RNA.

GENOME The set of chromosomes, or total amount of genetic material (DNA), in cells of any species.

GENUS A taxonomic category that ranks above species and below family, e.g. *Sturnus*.

GLIDING A form of aerial locomotion that does not require flapping flight.

GRANIVORE Feeding on a diet of plant seeds, grain and nuts.

HABITAT The natural home or environment occupied by an animal, plant or other organism.

HAEMOGLOBIN A red respiratory pigment or protein involved in the transport of oxygen by blood.

HOMEOSTASIS The maintenance of a constant internal, physiological environment (e.g. body temperature) despite marked changes in the external environment.

HOVERING A form of aerial locomotion where flapping wings direct thrust downwards to provide lift (e.g. as seen in hummingbirds).

HYPOTHALAMUS A structure at the base of the brain that transduces environmental or sensory information to chemical signals (hormones); part of the neuroendocrine system.

IMPRINTING The rapid learning process by which chicks develop recognition of, and attraction to, usually others of the same species, especially parents (or humans if they rear chicks).

INBREEDING Mating with genetically similar or closely related individuals, e.g. siblings; can cause inbreeding depression, a decrease in phenotypic quality.

INCUBATION Behaviours by which parents maintain eggs at temperatures required for embryo development (generally higher than 35°C).

INSECTIVORE Feeding on a diet of insects, arthropods or other invertebrates.

INTEGUMENT The body coverings of birds, including skin, scales, claws and feathers.

IRRUPTION A movement of birds, often in large numbers, beyond their normal range usually in response to low food availability (cf. migration, nomadism).

JUVENILE A young bird in the period between the chick stage (or fledging) and adulthood.

KEEL A bone extending outward from the sternum where the pectoral (flight) muscles attach.

LORE The area around the eye or beak.

MANDIBLE The lower half of the beak (cf. maxilla).

MAXILLA The upper half of the beak (cf. mandible).

MELANINS Endogenous pigments that generate a range of colours in birds, from black to reddish brown and yellow.

MELANOSOMES Cellular organelles that contain different forms of melanin pigments.

METABOLIC RATE The amount of energy used per unit time, measured as heat production or oxygen consumption.

MIGRATION The regular seasonal movements of birds between their breeding grounds and non-breeding (wintering) areas (cf. irruption, nomadism).

MONOGAMY Where one male and one female share a territory and cooperate to raise their offspring.

MOULT The process of shedding some or all feathers (plumage) and replacing them with newly grown feathers.

MUTUALISM An ecological relationship between multiple individuals, or different species, where both parties obtain benefits.

MYOGLOBIN An oxygen-binding protein, occurring mainly in muscle tissue, that increases oxygen storage in diving birds (e.g. penguins).

NARES Two openings of the nasal cavity located on either side of the upper beak.

NATURAL SELECTION The evolutionary process whereby some individuals in a population with favourable traits are more successful at reproduction or have better survival (i.e. they have higher fitness), leading to traits increasing over evolutionary time.

NICHE The place an organism occupies in an ecosystem, including its habitat, resource needs and the effect it has on other species and the environment.

NOMADISM An irregular movement of birds, usually in response to variable food availability (cf. irruption, migration).

OIL GLAND See 'preen gland'.

OMNIVORE Feeding on a diet that includes a variety of animal and plant material.

OSMOREGULATION Homeostasis of osmotic pressure in body fluids by regulating the uptake and loss of water and solutes (e.g. salts).

PARASITE An organism that obtains food or other resources from, and at the expense of, another organism (the host).

PHENOTYPE The characteristics of an individual (morphological, behavioural, physiological) resulting from the interaction of its genotype with the environment.

PHILOPATRY The tendency for individuals to return to the same breeding location, or nest site, in successive years.

PHOTOPERIOD The period of time each day during which an organism is exposed to light (i.e. daylength).

PHYLOGENY The evolutionary relationships or history of a set of organisms.

PINNATE Having branches on each side of an axis, like the vanes of a feather.

PISCIVORE Feeding on a diet of fish.

PITUITARY GLAND Part of the neuroendocrine system below, and connected to, the hypothalamus that secretes or releases hormones controlling the peripheral endocrine system.

POLYANDRY A mating system in which a single female raises offspring with more than one male.

POLYGYNY A mating system in which some males attract and raise offspring with two or more females.

PRECOCIAL A species where chicks hatch in a relatively well developed state, with feathers and ability to thermoregulate, walk and often self-feed.

PREDATOR Any animal that catches and kills other animals for food.

PREEN GLAND A gland near the upper base of the tail that secretes oil that the bird spreads over its feathers by preening; also called oil gland or uropygial gland.

PRIMARY FEATHERS The elongate flight feathers of a bird's wing (also called remiges); most birds have 10–12 primaries.

PROSPECTING The process whereby individual birds search for future breeding sites.

RACHIS The central midline of a feather off which barbs and barbules branch to form the feather vane.

RAPTORS Birds of prey that feed mainly on other vertebrates; includes eagles, hawks, falcons and owls.

RECTRICES Elongate flight feathers that form the tail, used in control, braking and steering during flight.

REMIGES Elongate flight feathers originating from the wing bones and forming the wing or aerofoil; include primary, secondary and (sometimes) tertiary feathers.

SALT GLANDS Specialised, paired osmoregulatory organs located just above the orbit of a bird's eye that extract and excrete excess salt from the blood.

SECONDARY FEATHERS A set of flight feathers making up the inner wing (inside the primaries) and attached to the ulna.

SEXUAL SELECTION A specific form of selection driven by access to mating opportunities, where males compete for females and females choose (good-quality) males; gives rise to exaggerated sexually selected traits when these traits increase the ability of individual males to compete for, and fertilise, more females, or if they are more attractive to females.

SOARING A form of gliding flight in which birds make use of rising air currents to provide sufficient lift to stay aloft for long durations without flapping.

SONG FLIGHT When a bird sings on the wing while in flight, often involves other behavioural displays such as 'parachuting' downwards.

SPECIES A fundamental unit of biological diversity; in general, a species comprises individuals that share many common traits, that can interbreed with members of the same species but rarely breed with members of other species, and that share a common evolutionary history.

STOPOVER SITES Specific locations where migratory birds land, transiently, to rest, eat or drink, before continuing their migratory flight.

STRUCTURAL COLOUR A colour produced by nanometre-scale light-refracting surfaces that selectively reflect certain wavelengths of light, producing blues, glossy blacks and bright, iridescent hues.

SYRINX The main sound-producing organ of a bird, located at the point where the trachea divides into two bronchi before entering the lungs.

TAXONOMY A system for classifying organisms within hierarchical levels or taxa (e.g. genera, families) based on the shared characteristics and relatedness to other taxa.

THERMOREGULATION The process by which animals regulate or maintain their internal body temperature.

TRAIT Any distinguishing morphological, behavioural or physiological characteristic of an individual or species.

UPSTROKE The recovery stroke during flapping flight, where upward motion of the wing is powered by contraction of the supracoracoideus flight muscle.

UROPYGIAL GLAND See 'preen gland'.

Further reading

Avery, M. 2014. *A Message from Martha: The Extinction of the Passenger Pigeon and its Relevance Today*. Bloomsbury Natural History.

Birkhead, T. 2012. *Bird Sense: What It's Like to Be a Bird*. Bloomsbury Publishing.

Birkhead, T. 2016. *The Most Perfect Thing: Inside (and Outside) a Bird's Egg*. Bloomsbury Publishing.

Birkhead, T., Wimpenny, J. and Montgomerie, B. 2014. *Thousand Birds: Ornithology Since Darwin*. Princeton University Press.

Chadd, R. W. and Taylor, M. 2016. *Birds: Myth, Lore and Legend*. Bloomsbury Publishing.

Lederer, R. J. 2019. *The Art of the Bird: The History of Ornithological Art Through Forty Artists*. University of Chicago Press.

Lovette, I. J. and Fitzpatrick, J. W. (*eds*). 2016. *Cornell Lab of Ornithology's Handbook of Bird Biology*. 3rd edn. Wiley.

Marra, P. P. and Santella, C. 2016. *Cat Wars: The Devastating Consequences of a Cuddly Killer*. Princeton University Press.

Morrison, M. L., Rodewald, A. D., Voelker, G., Colón, M. R. and Prather, J. F. (*eds*). 2018. *Ornithology: Foundation, Analysis, and Application*. Johns Hopkins University Press.

Root, D. B. 2018. The feather thief: beauty, obsession, and the natural history heist of the century. *Library Journal*, 143(3): 72.

Scanes G. C. (*ed*.). 2015. *Sturkie's Avian Physiology*. 6th edn. Elsevier.

Stutchbury, B. 2009. *Silence of the Songbirds*. Bloomsbury Publishing.

References

Included here are the papers and studies referred to in the main text.

Chapter 1

Clarke, J. A., Chatterjee, S., Li, Z., Riede, T., Agnolin, F., Goller, F., Isasi, M. P., Martinioni, D. R., Mussel, F. J. and Novas, F. E. 2016. A fossil avian vocal organ from the Mesozoic. *Nature*, 538: 502–505. doi: 10.1038/nature19852.

Hu, D., Clarke, J. A., Eliason, C. M., Qiu, R., Li, Q., Shawkey, M., Zhao, C., D'Alba, L., Jiang, J. and Xu, X. 2018. A bony-crested Jurassic dinosaur with iridescent plumage highlights complexity in early paravian evolution. *Nature Communications*, 9: 217.

Li, Z., Zhou, X. and Clarke, J. A. 2018. Convergent evolution of a mobile and protrusible bony tongue in flighted dinosaurs and pterosaurs. *PLOS One*, 13(6): e0198078. https://doi.org/10.1371/journal.pone.0198078

Chapter 2

Dove, C. J. 2000. *A Descriptive and Phylogenetic Analysis of Plumulaceous Feather Characters in Charadriiformes*. Ornithological Monographs 51. American Ornithologists' Union.

Dove, C. J. and Agreda, A. 2007. Differences in plumulaceous feather characters of dabbling and diving ducks. *Condor*, 109: 192–199.

Gomez, D. and Théry, M. 2004. Influence of ambient light on the evolution of colour signals: comparative analysis of a Neotropical rainforest bird community. *Ecology Letters*, 7: 279–284.

Igic, B., D'Alba, L. and Shawkey, M. D. 2016. Manakins can produce iridescent and bright feather colours without melanosomes. *Journal of Experimental Biology*, 219: 1851–1859.

McGraw, K. J. 2006. Mechanics of melanin-based coloration: mechanisms and measurements. *In*: Hill, G. E. and McGraw, K. J. (*eds*). *Bird Coloration. Vol. 1*. Harvard University Press. Pp. 243–294.

Price, T. D. and Pavelka, M. 1996. Evolution of a colour pattern: history, development, and selection. *Journal of Evolutionary Biology*, 9: 451–470.

Prum, R. O. 1990. Phylogenetic analysis of the evolution of display behavior in the Neotropical manakins (Aves: Pipridae). *Ethology*, 84(3): 202–231.

Prum, R. O. 2017. *The Evolution of Beauty: How Darwin's Forgotten Theory of Mate Choice Shapes the Animal World – and Us*. Anchor Press.

Shawkey, M. D. and D'Alba, L. 2017. Interactions between colour-producing mechanisms and their effects on the integumentary colour palette. *Philosophical Transactions of the Royal Society B*, 372(1724). doi: 10.1098/rstb.2016.0536

Chapter 4

McKechnie, A. E. and Wolf, B. O. 2004. The allometry of avian basal metabolic rate: Good predictions need good data. *Physiological and Biochemical Zoology*, 77: 502–521.

Chapter 5

Berthold, P. 1993. *Bird Migration: A General Survey*. Oxford University Press.

Cooper-Mullin, C. and McWilliams, S. R. 2016. Response of the antioxidant system to intense endurance exercise: lessons from migrating birds. *Journal of Experimental Biology*, 219: 3684–3695.

Ferretti, A., Rattenborg, N. C., Ruf, T., McWilliams, S. R., Cardinale, M. and Fusani, L. 2019. Sleeping unsafely tucked in to conserve energy in a nocturnal migratory songbird. *Current Biology*, 29: 2766–2772.

Heers, A. M., Baier, D. B., Jackson, B. E. and Dial, K. P. 2016. Flapping before flight: high resolution, three-dimensional skeletal kinematics of wings and legs during avian development. *PLOS One* 11(4): e0153446. https://doi.org/10.1371/journal.pone.0153446.

Lovette, I. J. and Fitzpatrick, J. W. (*eds*). 2016. *Cornell Lab of Ornithology's Handbook of Bird Biology*. 3rd edn. Wiley.

Newton, I. 2008. *The Migration Ecology of Birds*. Academic Press.

Rattenborg, N. C., Voirin, B., Cruz, S. M., Tisdale, R., Dell'Omo, G., Lipp, H.- P., Wikelski, M. *et al*. 2016. Evidence that birds sleep in mid-flight. *Nature Communications*, 7: 12468. doi: 10.1038/ncomms12468

Tobalske, B. W., Peacock, W. L. and Dial, K. P. 1999. Kinematics of flap-bounding flight in the zebra finch over a wide range of speeds. *Journal of Experimental Biology*, 202: 1725–1739.

Chapter 6

Berthold, P. 1993. *Bird Migration: A General Survey*. Oxford University Press.

Dingle, H. 2014. *Migration: The Biology of Life on the Move*. 2nd edn. Oxford University Press.

Greenberg, R. and Marra, P. (*eds*). 2005. *Birds of Two Worlds*. Johns Hopkins University Press.

Lovette, I. J. and Fitzpatrick, J. W. (*eds*). 2016. *Cornell Lab of Ornithology's Handbook of Bird Biology*. 3rd edn. Wiley.

McWilliams, S. R., Pierce, B. J., Guglielmo, C. and Klaassen, M. 2004. Flying, fasting, and feeding in birds during migration: a physiological ecology perspective. *Journal of Avian Biology*, 35: 377–393.

Newton, I. 2008. *The Migration Ecology of Birds*. Academic Press.

Skrip, M. and McWilliams, S.R. 2016. Oxidative balance in birds: an atoms-to-organisms-to-ecology primer for ornithologists. *Journal of Field Ornithology*, 87: 1–20.

Chapter 7

Nevitt, G. A. (1999) Foraging by seabirds on an olfactory landscape. *American Scientist*, 87: 46–53.

Nevitt, G. A., Veit, R. R. and Kareiva, P. (1995) Dimethyl sulphide as a foraging cue for Antarctic Procellariiform seabirds. *Nature*, 376, 680–682.

Chapter 8

Cornell, H. N., Marzluff, J. M. and Pecoraro, S. 2012. Social learning spreads knowledge about dangerous humans among American crows. *Proceedings of the Royal Society B*, 279: 499–508.

Chapter 10

Both, C., Bouwhuis, S., Lessells, C. M. and Visser, M. E. 2006. Climate change and population declines in a long-distance migratory bird. *Nature*, 441(7089): 81–83.

Brochet, A.-L., van den Bossche, W. and Kariuki Ndang'ang'a, P. 2016. Preliminary assessment of the scope and scale of illegal killing and taking of birds in the Mediterranean. *Bird Conservation International*, 26(1): 1–28.

Cochran, W. W., Mouritsen, H. and Wikelski, M. 2004. Migrating songbirds recalibrate their magnetic compass daily from twilight cues. *Science*, 304(5669): 405–408.

Cox, A. R., Robertson, R. J., Lendvai, Á. Z., Everitt, K. and Bonier, F. 2019. Rainy springs linked to poor nestling growth in a declining avian aerial insectivore (*Tachycineta bicolor*). *Proceedings of the Royal Society B*, 286(1898). https://doi.org/10.1098/rspb.2019.0018.

Eng, M. L., Stutchbury, B. J. and Morrissey, C. A. 2019. A neonicotinoid insecticide reduces fueling and delays migration in songbirds. *Science*, 365(6458): 1177–1180.

Freeman, B. G., Scholer, M. N., Ruiz-Gutierrez, V. and Fitzpatrick, J. W. 2018. Climate change causes upslope shifts and mountaintop extirpations in a tropical bird community. *Proceedings of the National Academy of Sciences*, 115(47): 11982–11987.

Gross, M. 2019. Hunting wildlife to extinction. *Current Biology*, 29(12): R551–R554.

Gruber, R., Schiestl, M., Boeckle, M., Frohnwieser, A., Miller, R., Gray, R. D., Clayton, N. S. and Taylor, A. H. 2019. New Caledonian crows use mental representations to solve metatool problems. *Current Biology*, 29(4): 686–692.

Hickey, J. J. and Anderson, D. W. 1968. Chlorinated hydrocarbons and eggshell changes in raptorial and fish-eating birds. *Science*, 162(3850): 271–273.

Laughlin, A. J., Taylor, C. M., Bradley, D. W., Leclair, D., Clark, R. C., Dawson, R. D., Dunn, P. O. *et al*. 2013. Integrating information from geolocators, weather radar, and citizen science to uncover a key stopover area of an aerial insectivore. *The Auk*, 130(2): 230–239.

Martin, P. R. and Bonier, F. 2018. Species interactions limit the occurrence of urban-adapted birds in cities. *Proceedings of the National Academy of Sciences*, 115(49): E11495–E11504.

Rosenberg, K. V., Dokter, A. M., Blancher, P. J., Sauer, J. R., Smith, A. C., Smith, P. A., Stanton, J. C. *et al*. 2019. Decline of the North American avifauna. *Science*, 366(6461): 120–124.

Index

Note: page numbers in **bold** refer to information contained in captions.

Picture credits

(T = top, M = middle, B = bottom, L = left, R = right)

Illustrations by Robert Brandt: 23 (redrawn from Sues, H. 2001. Ruffling feathers. *Nature* 410: 1036–1037), 32L & R, 33B, 47L (redrawn from Vardanis, Y., *et al.* 2011. Individuality in bird migration: routes and timing. *Biology Letters*, 7502–505), 61T & M, 62T, 95BR, 96B, 97B, 99R, 101L, 102B, 108, 112M, 128, 129T, 130, 132, 133, 136M, 138 (redrawn from Lovette, I. J. & Fitzpatrick, J. W. (eds). 2016. *The Cornell Lab of Ornithology Handbook of Bird Biology*, third edition), 139T, 140, 144, 145TL & BL, 146R, 149B, 151, 152B (redrawn from Williams, T. D. 2018. Avian Reproduction – Overview (wild birds). In: *Encyclopedia of Reproduction* (E. Knobil & J. D. Neill, eds.), *Vol. 6: Comparative Reproduction*, pp.595–601), 155T, 158, 164R, 166, 167T & B, 170, 175, (redrawn from Ferretti, A., *et al.* 2019. Sleeping unsafely tucked in to conserve energy in a nocturnal migratory songbird. *Current Biology*, 29: 2766–2772), 181R (redrawn from Butler, P. J. 2016. The physiological basis of bird flight. *Philosophical Transactions of the Royal Society*, B 371: 20150384), 188, 190, 191B, 192–3, (redrawn from Cohen, E. B., *et al.* 2018. Quantifying the strength of migratory connectivity. *Methods in Ecology and Evolution*, 9: 513–524), 193M, 197, 213 (redrawn with permission by Georgia Department of Natural Resources and the Center for Conservation Biology (at William and Mary)), 221 (redrawn from Rotenberry, J. T. 1980. Dietary relationships among shrub steppe passerine birds: Competition or opportunism in a variable environment? *Ecological Monographs*, 50: 93–110), 231 (redrawn from Wilson, R. 1995. Foraging ecology. In: Williams, T. D. *The Penguins*. pp.81–106), 242R, 273R (redrawn from Baker, M. C. 1982. Vocal dialect recognition and population genetic consequences. *American Zoologist*, 22: 561–569), 287L, 289T (redrawn from Goodfellow, D. J. 1988. Singing behaviour and its relationship to territoriality in the Chaffinch (*Fringilla coelebs* L.). D.Phil thesis, University of Sussex, Brighton), 302 (redrawn from Williams, T. D. 2012. *Physiological Adaptations for Breeding in Birds*), 305T (redrawn from Williams (2012)), 306T, 338 (CC BY-SA. Redrawn from original by H. Ritchie and M. Roser at OurWorldinData.org), 340 (redrawn using information from Freeman, B. G., *et al.* 2018. Climate change causes upslope shifts and mountaintop extirpations in a tropical bird community. *Proceedings of the National Academy of Sciences*, 115 (47): 11982–11987), 347L (redrawn from Nygård, T., *et al.* 2019. Population recovery of peregrine falcons in central Norway in the 4 decades since the DDT-ban. *Ecotoxicology*, 28: 1160–1168), 349 (redrawn from Brochet, A.-L., *et al.* 2016. Preliminary assessment of the scope and scale of illegal killing and taking of birds in the Mediterranean. *Bird Conservation International*, 26.1: 1–28).

Alamy: 24M Martin Shields; 36 Doug Allan/Nature Picture Library; 40T Larry Minden/Minden Pictures; 59T Amazon-Images; 81BL Ray Wilson; 106R Michael Stubblefield; 123T William Leaman; 126L Arco Images GmbH; 145TR Ivan Kuzmin; 146M Dorling Kindersley Ltd; 162B Andy Harmer; 172 Octavio Campos Salles; 182R Michael Nolan/robertharding; 183 Scotland: The Big Picture/Nature Picture Library; 185T Cordier Sylvain/hemis.fr; 226L Alan Williams/Nature Picture Library; 227BL blickwinkel/B. Zoller; 233 Prisma by Dukas Presseagentur GmbH; 237 Thomas P. Peschak/National Geographic Image Collection; 240 Nigel Blake; 245L D. Parer & E. Parer-Cook/Minden Pictures; 245R Jeff Mauritzen/National Geographic Image Collection; 253B blickwinkel/Agami/C. van Rijswijk; 254 Ashley Whitworth; 258 Dave Watts; 266 Richard Mittleman; 267T blickwinkel/McPHOTO/ODE; 274R Dorling Kindersley Ltd; 278 Jim Brandenburg/Minden Pictures; 288L blickwinkel/Agami/M. Guyt; 288R Octavio Campos Salles; 291 Barrie Britton/Nature Picture Library; 295L Wayne Lynch/All Canada Photos; 297B FLPA; 310–11 Accent Alaska.com; 311R Franz Wögerer/imageBROKER; 315T Joe Blossom; 317 blickwinkel/Hartl; 324 Douglas Lander; 335R National Geographic Image Collection; 342 Mohammad Adam; 343T David Tipling Photo Library.

Lucy Alpin 274L.

Staffan Andersson 115L & R.

Nyambayar Batbayar 169B.

Peter Beninger 241BL & BR (reproduced from Beninger, P. & Elner, R.W. 2020. On the tip of the tongue: natural history observations that transformed shorebird ecology. *Ecosphere* DOI 10.1002/ecs2.3133.)

Alan Benson 117TL, TM & TR, 222, 227R.

BirdLife International 189T.

BirdSafe@UWO team 210–11.

Peter Boesman 173B.

Xavi Bou 162T.

BTO 346B (reproduced from *Bird Atlas 2007–11*, which was a joint project between BTO, BirdWatch Ireland and the Scottish Ornithologists' Club, with permission from the British Trust for Ornithology).

CC A 4.0 International 149TR (reproduced from Leão, R. A. C. *et al.* 2017. *Arquivo Brasileiro de Medicina Veterinária e Zootecnia*. 69: 1529–1538. 10.1590/1678-4162-9233).

CC BY 4.0 159T (reproduced from Heers, A. M., *et al.* 2016, Flapping before flight: high resolution, three-dimensional skeletal kinematics of wings and legs during avian development. PLoS ONE 11(4): e0153446).

CC BY-SA 4.0 (reproduced from Grigg, N. P., *et al.* 2017. Anatomical evidence for scent guided foraging in the turkey vulture. *Scientific Reports* 7: 17408.) 101TR.

Christopher Clark 173T.

Julia Clarke 28TR.

Kimberley Collins (Flickr CC BY 2.0) 343B.

Joseph Corbo 112L.

Cornell University College of Veterinary Medicine 136L & R.

Dreamstime: 14B Alexander Potapov; 126R Khunaspix; 142–3 Simon Fletcher; 160L Martin Pelanek; 163 Mcasabar; 257 Luis Tejo; 282 Kim Nelson; 314 Adri Pendleton.

Chad Eliason 76L & R, 77BR, 81TL.

Feather Friendly Technologies Inc. 211T.

Getty Images: 28M Will Giles; 238–9 Frans Lemmens; 271T tracielouise; 325R Eye Ubiquitous/Universal Images Group.

Gavin Hunt, University of Auckland 275L & R.

Johnathan W. Hurst 327TR.

Erich D. Jarvis 100BL.

Alex Kirschel 25B.

Andrew Laughlin 352B (reproduced from Laughlin, A. J., *et al.* 2013. Integrating information from geolocators, weather radar, and citizen science to uncover a key stopover area of an aerial insectivore. *The Auk* 130(2): 230–239).

Carolyn Lehrke 264–5.

Scott MacDougall-Shackleton 122T.

Max-Planck-Institut für Verhaltensbiologie 353T.

Tony Mills 228–9.

Colin Miskelly/New Zealand Birds Online 301L.

© National Audubon Society 350B.

Nature Photographers Ltd (Paul Sterry) 48–9, 86R, 198–9, 312–13.

Nature Picture Library (Shattil & Rozinski) 298T.

Brian Peer 103TL & TR.

Jason Puddifoot 241ML & MR.

Science Photo Library: 20 Martin Shields; 26–27 Millard H. Sharp; 56L Kage Mikrofotografie GbR; 57TL Power and Syred; 66BR Power and Syred; 80L Dennis Kunkel Microscopy; 90L Steve Gschmeissner; 92L Dorling Kindersley/UIG; 92R D. Roberts; 131TL Steve Gschmeissner; 154 Dennis Kunkel Microscopy; 165 Jaime Chirinos; 225B Jim Zipp; 263R Cavallini James/BSIP; 336 Jaime Chirinos.

Shutterstock: 1 Vitaliy Kaplin; 4 Eric Isselee; 8 Daniel Danckwerts; 10 serkan mutan; 11 FotoRequest; 12 Erni; 14T schankz; 15M Gallinago_media; 15TL Eric Isselee; 15TR Martin Mecnarowski; 15ML pandapaw; 15MR Kotomiti Okuma; 15BL Christian Weber; 15BR Butterfly Hunter; 16 Arctium Lappa; 17 John Wollwerth; 18T Anna Filippenok; 19T Linda Bucklin; 19M Gregory Johnston; 19B Drakuliren; 21TL Vladimir Sviracevic; 21TR Nikita Konashenkov; 22 Warpaint; 24T Mike Truchon; 25T Catmando; 27B YuRi Photolife; 28TL Emre Terim; 29 sljones; 30 Eric Isselee; 33T MZPHOTO. CZ; 34T buteo; 34TM Gertjan Hooijer; 34M WaceQ; 34BM Imogen Warren;

34BR Sandra Standbridge; 35M zieusin; 35TL Gallinago_media; 35TR YK; 35ML Gallinago_media; 35MR LifetimeStock; 35BR AlessandroZocc; 37L Tony Brindley; 37R Dennis W Donohue; 40B Independent birds; 42T serkan mutan; 42BL Jarkko Jokelainen; 42ML Lost Mountain Studio; 42MR Karel Bock; 42BR szpeti; 43T Samantha Crimmin; 43B Bob Hilscher; 45T Dave Montreuil; 45BR Jukka Jantunen; 46T Gualberto Becerra; 46B Paul Reeves Photography; 47R Edwin Butter; 50L Kevin Sawford; 50R Estuary Pig; 51T JMx Images; 51B Syed F Abbas; 52 Ivan Kuzmin; 53R Erni; 54 buteo; 56–7 Eric Isselee; 57TR ThomasLENNE; 58B Ralph Eshelman; 59ML & R Gwoeii; 59M & BL phugunfire; 60 S.R. Maglione; 61B Marina Sliusarenko; 62B On the Wing Photography; 63 Iv Nikolny; 64T Lotus_studio; 64BL Siwakorn1933; 64BR Eric Isselee; 65T Andi111; 65BL Erni; 65BR HASPhotos; 66–7 BERNATSKAIA OKSANA; 66BL Lena Ivanova; 68 Dicky Asmoro; 70L Piotr Krzeslak; 70R Paul Alexander Watkins; 71TL Chursina Viktoriia; 71TR Karel Stepan; 71BL Jennifer Bosvert; 71BR Robert L Kothenbeutel; 72 Steve Byland; 73 oguenaydin; 74L Steven Russell Smith Ohio; 74R Jen DeVos; 75 GUDKOV ANDREY; 77T Agami Photo Agency; 77BL Feng Yu; 78T Randy Bjorklund; 78BL Agnieszka Bacal; 78BR Agami Photo Agency; 79 Tarpan Royalty; 80R Alice Hebb; 82 Alexey Seafarer; 83L Eric Isselee; 83R Karin Jaehne; 86L David James Chatterton; 87BR Eric Isselee; 88 Menno Schaefer; 90R Panaiotidi; 91TL, ML, MR & BR KUCO; 91M Marzolino; 93T Rob Brandt; 93B Nynke van Holten; 94–5 JayPierstorff; 95T Emre Terim; 96T Neal Cooper; 97TL Chris Watson; 97TR Eric Isselee; 98L Rudi van den Heever; 98R Chris de Blank; 99TL Krakenimages.com; 100TL Jesada Sabai; 100TR majivecka; 101BR danielecolombo; 102T Maslov Dmitry; 103B Paul Reeves Photography; 105 serkan mutan; 106L Agami Photo Agency; 107L Martin Fowler; 107R valipatov; 109T Lowie51; 109B Michael Shake; 110 Peter Krejzl; 111L Eric Isselee; 111R Wolfgang Kruck; 113BL Les Perysty; 113BR Vector Tradition; 114TL WildlifeWorld; 114TM Eric Isselee; 114ML Marina Maytin Maduro; 114M Eric Isselee; 114BR Les Perysty; 115M Eric Isselee; 116 Ysbrand Cosijn; 117BR Gallinago_media; 118–19 Morpheus Szeto; 120L tea maeklong; 120R Lopatin Anton; 121TL Morphart Creation; 121B Eric Isselee; 122B Charles Brutlag; 123B Lost Mountain Studio; 127T semue85; 127B Hennadii H; 129B jo Crebbin; 131BL Allgord; 131BR Nitin Chandra; 135TL plenoy m; 135TR Jubal Harshaw; 135BL Paul Reeves Photography; 137B Jamil Bin Mat Isa; 139B colacat; 141B Potapov Alexander; 145BR Brian A Wolf; 146L Valentina Proskurina; 147L fieldwork; 147R Volodymyr Goinyk; 149TL Sadie P Photography; 150B Khairil Azhar Junos; 152T Natapop; 152B (background) Koshevnyk; 153 Super Prin; 155B showcake; 156 Chesapeake Images; 159B Bonnie Taylor Barry; 160R Victor Tyakht; 161T tea maeklong; 161B Stubblefield Photography; 164L NaturesMomentsuk; 164 (background) oorka; 168T Wang LiQiang; 168B slowmotiongli; 171 mezzotint; 174 Toni Genes; 176T JayPierstorff; 176B MintImages; 177TL Ihnatovich Maryia; 177TR Piranjya; 177BL VladimirCeresnak; 177BR Roi and Roi; 180L Eric Isselee; 181TL John Carnemolla; 181R (background) Koshevnyk; 182L Matt Elliott; 184TL Jukka Jantunen; 184TR Jukka Jantunen; 184–5 Neil Bradfield; 186 Alyce Taylor; 189B photomaster; 191T clarst5; 192L Mike Truchon; 192R Rabbitti; 193BR Brian Lasenby; 195 forgetselfies; 196 Tim Zurowski; 200 Toni Genes; 201 (background) Iri_sha; 201 (inset) John L. Absher; 202 Rabbitti; 203 manfredxy; 204–5 valleyboi63; 205 AnastasyaM; 207 Vladimir Kogan Michael; 208–9 Jim Cumming; 209T Thomas Clode; 209M Linn Currie; 210 pio3; 212 Erik Abet Andrew Munot; 213 (overlay) aDam Wildlife; 214 Jesus Giraldo Gutierrez; 215 Martin Fowler; 216 Clayton Burne; 218 Brian Lasenby; 219M fewerton; 219TL AZ Outdoor Photography; 219TR LecartPhotos; 219TML Luca Nichetti; 219TMR Ian Maton; 219BML Massimiliano Paolino; 219BMR Gaston Piccinetti; 219BL Carolyn Smith1; 219BR Chandanaroy; 224 Arun RaJpuT6621; 225T Jim Cumming; 226R Karel Bock; 229B Samantha Crimmin; 232 Thorsten Spoerlein; 234T Dennis W Donohue; 234B OSTILL is Franck Camhi; 235 Marina Maytin Maduro; 236 Karel Bartik; 241TL Natalia Kuzmina; 241TR Ekaterina V. Borisova; 242L Andrew Burgess; 243T Agami Photo Agency; 243B Imogen Warren; 244 malewick; 246 Voodison328; 247 Hayley Crews; 248 clarst5; 250 MZPHOTO.CZ; 251T Kuttelvaserova Stuchelova; 251B Cathy Withers-Clarke; 252B Nowaczyk; 253T Tom Reichner; 255 cvrestan; 256 Lilly P. Green; 259L Imogen Warren; 259R Steve Byland; 262 Sue A Dunning; 263L David Roy Carson; 267B Marut Sayannikroth; 269T Sue Harper Photography; 269B Imre Tuvi; 270 BOONCHUAY PROMJIAM; 271B Ondrej Prosicky; 273L Frode Jacobsen; 276L Carol Kelpin; 276R Uwe Bergwitz; 277T Natalya Erofeeva; 277BL Jukka Jantunen; 277BR francesco de marco; 279T PHOTOCREO Michal Bednarek; 280TL royaltystockphoto.com; 280TR Roman Samokhin; 280BL Eleanor McDonie; 281T vagabond54; 281B Rocky Cranenburgh; 284 Rudmer Zwerver; 285T Petr Simon; 285B Gualberto Becerra; 286 Red ivory; 287BR hfuchs; 289B Bachkova Natalia; 290 Marty Oishi; 292L Rudmer Zwerver; 292R Rich Lindie; 293L Super Prin; 293R iamtui7; 294L Ondrej Prosicky; 294R Vinod Pillai; 295R Samantha Crimmin; 296T Petrova Maria; 296B Karel Bartik; 297T Ilmarin; 298B ledokolua; 299BR Parrotfish Journey; 300B Vishnevskiy Vasily; 303ML Cuhle-Fotos; 303BL flaviano fabrizi; 303R David Osborn; 304–5 JeremyRichards; 306M Agami Photo Agency; 306B anne-tipodees; 307T Francois Loubser; 307B Greg A Wilson; 308 John L. Absher; 309 Marina Sliusarenko; 315B Nick Pecker; 316L Vishnevskiy Vasily; 316R Vishnevskiy Vasily; 318 Streeteh; 319T Agami Photo Agency; 319B Agami Photo Agency; 320 De Visu; 322T chrisdorney; 323B Ondrej Prosicky; 326BL Pornanun K; 326BR Deborah Ferrin; 327TL VectorArtFactory; 327BL Karolis Kavolelis; 327BR aaltair; 328 Joule Sorubou; 329T Maren Winter; 329B Lauren Pretorius; 330 Ram Jagan; 331T LI KWONG-MAN; 331B Tom Reichner; 332L Mila Drumeva; 332M Eric Isselee; 332R photomaster; 333 dowraik; 334 PLOO Galary; 339T Erni; 339B Danny Ye; 341 Ryzhkov Sergey; 344 Natalya Erofeeva; 345TL & TR Oceloti; 345TM Rauf Aliyev; 345M Eric Isselee; 345MR kontur-vid; 345BL Piter Kidanchuk; 345BM KOKTARO; 345BR VikiVector; 346T xpixel; 347L (background) Koshevnyk; 347R Michael de Nysschen; 348 Philippe Clement; 350T Charles Brutlag; 351 (inset) Agami Photo Agency; 352T Erni; 353B Lowell Monke; 357 Pavel Kovaricek.

Alwyn Simple 206.

Miguel Siu Rios 44.

Dr Joel Slade 272.

Peter J. B. Slater 289TR.

Slater Museum of Natural History, University of Puget Sound. 148L & R.

Oliver Smart 124.

Claire Spottiswoode 279B.

Stamen Design, Courtesy of the National Audubon Society 39.

Hanna Støstad 150TL & TR.

Unsplash: 2 Alexander Sinn; 41 Ivan Botha; 261 Ian Parker.

USGS Patuxent Wildlife Research Center 351.

Velizar Simeonovski for UT Austin 18B.

Wellcome Collection, Attribution 4.0 International (CC BY 4.0): 38, 131TR, 137T.

From Wikimedia ordered by license:

CC BY 2.0: 134 Christopher Michel; 230 Derek Keats; 252T Andy Morffew; 193T Andrew C.

CC BY 3.0: 87BL Scottmliddell; 194 John Manger, CSIRO; 323T Eddo.

CC BY 4.0: 84–5 Marie-Lan Taÿ Pamart; 180R Judi Lapsley Miller; 268 Marie-Lan Taÿ Pamart.

CC BY-SA 2.0: 223 cuatrok77; 303TL nottsexminer.

CC BY-SA 2.5: 24B Steveoc 86; 53L Andreas Trepte.

CC BY-SA 3.0: 141T Ianaré Sévi; 113TL Olaf Oliviero Riemer; 322B CarstenNorgaard; 45BL DiverDave.

CC BY-SA 4.0: 299BL Roger Culos; 299BM Didier Descouens; 301R Andrew Shiva; 300T Roger Culos; 58TR Statistique.

CC0 1.0: 81BR Boris Smokrovic.

Paul Williams 178–9.

Tony D. Williams 87T, 152M.

Every effort has been made to credit the copyright holders of the images used in this book. We apologise for any unintentional omissions or errors, and will insert the appropriate acknowledgement to any companies or individuals in subsequent editions of the work.

Acknowledgements

Tony D. Williams: thanks to Sir Brian Follett FRS for reading the complete draft manuscript and providing many useful comments.

Scott R. McWilliams: thanks to the U. S. National Science Foundation, U. S. Department of Agriculture, University of Rhode Island College of Environment and Life Sciences, as well as TSM for financial and other forms of support.

Elizabeth MacDougall-Shackleton: thanks to Joel Slade for providing spectrograms in Chapter 8, page 272.

Frances Bonier: thanks to Paul Martin and Bob Montgomerie for feedback and assistance with drafts of Chapter 10.